NEWFOUNDLAND—LABRADOR

Outlines of the Geography, Life & Customs of

NEWFOUNDLAND—LABRADOR

(The Eastern Part of the Labrador Peninsula)

Based upon Observations made during the
Finland—Labrador Expedition in 1937, and the
Tanner Labrador Expedition in 1939, and upon
information available in the literature and cartography

Volume II

By **V. TANNER**, Professor in Geography at the
University of Helsingfors

With 342 maps, diagrams, and photographs

CAMBRIDGE: At the University Press, 1947

CAMBRIDGE
UNIVERSITY PRESS

University Printing House, Cambridge CB2 8BS, United Kingdom

Published in the United States of America by Cambridge University Press, New York

Cambridge University Press is part of the University of Cambridge.

It furthers the University's mission by disseminating knowledge in the pursuit of education, learning and research at the highest international levels of excellence.

www.cambridge.org
Information on this title: www.cambridge.org/9781107692916

First issued in Acta Geographica (T. 8, No. 1) 1944
First published 1947
First paperback edition 2014

A catalogue record for this publication is available from the British Library

ISBN 978-1-107-69291-6 Paperback

Part VI. Human life.

Some general considerations.

THE GEOGRAPHICAL DISTRIBUTION OF THE POPULATION.

It has been said that Labrador is a little world of its own, and there is some good reason for this idea.

According to the official census this little world comprised a population of 4716 in 1935 (222). The figure however is somewhat misleading, for during at least nine months of the year, Indian hunting nomads, the so-called *Canadian Indians* (p. 609), to the number of some 280 are at work in the southern parts of the forested areas. If this number is added to the population figure for the whole of Newfoundland-Labrador the density is 0.01 persons per 1 km². These figures are of course fictitious for it would never occur to anyone to settle down for instance in the Torngak Mountains. The inhabited area is mainly restricted to the more or less forested district and the total population figure distributed over these parts indicates an area of 46 km² for each individual; but even this figure lacks reality, for a considerable number of the people live outside the forested area.

In Fig. 210 is given the approximate geographical extension of the population mentioned in the statistical reports; a dot indicates five Eskimoes or whites, a cross indicates five Indians. As it is so difficult to localise the Indians and the Eskimoes the map must be regarded as merely schematic as regards the northern parts. The permanent habitations are always at the seashore and no dwellings have been set up out of sight of the high-tide boundary. More than half of the population are found to the south of lat. 54°; cf. the notes on the map.

THE ENDEMIC GROUPINGS.

Still in our day Newfoundland-Labrador is a primeval wilderness. With the exception of the widely scattered camping-grounds of the interior and the few inhabited spots on the coast, there are very few indications that human beings live in the country. For seven or eight months of the year a belt of coastal ice

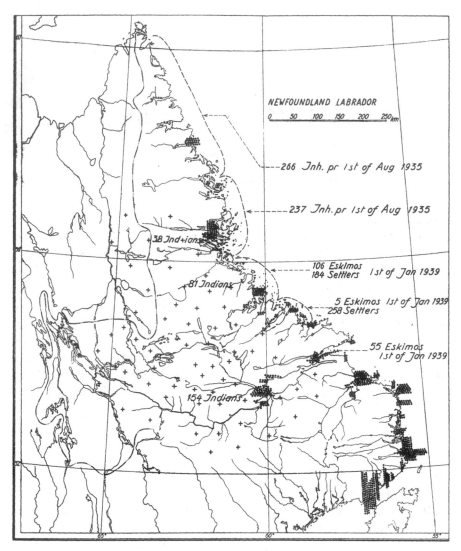

Fig. 210. Schematic sketch indicating the distribution of man in Newfoundland-Labrador according to the census of 1935 (cf. 222). A point indicates five white men or Eskimoes, a cross five Indians.

encloses the deserted land apparently almost completely separating it from the rest of the world. The people in this wilderness must then secure a livelihood from the resources of the country alone. But the land is barren and stingy; the people must strain their utmost even to exist, and this they can do only if their way of living is in agreement with the resources of the country. But these vary from district to district. Moreover the annual rhythm of nature determines the possibilities of exploiting them. The result of this is that in time different types of population have become segregated, primitive societies all of which bear the marks of endemicity. Inversely these societies clearly show the natural presumptions for their rise and the conditions for their continued existence. They show us how small groups of people can manage to live in the world-wide, subarctic land-outskirts and therefore are of important ethnological interest. They really form a little world in themselves.

THE TRANSITIONAL STATE OF THE POPULATION.

In the isolated milieu the life of these small societies has thus become set in forms determined by nature. For decades and centuries they have lived on the same lines. All fruitful impulses from the outer world have long been intuitively repulsed. The stock has also of late years remained unchanged in the main; the figures rising and falling slightly (141, 163—167, 222, 675 b, 676, 1122):

1857	1869	1874	1884	1891	1901	1911	1921	1935
1650	2150	2416	4211	4106	3947	3949	3774	4716

However, chiefly during the last hundred years stimulating impulses have found their way even to the most distant and most repulsive societies of the wilderness and contaminated them. The old societies in the country under consideration began to ferment and soon the traditional inhibitions to development were broken down one after the other; the development of the people entered upon new and unknown ways and it is still difficult to say whither they will lead if the people are left to themselves: deterioration? improvement? Labrador's people now stand at the cross-roads of good and evil.

It can be expected that the different groups will react differently to the new impulses because of the somewhat differently composed psychological structure and attitude to the tasks of life of the individuals. A considerable difference in this respect can be noticed between the three main ethnic groups of the population: the Eskimoes, the Indians and the White People. For this reason we will now consider these three elements separately. For historical reasons it is justifiable to study first the aboriginal inhabitants.

The aborigines.

The culture of primitive man is always rooted in natural conditions, and this is unusually clear in Labrador. We have for this purpose only to glance at the distribution of the ethnic groupings in the area under consideration in relation to the varying environments.

Farthest north lies a greatly cut up track of high mountains (cf. p. 330). From this m o u n t a i n t u n d r a a r e a a small strip of forestless c o a s t a l t u n d r a stretches southward along the coast towards Pinware Bay on the northern shore of the Straits of Belle Isle. Looked at from the sea this coastal tundra gives the peninsula such a poor appearance that one of the first men to sail these waters, the Frenchman JACQUES CARTIER in 1534, gave Labrador the opprobrious designation: 'the land that God gave to Cain'. With that was born the misleading conception — so difficult to eliminate — of the extreme barrenness of the whole of Labrador. The two tundra types occupy larger and larger areas of the country towards the north so that finally farthest away, beyond Ungava Bay, the forest only forms small strips on the valley bottoms of the great rivers and patches here and there in sheltered hollows.

Within the coastal tundra runs the endless t a i g a with its in parts nearly impenetrable, primeval woods of black spruce, its lakes, marshes, and heights with small bare summits. Cf. Fig. 156.

The two different habitats of Eskimoes and Indians.

The Eskimoes occupy their natural home in the bleak and rockbound coastal tundra and the neighbouring strips of wood, in the islands, the fiords, and the ocean shores with their thundering surf, their drift-ice and icebergs in the summer, and their fringe of ice where the seals sport in winter. From time immemorial the slanting-eyed 'huskies' have sat there in their igloos and gorged themselves upon seal flesh; see Fig. 211.

The Indians again harbour in the endless, soughing forest lands with their glittering labyrinths of lakes and network of rivers, their dim toned mountains and hills, the mirelands and the scattered patches of 'land without trees' as the tundra is called by the aborigines. In the great forests and on the bare crests in the woods the sharp-eyed Indian hunter is on the look-out for game; see Fig. 212.

The feud between Eskimoes and Indians.

It must not however be thought that the natural limits have been in any way respected as a fixed separation between the spheres of interests of the two races. The Indians have a tradition that when *Tcè mentu*, the Great Spirit, (a conception which the Indians probably incorporated in their spiritual ideas after the arrival of the Catholic Mission), wished to give people a clear sign of what is right in Labrador he created the two different natural areas: the forest and the coastal tundra. He decided then that all the forest land should belong to the Indians, while the Eskimoes should keep out on the barren tundra strips along the coast. But the great spirit of the Eskimoes, *tornarsuk*, did not speak the same language; he told the Eskimoes that they could take what they considered they needed. For this reason it has come about that also in these endless wildernesses the boundaries were regulated in accordance with the interests of the stronger. Every attempt made by the Eskimo to force their way into the Indians' hunting-ground in the taiga was considered a grave violation of a holy commandment which demanded restitution, started feelings of revenge in the Indians which had their natural consequences among members of the other race. This tradition is probably artificial and scarcely older than the last Eskimo invasion (see farther on), but in any case it throws light upon a condition which at first glance is astonishing, that even in these deserted places where mankind is a rarity, racial enmity to the death has prevailed for a long time between the Indian and the Eskimo, an animal, inherent hate, originating from a primitive instinct just as that between the wolf and the lynx in the same country; wherever an Indian met an Eskimo the Great Spirit bade the one kill the other.

The spear many a time decided where the boundary should be, and in the course of time streams of blood have flowed here. When ROGER CURTIS in 1772—73 travelled along the Atlantic coast he was struck by the irony of the fact that the comparatively few tribes living there should be so set upon exterminating each other (196); they were at perpetual war with each other. Between the Mountaineers and the Esquimaux there subsists an unconquerable aversion; the Esquimaux live always upon the sea-shores, from their dread of the Mountaineers. Conditions corresponded to those of our day in northern Canada where it is said that a redskin does not dare to enter the *huskies'*, i.e. the Eskimo's land. At the beginning of the 19th century the Indians and the Eskimoes south of Hopedale sometimes meet, reports KOHLMEISTER (532), but as the Hopedale Eskimo seek to cultivate their friendship, quarrels and bloodshed never occur. In Ungava, on the other hand, though they often exchange tokens of friendship they were apt to give way to their natural jealousies, and provocations being aggravated their meetings terminated in murder. The

Eskimoes were much afraid of the Indians, who are a more nimble and active race. When Hudson's Bay Company's factory Fort Chimo was set up in 1831 it was considered advisable to surround the post with a defensive palisade to prevent surprise if the clients who met there should begin to fight, as might be expected at any minute seeing that war had so recently ceased and the feelings of hate had not yet died away (963).

Pacification has proceeded very slowly. The open war between the Indians and the Eskimo ceased near the Atlantic coast for a reason which at first

Fig 211.   The land of the Eskimoes. Looking over Paul Island in the Nain archipelago. In the background is the open ocean, in the foreground rocky islands are seen with some small patches of black spruce forest, mostly only brushwood, in valleys and sheltered clefts. Aerophoto the FORBES-GRENFELL EXPEDITION, 1931.

seems rather surprising. The practical British traders succeeded in gradually bringing about a kind of peace, at least on the business premises. As fighting was injurious to the commercial interests the managers at the Hudson's Bay Company's posts made the enemies understand that the posts would be withdrawn and the natives lose the advantages the trade gave them if they did not make peace. And this the latter soon grasped, finding it better to catch fur-bearing animals and exchange their skins for all kinds of useful products than constantly to destroy their possessions and their families by continuing the blood feuds. Further the Indians probably realized that the Eskimo, if necessary, could now depend upon the moral support of the white trappers who had married Eskimo women; this was a handicap for the Indians.

The Moravian Mission also worked hard to create peace and have earned the thanks of many by modifying the racial hate of the Eskimoes for the Indians. The first most remarkable sign of appeasement between the rival parties

Fig. 212. The land of the Indians. Looking south-east over the wilderness of forest on the southern part of the Labrador Lake Plateau. Aerophoto the author, Aug. 2nd 1937.

occurred as far as is known in 1843 when a starving band of Montagnais appeared at Hopedale. Some decades earlier the Eskimoes had completely exterminated them, but now they received them with friendliness, took them into their huts and supplied them with food (325). In 1855—60 the Indians suffered terribly from lack of food. Many parties of them were brought by the Eskimoes to the Mission Stations where they were helped. On an other occasion Indians rescued a group of Eskimoes, who having been unsuccessful in their salmon-fishing, were dying of starvation (325).

However, still in our days certain reflexes from the old fighting period are visible in the distant parts. FLAHERTY for example says that *Adlit*, the

Indian, even to this day is dreaded by the Eskimo in the country north-west from Ungava Bay. HUTTON tells that Eskimoes still in his day, at the beginning of our century, were scared at the word 'Indian'; when he once travelled southward from Okkak his drivers asked him in awestruck voices: 'Shall we see the Allat (= Indians) (476)?

It should be specially remembered that the white men have also brought a blessing to these primitive peoples, for it was their common sense which stopped the meaningless feuds in Labrador.

At the present time the Eskimoes and the Indians live side by side in perfect peace. There is, however, no attraction between the both races and mostly no open intercourse. The Indians say that they cannot bear the smell of the Eskimo; on the other hand the Indians themselves are so impregnated with the smell of the camping fires that the Eskimoes cannot bear them! Yet, the Indians do sometimes take a steam bath, whereas an Eskimo never in his life time gets rid of his bodily dirt knowingly and willingly. Yet they do not love each other. It is said that they scarcely ever intermarry. The Rev. PAUL HETTASCH, the Moravian missionary, which lived in the country for fourty years told me that he knew of only one mixed marriage, namely the parents of EDWARD RICH whose father was a Naskaupee Indian and his mother an Eskimo; the Riches now form a kind of clan which usually lives apart from the others. On the other hand the aboriginal women are not unwilling to marry the white men.

Their very way of living, their thirst for battle and their dislike of each other show that the aborigenes have had a vague apprehension of their ethnological differences and behaved as civilized people are in the habit of doing in the present period of nationalism. Their somatic characters emphasize the differences.

Anthropometrical characteristics of Eskimoes and Indians

Thanks to different measurements dating back to 1880 and especially to HALLOWELL's pioneer study of 1929 and the measurements secured by W. D. STRONG during the Rawson-MacMillan Subarctic Expedition of Field Museum in 1927—28 among the Eskimo and a small group of Indians living on the north-east coast of Labrador, relatively reliable knowledge of the *living* aborigines is available (cf. especially 86 a, 823 a, 694 a, 152, 230, 231, 752, 897 a, 982, 963, 965, 80a, 906).

Anthropometric characteristics of the Newfoundland-Labrador Aborigines:

Eskimoes		Indians	
male	female	male	female

S t a t u r e:

(58) 158.3 [St]	(78) 148.3 [St]	(10) 164.6 [St]	(7) 153.3 [St]
(11) 157.7 $\begin{bmatrix}\text{L, So,}\\\text{P, V}\end{bmatrix}$	(10) 149.7 $\begin{bmatrix}\text{L, So,}\\\text{P, V}\end{bmatrix}$	(41) 166.2 [H]	(29) 154.6 [H]
(37) 157.0 [D]	(22) 150.5 [D]		

Head Length in mm:

(58) 192.17 [St]	(79) 185.04 [St]	(11) 189.4 [St]	(7) 184.1 [St]
(37) 192.89 $\begin{bmatrix}\text{L, So,}\\\text{P, V}\end{bmatrix}$	(21) 189.48 $\begin{bmatrix}\text{L, So,}\\\text{P, V}\end{bmatrix}$	(67) 194.5 [H]	(54) 187.3 [H]
(11) 191.2 [D]	(10) 190.2 [D]		

Head Breadth mm:

(58) 148.31 [St]	(79) 142.26 [St]	(11) 144.1 [St]	(7) 144.0 [St]
(37) 151.49 $\begin{bmatrix}\text{L, So,}\\\text{P, V}\end{bmatrix}$	(21) 143.72 $\begin{bmatrix}\text{L, So,}\\\text{P, V}\end{bmatrix}$	(67) 156.8 [H]	(54) 151.7 [H]
(11) 147.6 [D]	(10) 141.8 [D]		

Cephalic Index:

(58) 77.28 [St]	(79) 76.94 [St]	(11) 76.1 [St]	(7) 78.2 [St]
(37) 78.61 $\begin{bmatrix}\text{L, So,}\\\text{P, V}\end{bmatrix}$	(21) 75.88 $\begin{bmatrix}\text{L, So,}\\\text{P, V}\end{bmatrix}$	(74) 80.60 [H]	(58) 81.10 [H]
(11) 77.0 [D]	(10) 74.5 [D]	(79) 81.43 [B]	

Cephalo-Facial Index:

(58) 95.66 [St]	(79) 93.78 [St]	(11) 98.6 [St]	(7) 93.2 [St]
(37) 95.77 $\begin{bmatrix}\text{L, So,}\\\text{P, V}\end{bmatrix}$	(21) 94.35 $\begin{bmatrix}\text{L, So,}\\\text{P, V}\end{bmatrix}$	(67) 94.0 [H]	(54) 92.50 [H]
(11) 96.3 [D]	(10) 96.3 [D]		

Nasal Index:

(58) 66.98 [St]	(79) 62.54 [St]	(10) 68.9 [St]	(7) 68.6 [St]
(37) 73.81 $\begin{bmatrix}\text{L, So,}\\\text{P, V}\end{bmatrix}$	(22) 72.77 $\begin{bmatrix}\text{L, So,}\\\text{P, V}\end{bmatrix}$	(44) 73.0 [H]	(29) 74.8 [H]
(10) 64.1 [D]	(4) 62.4 [D]		

Skin Color (v. LUSCHAN scale):

(52) 3—9 30.8 % [St]	(78) 3—9 2.6 % [St]	
10—13 63.5 % »	10—13 92.3 % »	
14—17 5.7 % »	14—17 5.1 % »	*Skin Color*

(10) 3—9 0 % [St]	(7) 0 % [St]
10—13 50 % »	57.1 % »
14—17 50 % »	42.9 % »
(53) 3—9 11.3 % [H]	(35) 2.8 % [H]
10—13 62.3 % »	68.6 % »
14—17 26.4 % »	28.6 % »

The figures in parenthesis indicate the number of measured individuals. Abreviations: D = Duckworth (230, 231), H = Hallowell (394), L = Lee (see 152), P = Pittard (752), So = Sornberger (897 a), St = Strong (897 a), V = Virchow (982).

In v. Luschan's scale 1—2 signifies light; 3—9 medium, 10—13 dark, 14 and all thereafter still darker.

As these measurements were collected from only a few individuals they are not fully reliable scientifically, for a few large families may easily have a greater influence on the average values than they really should have. Yet the above figures and general experience go to show that the Eskimo are not so tall as the whites and the Indians. The Eskimoes are somewhat more dolico-cephalic than the Indians. Both are coloured but the complexion of the Indian is always darker. In general it is not difficult to know at first glance whether one is dealing with an Eskimo, Fig. 213, or an Indian, Fig. 214. Cf. (230, 231, 262 a, 394, 651 a, 694 a, 752, 823 a, 829 a, 849, 897 a—897 c, 910, 982, 1050).

Fig. 213. The Labrador Eskimo type. Photo the author at Hebron, July 1937.

CHANGE OF THE ESKIMO TYPE.

It is an interesting fact that a change must have occurred in the Eskimo type (80 a). The bearers of the old Thule culture (see later on) at Naujan on Repulse Bay had a height of 161—162 cm and a cranial index of 72.5 (262 a), whereas this index at Southampton Island was found to be 74.2; evidently the material from this latter place is too slender either for reliable calculations or for chronological determinations (80 a). The average index of 21 skulls from old graves is put at 71.8, whereas of 12 skulls from more recent graves is 72.6 (978 a). This indicates that the old Eskimoes of the eastern area (including perhaps Labrador) were decidedly dolichocranic.

The cause of the differences between past and present indices is uncertain. Variations often come about with hygienic or economic variations. STEWART (897 a) attributes the change in head shape with a possible decrease in stature to a change in diet. There is a growing body of evidence from other racial groups showing that both these characteristics are rather easily changed when environment, and especially nutrition, is altered. KROGMAN (1938), who summarized the literature on this subject, says that attention must be paid to a factor or a set of factors that is as difficult to evaluate as it is to describe: the environment, whatever connotation this term may have. The combined effects of disease and under-nourishment may result in a stunting of the presumably 'racial' growth pattern. Here many important problems present themselves.

The psychological characteristics.

Psychological differences are very soon noticed between the two peoples. The Eskimoes have developed far more in this respect than the Indians; their intellect is active, their receptivity great; they are lively, and their behaviour is friendly; at our first meeting I spontaneously called them *the smiling people*, and I still find them so even after the first enchantment has faded; see Fig. 227.

The Indians on the other hand are mostly sullen and sluggish, either reserved or noisy, and impress one as being less accessible and 'stiffer' than the Eskimoes; they are often very reserved towards strangers; the Indian only reluctantly approaches a newcomer, while the Eskimo gladly does so. No scientific research regarding the intelligence stage of the natives of Labrador has been made, as for instance in Alaska (cf. H. DEVEY ANDERSON and W. CROSBY EELLS: Alaskan Natives. Stanford Univ. Press 1935). However, it seemed to me easier to understand many of the characteristics of their mental development if one imagines that the Indian on an average is at the stage

Fig. 214. The Labrador Indian type. Photo E. H. KRANCK, at North West River, July 1937.

of a north European youth at fourteen, the Eskimo at eighteen in the best instances.

The distinctive race characteristics of the two peoples and the persistent differences in their way of living, their equipment and their language indicate an ethnic differentiation at a very distant period. Where and how this took place is still a mystery, but in any case their common home must be sought outside Labrador.

The origin of the aborigines of Labrador.

Many points about the Eskimo make them unique as a people. They are the only permanent conquerors of the Arctic. The most widely-spread race in the world, they are scattered along some five thousand miles of ice-bound coast, and everywhere they use practically the same language, have practically the same customs and, until quite recently hunted the same quarry with similar weapons; this they have done from prehistoric times. Their dress, food, dwelling-places, weapons, and habits are the best that could be evolved for their environment. It is evident that long ages were required to develop the peculiarities of their culture and accomodate them to their habitat. No wonder the question of their origin has so long remained a problem; only in the last decades has it been possible to find some keys.

It is not the right place here to go more deeply into these intricate questions, and the extensive literature which has been published on the subject. Yet considering the important interest which the aborigines in Labrador — so to say at the outer edge of the continent's periphery — have for ethnology it may be useful for future explorers here to review shortly some of the conclusions drawn from these people's anthropological indices in conjunction with the remains of their material culture found on the peninsula and the surrounding country. In this way it may be also possible to give the Labrador peoples their place in the scientific framework of today.

ON THE ORIGIN OF THE ESKIMOES.

Some resemblances have been found between the racial characteristics and the material culture of the late-Palaeolithic reindeer hunters of Central Europe and of the Eskimo (cf. SOLLAS, BOYD DAWKINS, KNUD RASMUSSEN, BIRKET-SMITH). From this the supposition developed at an early date that when the European ice-caps began to withdraw and the inland-ice margin slowly receded towards the north and the wild reindeer followed it, the Magdalenian hunters were compelled to follow their principal game towards the north. In this way they arrived at the coast of the Arctic Ocean and there spread along its shores eastward, finally arriving at Arctic America and Greenland. According to this supposition the Eskimoes are descended from the late Palaeolithic hunters who once lived between the Nordic and Alpine ice-sheets in Europe (cf. 87).

This theory has had to be abandoned; no unbroken connection can be proved in north Eurasia between the Cromagnon race to which these hunters belonged and the Eskimoes. On the Asiatic side of the Bering Strait — e.g. in the territory where the Chukchi live —, conditions do not confirm the assumption that the anthropological relics there must belong to the rearguard of the Eskimo race as supposed by MONTANDON. They may as well represent a local and temporary expansion of the Eskimo area in America (80 a).

On the other hand, some distant relationship must undoubtedly be recognized between the Eskimoes' material culture and that of the bearer of the so-called 'Arctic stone age' culture in northern Fennoscandia (915). Both these cultures are distinctly coastal and have similar types of tools and dwellings. I imagine provisionally that they have radiated from an unknown centre which now lies drowned on the shelf north of Asia.

When we speak of the Eskimoes we unconsciously think of three characteristics: race, culture, and language.

Anthropological indications.

On a purely anthropological basis the present day tendency is to see the
original home of the Eskimo in America. Close conformity has been found, for
instance, between the Eskimoes and the Indians living near Lake Athapasca. It is
further undeniable that anthropological similarities are remarkably great between the Eskimoes on the one hand and both the Cree and the Chipewyan Indians
on the other (80 a). Now it is uncertain whether both these groups are parts
of a great racial unit of the northern woodland Indian people, or a local Indian-
ised Eskimo survival.

This supposition requires for confirmation some correlation between the
somatic types in the Eskimo area, but according to BIRKET-SMITH none has yet
been confirmed (80 a). Yet, disregarding all slightly deviating details, he found
for example that the stature decreases from the Mackenzie 'hyper'-Eskimo type
both towards the east and the west. As regards the Eskimo of earlier times, the
somewhat scanty skeletal material available leads one to suppose that the
Thule Culture Eskimoes at Naujan rather closely resembled the present population at Point Barrow. This fact agrees with the similarity between the ancient
Thule Culture and the present culture in northern Alaska. If these similarities
can be confirmed it means that on the whole we can point to no essential differences between the bearers of the Thule Culture and the basic Eskimo type.

There is undoubtedly an old Europid long-headed type in America, and its
occurrence in the Algonkian stock has special significance when the likeness
between the Eskimo and the Cree are considered. The Eskimo race is also
thought to have arisen from the interbreeding of a long-headed race and the
brachycephalic Mongolid which later immigrated from Aisa. This possibility
cannot be dismissed, but it has still to be proved (871, 80a).

Many unknown factors have contributed to make-up the Eskimo tribes and
it will be an interesting task for the future to distinguish their various racial
types and prove to which one the Labrador Eskimo belong. It may be that such
types will reveal something about their tribal migrations, etc. too. Yet there seem
to be only slight differences within the eastern groups. I think we must agree
with BIRKET-SMITH, that it will be wiser to ignore speculations and merely say
that the basic type of the western, central and eastern Eskimo is the same, apart
from the changes wrought by the varying amount of admixture with foreign
elements.

Cultural indications.

Excellent descriptions of the distinctive material culture of the Labrador
Eskimo about the turn of the century in the Ungava district are given by TURNER
(963) and for the Atlantic Coast by HAWKES (424). CARTWRIGHT (134) tells of
the conditions at the end of the 18th century (cf. the Bibliography). On the
other hand little is known of the development of the material culture and the

29

waves of people who continually brought new cultural additions to Labrador. Few archeologists have done research work in the country and therefore I can only here present some main characteristics.

Recent Eskimo culture seems to have reached the north-east part of the country in almost the fully developed form revealed by the 18th century ruins. Considering that in all the sites investigated European objects are found (70 a), it appears that the Eskimo have not been there more than some 400 years. Although it must be borne in mind that iron may have been furnished by the Norsemen as early as 1000 A.D., or by the inter-tribal trade of drift-objects (800, 391, 387 a, 210).

The recent Eskimo culture.

Now, this recent Labrador culture has one very conspicuous trait. Before it was altered by contact with the civilization of the whites, it had been fairly closely related to that distributed over the central Arctic of America (635, 897 a). More of it seems to have been preserved in Labrador than in the central region proper, (dwellings, snow house, sledge, hunting implements, clothing, lamps, square cooking pots). Some remains (whaling-harpoons, woman's boats, etc.) show that the Labrador Eskimo inherited a good deal more from the Thule culture (see farther on) than their western neighbours. In addition they are coast dwellers and their life is less nomadic than that of the central Eskimo; yet they remain more closely related to the latter than to the Thule culture (635).

The Thule culture.

The original T h u l e c u l t u r e was once spread over the arctic of America, having its centre in the Hudson Bay region and stretching eastward into Greenland. In Labrador it has lasted down to modern times (920) and it appears probable that its bearers were the so-called *Tunnit*. On Southampton Island in Hudson Bay it is known to have survived even as late as 1902 (635). MATHIAS-SEN has described this culture in detail, showing that it has all the fundamental traits of the recent Eskimo material culture, and some explorers have considered it the original Eskimo culture of the eastern Arctic of America (635).

The Dorset culture.

Remains have also been found of another culture, but this disappeared long ago: the Cape D o r s e t C u l t u r e. JENNESS first discovered and described this culture (489); later it was found in scattered districts throughout the eastern Arctic (490). On the southern coast of Labrador some remains were found by WINTEMBERG in the district of Blanc Sablon-Bradore (1041, 1042, 1044). It may be an early phase of Eskimo culture. On the southern sites it was found that relics of the Thule culture were conspicuously absent and it may therefore be considered quite independent.

Yet in many of the Dorset sites a relationship can be discovered between it and those of the Indians and Eskimoes. JENNESS has pointed out that the Dorset culture shows unmistakable Indian affinities, particularly with the Beothuks of Newfoundland and the prehistoric 'Red Paint' (490), and MATHIASSEN (640 a) and COLLINS (1937) have suggested that the Dorset group may have been of Indian origin. The correct explanation may be that this Dorset culture worked northward and became at a later stage still more influenced by the Eskimo. The Thule culture is in this way in some places contemporaneous with the Dorset culture, but in others succeeds it and has probably extinguished it. It is easy. to understand that with the advent of the better equipped and more aggressive bearers of the Thule culture the Dorset people were forced to give way and gradually succumbed (897 a).

STRONG's stone culture sites in eastern Labrador.

Yet there is in Newfoundland-Labrador evidence of a still older culture. Between Hopedale and Nain, STRONG investigated three old sites (906, 1043) of which the stone tools (large flints, arrow-heads, chipped scrapers, crescent-shaped knives) differ from both the Eskimo relics and the Naskaupee Indians' bone and wood culture. They also possess some unique artefact types: gouges, a ground chisel and oval celts. Some isolated finds of stone adzes are also conspicuous. The stock of tools gives the impression of Indian affiliation, and STRONG considered them to be the remains of an old Indian population of long ago — long before both the Eskimoes and the Naskaupee; two of the sites he assumed were of Pleistocene age. Observations which I myself made on Satosuakh in the Nain archipelago seem to agree very closely with this assumption; the last mentioned site being of early post-Glacial age.

There is very little probability that these sites have anything in common with the Dorset culture or that of the Eskimoes, for they contain scarcely anything of the materials characterizing the Labrador Eskimo sites and the Thule culture: artefacts of bone, antler, ivory, or steatite. Moreover these old stone culture sites lack entirely the surface indications or the abundant bone debris that mark the Eskimo remains. It seems therefore most probable that this complex represents an old Indian culture.

It is moreover impossible to decide whether this culture has anything in common with the remains of an old Indian culture in the southern part of Labrador: the clusters of chipped points and flakes at the head of l'Anse du Diable (584), numerous spear and arrow-points and parts of highly polished stone gouges and sherds of pottery with roulette decorations like those in the Algonkin sites in Ontario discovered by WINTEMBERG, near Blanc Sablon and Bradore, and the workshop with strongly patinated flakes but no pottery or bone near Tadousac which SPECK believes to be very old, proto-Algonkian or possibly related to the Beothuks.

In the present state of our knowledge one is disposed to conclude that in Labrador there have been different waves of invasion: (1) a very old Indian [?] layer, (2) the Dorset culture, (3) the Thule culture (4) the recent Eskimo culture, and in very recent times (5) that of the Naskaupee and the Montagnais Indians.

Theories concerning the origin and migrations of the Eskimoes.

Several theories regarding the origin of the Eskimoes and their migrations have been put forward on the basis of the archaeological and ethnological observations made by American and Danish explorers. Space permits of the mention of only some of their salient features; the more so as the stage of speculation has not yet been left behind; the deeper we delve into Eskimo race history the clearer becomes our ignorance, as BIRKET-SMITH, one of the most prominent Eskimo experts, stated even in 1940.

Most explorers are of the opinion that the Eskimo culture rose in the regions round about the great lakes and rivers of northern Canada. Not very long ago, having detached themselves from the Indians whose development led them to the wooded lands, the Eskimoes occupied land on the barren tundra. Later on they left that part in the wake of the enormous spring trek of the caribou and travelled until they arrived at the ice-bound storm-beaten coasts of the North-West Passage. Here, according to KNUD RASMUSSEN for example, the whole of their material culture became gradually adapted to the ice of the Polar Sea and the big game of the coastal waters. This meant a revolution. The salmon spear became a seal harpoon, the open canoe of the rivers was transformed into the covered kayak, the stout one-man craft which sails the open sea in every kind of weather. They built the umiak for their hunting trips, a strong transport boat with a wonderful capacity. They made winter houses of stone, turf, and whalebone or of wind-blown snow. From the bones of marine animals, the tusks of the walrus and the narwhal and the antlers of the caribou, they worked out ingenious hunting weapons (cf. 635, 79, 489). The seal provided them with food, clothing, and coverings for boats and tents, and its fat was converted into light and heat for the winter-houses. The bearers of the Thule culture lived in small settlements forming communistic societies without chiefs, even though they recognised the advantage of bowing to the word of the best hunter or the strongest man (779).

After their arrival at the ocean shore the tribes moved slowly westward as far as the shores of the Pacific. In course of time the Alaskan group gave rise in its turn to a new type of culture: the whale-hunting or Thule culture. Later this spread eastward, all over Arctic America into Labrador and elsewhere. When Eric the Red arrived in Greenland five hundred years before Columbus went to America, that country had been visited and inhabited long before by the bearers of the Thule culture.

Later, according to this theory, a second group moved out of the central region to overcome the Thule people and to become the present-day Eskimo, the 'Eschato-Eskimo' (75, 76, 77, 78, 79, 80).

According to BIRKET-SMITH the small remnant of the Caribou Eskimoes (70 a, 74 a) in the interior westward from Hudson Bay is a relatively unchanged group of the population from which all the other Eskimoes arose.

MATHIASSEN (635, 639, 64 a) on the other hand, regards the Thule culture as the original Eskimo culture, the first to be spread eastward over the Arctic coast of America. To him the Caribou Eskimo are descendants of Thule people who went into the interior and gave up many of their former customs.

The most modern theory of this problem has been put forward by JENNESS (493, 494 a). Some time apparently in the first millenium A.D., Eskimo tribes from Arctic Alaska began to spread eastward, dropping settlers all along their route. Some families hugged the mainland and continued to Hudson's Bay; others scattered over the islands to the north and eventually reached Greenland.

Meanwhile, other and more primitive Eskimo roaming the hinterland of Hudson Bay sent out colonists to the coast of the eastern Arctic: Ellesmere Island, Greenland, the coast and islands of Hudson Strait and Labrador, or they traversed the heart of the Labrador peninsula and took possession of the northern headland of Newfoundland. Whether this movement preceded or coincided with the eastward movement of the Alaskan Eskimo is not settled: JENNESS suspected that it started several centuries earlier.

In places where the two peoples subsequently clashed the western Eskimo gained the mastery. There is reason to believe that they differed not only in culture but in physical type from the eastern Eskimo, both those who remained inland and the 'Dorset people' who settled on the coast. The eastern natives seem to have acquired the features of the neighbouring Algonkian people with whom they intermarried for many centuries.

About A.D. 1200 a new impulse can be detected surging through the Arctic. The Indian-like Eskimoes behind Hudson Bay began to stream seaward, to the Arctic coast northward and eastward. Gradually these newcomers swamped the inhabitants — both the Thule people and their own kinsmen of the Dorset culture — until they held undisputed sway from Coronation Gulf to Labrador; only a few Thule people managed to survive until the beginning of our century on Southampton Island. The Dorset culture disappeared before the arrival of Europeans. The majority of the seal- and whale-hunting population on the rising islands in the far north made their way to Greenland, where they may have helped in overwhelming the Norse settlements; others retreated to the mainland, only to be submerged by the tide of Eskimoes from the interior.

JENNESS concludes: we are now in a position to understand why the present-day Eskimoes of Canada fall naturally into three divisions. (1) The natives of the Mackenzie River delta (and also, until 1902, the inhabitants of Southampton Islands) are descended from some of the old Thule people. (2) The 'Caribou Eskimo' on the Barren Grounds behind Hudson Bay represent the survivors of the second great reservoir of the race, the inland Eskimo, now dwindled to a fast vanishing pool. (3) The Eskimo who flowed out of this inland reservoir about A.D. 1200 and overwhelmed the earlier coast dwellers now occupy the whole coast-line from Coronation Gulf to Labrador and in their new environment have gained a fresh lease of life and vigour.

JENNESS' theory applies especially to the eastern Arctic of America. In Labrador we still find Group 3, whereas Group 2, which perhaps represented the so-called *Tunnit*, died out a few centuries ago.

Each new discovery has generally proved to introduce modifications of the still comparatively young theories of the origin of the Eskimoes based upon archeological findings. It would therefore be surprising if the last word has yet been said on this question. For example does not the I p i u t a k c u l t u r e — at least 2000 years old — and the skeletons of its bearers discovered at Point Hope in Alaska give promise of quite new points of view in this regard. It will be best to await the results of HELGE LARSEN's investigations before we decide to support the one or the other theory. Cf. (492, 493, 516, 638, 639, 640, 829, 886, 887).

THE ORIGIN OF THE INDIANS

still lies in complete darkness and there is no possibility of systematically dealing with this question here. STRONG's discovery (p. 451) certainly appears to show that Labrador had an (Indian?) population already before there were any Eskimo in the country. This is also indicated by my observations on Satosuak in the Nain archipelago where the findings are probably derived from a period several thousands of years before our era. We shall later on return to the spread of the Labrador Indians in historic times (p. 578 sqqq.) and touch upon some theories regarding their separating out from the great body of prehistoric population of America (p. 697 sqqq.).

The study of the Labrador Indians offers undoubtedly one of the most grateful and important tasks which the peninsula has for a trained ethnologist.

The Eskimoes or Innuit.

T h e p r o v e n a n c e o f t h e d e n o m i n a t i o n. The Eskimoes never call themselves by this name; they use the designation i n n u i t (plur. of i n n u k = person), which simply means 'people', or 'men', calling the rest of mankind 'the foreigners', 'Kablunat', 'Kablunael' or 'Kablunak' (936) which is untranslatable. Translations such as 'inferior beings' (387 a) or 'sons of dogs' (761) are not exact; only later was this applied to white men following BOAS. HAKLUYT in 1584, in his *Discourse of Western Planting*' seems to have been the first European writer to use the name Eskimo in the literature; his form was »E s q u i m a w e s». *Ash'ki'mai* is the Cree Indians nickname for them and signifies: raw flesh eater; the Montagnais Indians of Labrador call them

Iees'te'mu which means the same thing. One point to explain why the Eskimo will eat raw meat while the Indian never does, is cited by SPECK (871): The former is sometimes prevented by nature from preparing it by means of fire, as e.g. when game has been killed on the sea ice, whereas the Indian game inhabits the timber, where fuel is obtainable.

The Newfoundland fishermen call them H u s k i e s or H u s k i m a w s. They also bear the name of *Carolit, Karalit* (434, 325) in older documents; this is used in part of Greenland and has perhaps been transferred by the Moravian Missionaries to Labrador; it has been argued that it has had something to do with the term *skræling* in the Icelandic sagas. We use the name Eskimo here because it is the most general.

The Census of the Eskimoes in Newfoundland-Labrador.

As the area inhabited by the Eskimo has been considerably reduced in historical times (p. 479) it is psychologically comprehensible that the opinion general both among the learned and the laity likes to assume that the number of them has also declined. PACKARD (716) calls the Labrador Eskimo a doomed race and this has been stated by many others (e.g. 364, 321). KOCH points out that while the numbers of the settlers is increasing the Eskimoes are decreasing because of infant mortality; the settlers are moving farther and farther to the north (528). STEWART notices (897a) that in spite of fluctuations the general trend of the Labrador Eskimo population is clearly downwards. Dr PADDON writes (731) that although there has been a slight rise in number since the great 'Spanish Flu' epidemic in 1918 (cf. table, p. 458), it is doubtful if there is really enough healthy bloodstock to recuperate the race. However, research work on this question has not been very satisfactory. The population figures obtainable do not go very far back in time, and added to that it is impossible to judge of the reliability of the older ones.

No trustworthy evidence is obtainable as to the number of Eskimoes in olden times. The Moravian Mission estimated their numbers at 3000 in 1763, CURTIS in 1773 from Straits of Belle Isle northwards to Ungava Bay at 1625 persons. What figures are obtainable have been summarized in the following table and given here for what they are worth, but they do not allow the question to be settled. As to the variation of the population the curve shows slightly erratic movements for the individual years.

The most reliable figures I know regarding the Mission's stations have been collected with great trouble by the Rev. PAUL HETTASCH who was

Statistics of the Eskimo Population in Newfoundland-Labrador.

Geographical groupings	1763[3]	1764[33]	1773[1]	c:a 1800[23]	1813[2]	1821[33]	1828[33]	1835[33]	1836	1840[2]	1850[2]	1856[2]	1857	1860[2]	1863[32]	1866[10]	1870[10]	1871[14]
From Nachvak north into Ungava Bay	210																	
Killinek (founded 1904)																		
Nachvak [Niuchvak]	60																	
Ramah [Noolaktucktoke] (founded) 1871, abandoned 1890)	30																	
Lamson Bay [Chukleluit]	40																	
Saglek Bay [? = Chuckbuk]	140																	
Hebron [Kanmuklookthuok] (founded 1830)	345							148	148	179	346			306	313			
Napawktoot	70																	
Okkak [? = Kunedlooke] (founded 1776)	360			48[35]	394	255	394	251		352	408			314	327			
Nain [? = Nanyoki] (founded 1771)	100			63[36]	168	168	232	278		298	314			277	275			
Zoar (founded 1865, abandoned 1890)																		
Hopedale (founded 1782)				51[37]	149	149	176	194		205	229			241	248			
Makkovik (founded 1896)																		
Rigolet [Karwalla]													10 f.					
Belle Isle [Obuctike]	270																	
Total	1625	c:a 400	c:a 3000		×711					1034[4]	1297[5]	c:a 1300[31] c:a 1200[36]		1138[6]	1163[14]	1048[11]	c:a 1200	1324

Geographical groupings	1874[10]	1880[2]	1881[20]	1884	1886[15]	1887[16]	1890[2]	1895[3]	1904[10]	1904/05[19]	1905[20]	1906/07[18]	1908[24]	1912[12]	1916[13]	1926[25]	1933[22]
From Nachvak north into Ungava Bay												87[40]					
Killinek (founded 1904)										44	48	27[41]					
Nachvak [Nuchvak]											c:a 30	69[41]					
Ramah [Noolaktucktoke] (founded) 1871, abandoned 1890)		44			71		59										
Lamson Bay [Chukleluit]																	
Saglek Bay [? = Chukleluit]																	
Hebron [Kanmuklookthuok] (founded 1830)		202			207						183	180[41]					
Napawktoot																	
Okkak [? = Kunedlooke] (founded 1776)		329			308		350[33]				350	329[41]					
Nain [? = Nanyoki] (founded 1771)		282			214		263				270						
Zoar (founded 1865, abandoned 1890)		315			90		89										
Hopedale (founded 1782)		130			160		331				250						
Makkovik (founded 1896)											150				35		
Rigolet [Karwalla]																	
Belle Isle [Obuctike]																	
Total	1176[13]	1302[7]	1150	1365[27]	1050	×1500	1092	1290[9]	1450[17] 1330[13]	34	1281[31]	34–40	1200[34]	1216[34]		1300[34]	c:a 1300[34]

kind enough to allow me to use them here (Tab. p. 458). It must be remembered that the Karwalla or Rigolet Eskimoes (about 50 individuals) are not included.

It is not easy to judge from these figures whether the number of Eskimoes in Newfoundland-Labrador has any real, intrinsic tendency to decline as is mostly believed. When severe epidemics rage these people, like other aborigines, can decline in numbers by 30 %, as they did after the 'Spanish sickness' (influenza) ravages of 1918—1919 in Labrador; of 310 in Okkak 216 died in a few days. But such sudden drops are usually soon compensated by nature herself and this seems rather to be confirmed by the curve. BELL's observation that the Christian Eskimo could not be compared with the others is easily understood: the children of the igloo are differentiated at an early stage by the hand of death so that only the strongest survive the age limit of infant mortality. — Yet so many different factors come into play here that each of them must be specially judged before any reliable answer can be given to the question:

Are the Labrador Eskimoes a vanishing race?

The history of the primitive savages has mostly been: slowly retreating from the outposts of civilization and dying from its diseases (614).

When we look at the tables above it is evident, even taking into account some uncertainty in the primary figures, that the stock of the Eskimoes as a whole has considerably diminished during historical times, beginning about 1775. From this it has been inferred that they are a vanishing race: under civiliza-

---

1) CURTIS. 2) Moravian Mission's estimate (Hawkes). 3) The total population of the coast. 4) North of Hebron emigration towards Ungava. 5) In 1842 influenza; seal scarce; 59 persons died from epidemics and starvation in 1855. 6) In 1857 dogs and wild game infected by mysterious disease, animals died in large numbers. 7) Eskimoes and settlers. In 1876 whooping-cough, over 100 died. 8) BELL. 9) Calculated at 5 persons per family; including also the ca. 35 Karwalla Eskimoes and south of Hamilton Inlet some 5 Eskimoes. 10 MAC GREGOR. 11) practically no settlers. 12) + 115 settlers; PACKARD following HATTON and HARWEY noted 1700 Eskimoes. 13 1018 Christians + 31 pagans (1 in Hebron, 30 in Eclipse Bay +280 settlers); in Nain in 1900 80 of the 350 Eskimoes died of typhoid fever introduced from Chicago. 14) BELL: in the 1860's 1400 Eskimos at the Moravian stations. 15) BELL; PACKARD following LA TROBE. 16) Between Hamilton Inlet and Ungava, PACKARD. 17) BELL: at the mission stations 1400 + northwards to Chidley 50. 18) HANTSCH. 19) All appertaining to the Ungava group. 20) MC GREGOR: tendency to settle southwards, lack of firewood. 21) In Ckkak influenza killed 65 Eskimoes in 1904; the stock still increasing compared with 1886. 22) The Roy. Commission. 23) HAWKES. 24) Moravian Mission Atlas. 25) Newfoundland and Labrador Pilot 1929. 26) About 1200, PACKARD following LA TROBE. 27) CURLING stated: from Blanc Sablon to Cape Chidley in 1884 4211 persons, between Hamilton Inlet and C. Chidley 1425 person of whom only 50 white people. 28) REICHEL: at the mission stations 1124 + north of Hebron 200 pagans. 29) REICHEL in 787 a. 30) v. DEWITZ: 1100 at the stations + about 50 in Aiviktok (Hamilton Inlet). 31) 1204 Christians. 32) With Ungava Bay and Hudson Bay the total number of Eskimos in Labrador was about 2500 in 1895. 33) In Okkak in 1892 about 400. 34) For the time 1901—1936 see special table to follow. 35) GOSLING: 22 professed Christianity. 36) GOSLING: 30 professed Christianity. 37) GOSLING: 33 professed Christianity. 38) GOSLING. 39) GOSLING: PALLISER made peace with 400 Eskimoes at Chateau Bay in 1764. 40) DILLON WALLACE noted 1127 fullblooded Eskimoes. 41) According to DILLON WALLACE spring 1906.

Year	1901	Individuals	1283		År	1918	Indiv.	875	Influenza, small-
»	1902	»	1273		»	1919	»	860	pox and measles
»	1903	»	1296		»	1920	»	867	
»	1904	»	1298		»	1921	»	859	
»	1905	»	1321		»	1922	»	891	
					»	1923	»	895	
»	1906	»	1309 (Flue epidemic) in Okkak		»	1924	»	919	
»	1907	»	1304 » » Hopedale		»	1925	»	904	
»	1908	»	1253 » » Nain		»	1926	»	937	
»	1909	»	1270		»	1927	»	948	
»	1910	»	1229		»	1928	»	968	
»	1911	»	1250		»	1929	»	975	
»	1912	»	1272		»	1930	»	987	
»	1913	»	1261		»	1931	»	1006	
»	1914	»	1276		»	1932	»	1017	
»	1915	»	1274		»	1933	»	1054	
»	1916	»	1244 measles in		»	1934	»	1051	
»	1917	»	1271 Okkak		»	1935	»	1092	
					»	1936	»	1135	

These figures have been plotted in Fig. 215.

tion with the white man's food and dress they must disappear from those northern confines of the earth which they have made their own and in which no other branch of the human family can or will succeed them (325).

Let us consider the question more closely, as far as the scattered information allows; most valuable are Dr. HUTTON's expositions which we will follow here.

The individuals are mostly strong and healthy.

JOHN DAVIS (209) wrote in the 1580's of the Labrador Eskimoes: The people are of good stature, well in body proportioned, with small slender hands and feet, with broad visages, and small eyes, wide mouths, the most part unbearded, great lips and close throat. — This also holds good to-day.

As if sprung from the earth the Eskimo suddenly stands before one like a ghostly shape; a small shock-headed man; he is a burly, muscular fellow, sturdy and squat, with a magnificent pair of shoulders, in every respect a normally developed type. The face is broad, ruddy and stolid, his keen brown eyes gleaming and twinkling under a great mane of shaggy hair. The skin is somewhat darkly pigmented and browned with the winter air and the frost. The face is always smiling, often shining because smeared with grease. He wears his hair long because he usually walks about with uncovered head

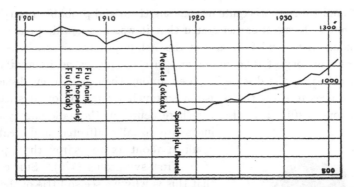

Fig. 215. The influence of the 'Spanish Flu' epidemic in 1918 on the numbers of the Eskimo population of Newfoundland-Labrador. Cf. table p. 458. The Lake Melville Eskimoes are not included.

even on the most bitter snowy day, cf. Figs. 216, 217 and 218. Even when his days as a hunter are at an end there remains a hint of bygone power and prowess in eye, arm, and shoulder, and great dignity in the rugged face under the few wisps of the grey, straggling hair of the stout men and the buxom mothers; Fig. 219. Beauty cannot be said to exist among them, but there is something very attractive in the young people, cf. Figs. 220 and 221. Most of the worthy matrons on the contrary are anything but lovely with a pipe in the fat face. It is better not to notice the malodorous smell they spread which is characteristic of them all: it is the smell of seal-oil sodden boots and harness, cooking and bedclothes, — it is a natural thing, partly the effect of the diet. On the other hand the Eskimo says: we can always tell if some of you have been in the house, you Kablunaks have a smell of your own (cf. 472, 474, 476, 478).

These people are unusually powerful physically even in our days; they can endure what would kill one of us outright; hunger and exposure belong to the very existence of the hunter and only seem to harden him. HUTTON, their surgeon for five years, states that their powers of recovery after injury are marvellous.

Pathological conditions.

We must not, however, forget that the white people have brought among them diseases previously unknown there: the gravest are tuberculosis and syphilis. It is not possible for me to estimate the degree of the influence of these scourges on the Eskimo stock; for details of these the reader is referred to HUTTON (475) and SUK (910) (cf. 897 a; 17, 116, 909; 440, 824).

SUK found tuberculosis quite new among them and probably brought there by the whites. HUTTON earlier stated that it was not a widespread cause of death among them; and SUK says of the Eskimo of the coast: it is a stock without any history of tuberculosis and in consequence no immunity at all. It is probable that the disease has been among them only a few decades and everything goes to show that it is not so much a question of constitution as of the history (910). This disease is here acute, of the galloping type; both lungs are usually affected and death may result in about six weeks from the appearance of the first symptoms (475). SUK considers that this is why we see so little of the disease among the Eskimo.

Fig. 216. Eskimo girl from Cut-throat. Photo the author, 1939.

SUK's investigations have shown further that an overwhelming majority of the infected cases are to be found among the mixed breeds, especially those who live on European food and often in not sufficient quantities. This is not very surprising; GRENFELL calculates that one-third of the deaths among the settlers on the south coast of Labrador are due to tuberculosis.

On the other hand those pure Eskimo who, in spite of the proximity of the white man, still maintain their old dietary are particularly free from tuberculosis, whereas the settlers, who live on the white man's food and have very little of it, lacking especially fresh meat and therefore resorting to tinned meat, are in a different position. It is no wonder that once having settled there tuberculosis takes its greatest toll among these quiet, friendly people (364). The same is to be said of beri-beri, scorbutus, and other avitaminoses which do not seem to occur among the Eskimoes, but certainly among the settlers.

The northern coasts of Newfoundland-Labrador long remained free from syphilis; in the so-called heathen graves for instance SUK did not find any trace of osseous syphilis (910). The disease began in 1902 with some Eskimoes who had visited an exhibiton in the States, and it spread rapidly but insidiously (475, cf. p. 472). Few of this party returned to Hebron, and they were enriched only by these heirlooms of civilization, the germs of which most effectively put a stop to the growth of the community and left people who were a constant danger to every Innuit on the coast, says GRENFELL (364). In 1909 the disease was also noticed among an isolated tribe in the north; the most likely origin was the visit of a whaling vessel (475); in another case it was traced to a vessel which used to visit some of the northern stations (910). HUTTON, who left his medical work on the northern coast in 1912, had not noticed a single case in the tertiary stage.

Among the Eskimoes and the mixed breeds on the northern coast the disease runs about the same course as in most parts of the world. SUK stated in 1926 that the future outlook is undoubtedly grave, for the plague is increasing. He met not a few women each of whom had several abortions on record, most of which he had sufficient evidence to ascribe to syphilis. He concludes that the future of this group is not altogether cheerful.

Fig. 217. Eskimo boy at Nain. Photo the author, 1939.

PADDON stated in 1938 that two sinister pairs of initials, 'T. B.' and 'V. D.' are strong arguments against the recovery of the Eskimo race. There is the menace, a health and social problem of the first magnitude, and it is not overdrawn. It cannot be handled by sporadic visits of a patrol boat operating over five hundred miles of coast. It demands that a trained worker with adequate equipment spend the necessary time in each of the chief centres. If anything is to be done, there is no time to lose.

Marriages.

The Eskimoes love children.' If they have none of their own their desire to adopt is very strong, for they require some one to look after them when they grow old as their own children would have done. A home need not remain childless; an orphan never needs to lack foster parents; someone always comes forward with the offer of a home. Adoption is also very common; families

Fig. 218. Eskimoes working temporarily for Hudson's Bay Company in Nain. Photo
C.-G. WENNER, 1939.

sometimes exchange a child or two; a family wanting a boy hands over a
superfluous girl (476).

The Eskimoes marry early, especially the young women. The younger
widowers were readily consoled in earlier times; the men only reluctantly wait
the six weeks insisted upon by the missionaries before they marry again
(761). Their excuse is that they need help in the home.

The number of marriages is however not so great as is desirable. The tragic
fate of this race is their lack of women; but unfortunately I have no exact

figures for Labrador, though the preponderance of males there is not so great as for instance among the Netsilik Eskimoes where BIRKET-SMITH in 1921 found 247 males to 176 females (80 a). This was found to be the result of the frequent practise of killing new-born girls. But this would never happen in Labrador. The birth-rate of girls among the Eskimo of Labrador is meanwhile much lower than of boys, and therefore the many young men have no possibility of marrying. This is the more to be deplored as even the widows do not in these days seem willing to remarry, possibly because the life of an Eskimo wife and mother is still extremely hard. When a hunter wants a wife for company and to perform the duties of the home, — drying the meat, sewing the boots, cleaning the house, mending the clothes — it may be that he must resign himself to going without for there is no girl left for him (475). This may be the explanation of the fact that in bygone times the girls were often taken as wives before they had attained puberty. This is also said to be the reason why they seldom have large families, 2, 3, or 4 children being the usual number, says TURNER (183). A boy was not considered a grown-up hunter able to marry until he had caught his first seal.

In former days when the wife of his youth grew too old and feeble to do all the work of his house the hunter would take another wife. Polygamy was a normal institution, and the lack of women was then still more disadvantageous for prolificness as the young men were not able to find wives. It was also the custom then for a young man to buy his wife, his parents offering seals on his behalf; if it seemed good the bargain was struck and the delighted bridegroom led his purchase home; the most important of all is said to have been that she should be a good bootmaker (476)! It can be said that it is very nearly a century since polygamy was abandoned.

Birth-rate and infant mortality.

The nativity may be considered normal for the marriages but the infant mortality is terribly high. The following percentages give a very instructive comparison (897 a):

		1901	1902	1903	1904	1905	1906	1907
Eskimoes'	Births	4.4	4.5	6.0	4.3	5.8	4.3	4.2
»	Deaths	4.6	6.2	5.8	9.6	7.8	6.5	6.6
		— 0.2	— 1.7	+ 0.2	— 5.3	— 2.0	— 2.2	— 2.4
Settlers'	Births	3.2	2.8	4.4	2.7	4.0	3.0	2.7
»	Deaths	2.1	2.8	1.0	2.3	1.4	2.0	2.3
		+ 1.1	± 0.0	+ 3.4	+ 0.4	+ 2.6	+ 1.0	+ 0.4

From this it appears that the birth-rate is higher among the Eskimo than among the settlers (see below), but the death-rate is also much higher. While there is an excess of births over deaths among the settlers, the opposite is often the case among the Eskimo. We must however remember that only a few years are included in the table.

It is a generally known fact that infant mortality among the Labrador Eskimoes is distressingly high. What is the cause? The unhygienic conditions. *Horribile dictu*, SUK says, even among the Eskimo women bottle-feeding is a fashion, that in practising it one has to resort to condensed milk and that it is hardly possible to induce the mothers to keep the feeding bottles clean so that the mortality of such infants is considerable.

Fig. 219. Two happy Eskimo couples at Nain. Photo the author, 1939.

The idea has been put forward that the frequent marriage of near relatives is against their chances of large families. These little people have gone on for years marrying and intermarrying until it is hard to find a family that cannot claim kinship of a sort with the greater part of the village (476). However, one does not get the impression that either the physical or mental qualities have suffered from this inter-marriage; I do not know of any case of imbecility among the Eskimo; but things are otherwise among the white settlers.

Longevity.

The relative rarity of aged people soon strikes the stranger (761). In olden times it was customary to kill the aged and infirm (cf. p. 551). To-day they patiently and devotedly care for their aged ones; the old parents are sure of a home with some relative, and if they have none the hospitality in the hearts of the people makes them open their homes to the needy.

These people undoubtedly grow old quickly, subject as they are to the rigorous environment of the Labradorean Atlantic coast. A man is old at forty-five, the hunters being by that time worn out by the hardships of the autumn seal-fishing. The face of a girl about twelve years old would pass as belonging to a woman of thirty (604). After fifty a man begins to decline and few live long after sixty, writes HUTTON (472); he had known a few over seventy and with astonishment one heard of a woman who lived to be eighty-two. It must be unique to meet a greatgrandmother. The very old people are always those who have clung closely to their native food and can speak of having been mighty hunters once upon a time, adds this surgeon. The census of 1935 lists fifteen individuals, mostly females, of seventy years or over at Hopedale, Nain and Hebron (897 a).

The general absence of longevity is therefore not a sign of lack of vitality; it is due to the more or less hard life in the struggle for existence. Yet there are among the Labrador Eskimoes old people who have lived till over eighty and worked regularly until the last. It can truly be said that they die in harness, dropping out as if too tired to go any further and passing away without illness or suffering (476).

Fig. 220. Eskimo children at Nain. Photo the author, 1939.

Liability to accidents.

The high death rate among little boys HUTTON thinks is very greatly due to the laxity of their upbringing, as the parents refrain from wisely controlling them. The children which have survived the age of infant mortality are well fed and well clothed. They are left a great deal to themselves and this helps them to grow up strong and hardy: the perky little rascals shall one day be big hunters! Yet not one of them has learnt to swim. It is risky to venture in a small boat on the fringe of the Ocean, or to try to sail on an ice-floe even on calm days. If they fall into the water it is not easy to get a hold on the slippery kelp-clad rocks of the beach, the suction of the retreating water will bear them down again. Many a fine and promising lad and many a staid man in the prime of life have been the victims of the merciless elements of snow or storm (478).

Later on the life of a grown-up Eskimo seems very matter-of-fact: he goes hunting, he drives his dog-team, he tramps away on his snow-shoes to set his fur-traps, but, says HUTTON, one must live among them before one can guess at the risks they run and the dangers they meet in their everyday life.

None of the circumstances mentioned above may be the decisive cause of the slow growth of the Eskimo population in the last hundred years. A much more potent obstructive element has been, it seems to me, the

Intermixture with white men.

The immigrants of course needed wives to manage their houses. Consequently the intermixture started early. THOMAS HICKSON (1063) recorded as early as 1824 from Hamilton Inlet that the probable number of the inhabitants there was as follows:

Pure Eskimo adults	100	
Pure Eskimo children	60	160
Half-Eskimoes	60	60
European settlers	90	
Canadian settlers	16	106

HICKSON tells of Eskimo women who lived in a state of concubinage with their European partners, a practice very common in this part of the world. It was however indispensable for the early white immigrants to have wives, and thus the habit developed of the white men taking Eskimo wives. In this way the s e t t l e r type arose, i.e. white and Eskimo mixed breed in various shades and degrees. The Eskimo women were not unwilling to marry the whites, and the whites have found that the attractive Eskimo girls are splendid and faithful wives. It seems that the talents of the Eskimo in many cases have a good opportunity to develop through marriage with the whites; the second and third generations are considered very gifted.

A second source of admixture has been the white Labrador fishermen (716). Until 1864 few of them visited the coast north of Hopedale, and even in 1900 very few fishermen went beyond Nain or Port Manvers (218). In any case many of these foreign fishermen are said to have behaved most outrageously. The missionaries (604) say that they not infrequently interfere in the family affairs of the Eskimo even to-day.

The assimilation with the whites has of course checked the increase of the pure Eskimo population. There are certainly no settlers north of Makkovik without some amount of Eskimo blood in their veins. Another opinion which seems to prevail is that the danger of the Labrador Eskimoes being exterminated by intermixture with the whites has passed. However, such mixed marriages are still not uncommon to-day.

Compared with the full-blooded whites, the half-caste settlers here are for a great part healthy and hard-working. Specialists who know about these

relationships are of the opinion that the intermixture with the Eskimo disposition has been of great advantage: the larger the percentage of Eskimo blood,
the better adapted to the milieu are the mixed people. It is this mixed breed
part of the population, say those who best know the conditions (GRENFELL,
PADDON), who will form Labrador's new population, in the same way as the
so-called Greenlanders in Greenland. CAREGA (126) expressed the opinion that
it was just the intermixture of the white blood that increased the Eskimo's
physical powers and reproduction capacity. I myself am not quite sure about
this conclusion. CAREGA's opinion has also been opposed by JENNESS (491)
who stated that the whites have brought with them numerous diseases of
which the influence on the health and fertility of the natives has not yet been
elucidated. In any case it is certain that this blood-letting of the Eskimo stock
is inimical to its reproductivity.

Very few somewhat reliable figures are obtainable of the proportions of
full-blooded Eskimoes, mixed bloods, and whites constituting the population
of the Labrador coast. It seems probable (897 a) that miscegenation, starting
earlier in the south, reaches a peak around Hamilton Inlet. It seems to me
not impossible that in earlier times considerable numbers of mixed bloods from
the south were also introduced into the northern communities, in spite of
the shift of population which took place in the last half-century from the
northern to the southern stations (604, 897 a).

Report gives the following details for the coast (603, 604):
in 1856 practically no half-castes
in 1874   115 settlers
in 1904   280   »

On December 31st 1905 (603, 604):

Killinek	78 Eskimo	3 settlers
Ramah	75 »	4 »
Hebron	166 »	8 »
Okkak	329 »	0 »
Nain	233 »	54 »
Hopedale	123 »	110 »
Makkovik		138 ['nearly all settlers']

In 1926 SUK (910) was of the opinion that at Hebron the people were almost
entirely pure Eskimoes but for five or six of mixed breed. At Hopedale, Nain
and Okkak pure Eskimoes were in the majority, but at Makkovik they were
mostly settlers. It is not clear to me how SUK determined the degree of
hybridism.

On January 1st 1939 there were large majorities of settlers in Makkovik and Hopedale; cf. Fig. 210. Most of the Newfoundland-Labrador Eskimoes are probably more or less mixed with white men's blood.

The Eskimoes degenerate when they take white men's food.

In the olden days the Eskimoes' life was simple and healthy, PRICHARD writes. The struggle for existence has certainly always been hard. But they lived long and contentedly. Death presented himself to them under the form of hunger or accident at sea or when the old people so to say expired by falling asleep for ever. Not until they met the whites did death wear the face of disease.

They have a hand-to-mouth existence, sometimes surrounded by plenty, sometimes in want, but there is always food for those who care to get it. Among no other people can be met such a large number of well-nourished and fat faces, slim types are rare. This is due to their open-air life and very nourishing food

Fig. 221. Eskimo children at Hebron. Photo
E. H. KRANCK, 1937.

— animal flesh and fish. Little children are so fat that the eyes are almost closed and the nose is almost hidden in the cheeks; they are a perfect picture of health and well-being (647). The autumn seal hunt gives them their best and most fattening food just when the cold weather begins and they have thus a fine natural protection against the cold. They need no fire to warm them; their physique has been altered century by century to suit the surroundings as was the case with the cave man (761). Dr HUTTON who studied the Eskimoes of the coast while practising as their surgeon at Okkak, is of the opinion that the Eskimo phenotype has got in to a stage of transmutation. In the north generally the Eskimoes are broad and plump with fat, flat faces and sunken noses. Further south HUTTON met lean, sharp-faced Eskimoes with bony limbs and pointed noses. He thinks that they, too, are pure-blooded and that the change is due to their altered food and habits. The Eskimoes are born imitators, he says, and in the southern stations they have had more contact with the English-speaking settlers who were there for cod-fishing and fur-trapping. It is little wonder that they assumed the settlers' habits regarding food and clothing. Cloth garments took the place of the sealskin, bread and tea and cooked meats replaced the raw meat and blubber. But the result is that the southern Eskimoes are less hardy now; they cannot bear cold well. The children are puny; they can dwindle and die off. These conditions result in a real danger; the natives — the Eskimoes and Indians — degenerate when they take to the white man's food. The Eskimo loses his natural coating of fat to a great extent and needs more clothing, he becomes less able to endure fatigue and the children are often small and weak; so say people who know the Labrador Eskimoes well.

The influence of the altered diet on the type and health of the Newfoundland-Labrador Eskimo was strongly emphasized in my conversations with the missionaries. Many scientists are, as we have seen, of the same opinion, for instance, SUK and STEWART.

Primitive hygiene and epidemics.

The worst scourge of the Eskimo and the Indian is their primitive hygiene and disease.

When the missionaries began their civilizing work among them the Eskimoes were dirty because they knew no better and were content to remain so (cf. p. 516). The filthy floor and the horrible smell in their subterranean winter dwellings must have been unbelievable. Nowadays the schoolmistress has taught the children to be neat and clean, but when they grow up their keenness dies away in many cases so that the race as a whole remains far behind the white people in regard to habits of cleanliness and sanitation. Meanwhile there seems to be a difference in this respect between the southern and the northern Eskimoes; the former who touch the fringe of civilization are far cleaner than the latter. Yet even on the northern coast there are Eskimoes who look after themselves fairly well and they are in their way clean, do not have vermin, and on the whole smell less badly than primitive people in general. When I quite unexpectedly entered the salmon-fishing hut of the Eskimo, Peter Adams, at The Narrows, I found perfect order, and the air was no worse than that of similar huts among the whites. My friend and pilot Julius Nathanael was distinguished by his neat appearance, -- and had such good table manners that I should have been glad to show them to some of my students as an example.

However, indifference to cleanliness can hardly be so harmful in the pure atmosphere of Labrador as it would be in the centres of the civilized world where the air is filled with miasma.

It is also remarkable that the Eskimo use no drugs since the law against the import of alcoholic drinks was put into force; drunkeness has been rare. There was a fatal time when the Eskimo learned to make beer with molasses and a furtive concoction of hard biscuits. The drink was not very strong but the brain of the Eskimo easily became inflamed, and in consequence their evil passions were loosed, women were beaten and various domestic disturbances arose. So far as I could find out this practice has now disappeared completely, and the Labrador Eskimoes may be said to be a teetotal race; when they see a drunk person they keep at a safe distance thinking 'the man has a devil' (475).

Men and women alike smoke very much, but it can scarcely be very danger-ous for their health. This habit they acquired from passing vessels, because they know nothing of 'the pipe of peace' (476).

One very distressing fact is the difficulty of rooting out their old habit of employing the native doctors to whom the Eskimo still likes to entrust his physical troubles. Nowadays, it is true, the work of these doctors is quite innocent; all witchcraft, sorcery, magic and the like have passed away in general with other relics of heathen times, Dr HUTTON was able to state. But if the treatment is not dangerous, it is also not wholesome. If an Eskimo has pain in some part of his body, that part is, to his method of thought, broken; and on this idea the native doctors work: something is broken and must be mended. Several of these 'doctors' have gained a reputation for unusual skill in dealing with illness. They use several plants (518) as medicines as tea: the onions which grow on the sea shore are good for scurvy, *Ledum latifolium* is good for fever because it lowers the temperature, *Ledum palustre* is good for chills, tansy for colds, seaweed is a cure for skin disease, dandelion is used to lighten the meat diet. Raw seal's liver is a universal remedy among them. They also make a stew of rosemary twigs as well as of the brain of the cod. A treatment is carried out with lengthy and mysterious manipulations. Physiology is beyond the grasp of their child minds. The native doctor is always shy of letting a foreigner see him at his work, and no doubt he owes his popularity to the old Eskimo conservatism (476).

I was able to note that their powers of recovery after injury are marvellous. But disease is another thing, their power of resistance is low and they are soon prostrate like all primitive people (476). They are extremely liable to infection, those of Killinek more than those of Hebron and Nain who have preserved some degree of immunity. Comparatively harmless diseases like measles or smallpox have been terrible scourges when brought for the first time to the

remote tribes of Labrador (cf. the tragic descriptions left by CARTWRIGHT
(134) and DAVEY (205). We read repeatedly in the missionary diaries of other
epidemics among the Eskimoes, especially influenza. A form of this last illness
visits the villages practically every year, but mostly in a mild form. Yet at
intervals it has descended upon them with virulent force and usually caused
a number of deaths: in 1842 severe influenza, 1855 starvation and disease,
1857 disease amongst the dogs infected the game, 1876 severe whooping-
cough (HAWKES); in 1900 typhoid fever, in 1904 in Okkak influenza (Mc
GREGOR, HUTTON); influenza raged in Okkak in 1906, in Hopedale 1907,
in Nain 1908, measles in Okkak 1906, 1918—1919 influenza, smallpox and
measles, in Okkak 216 died out of 310, in Hebron 140 in a few days (HET-
TASCH: see table above). The records of epidemics are usually imperfect and
the causative agents are often uncertain; but the results are reflected in the
mission's census records, and with each epidemic the population declines
(cf. tables, pp. 456, 458).

When an epidemic starts in a hut it soon spreads because of much visiting.
Fatalism is deeply rooted in the Eskimo nature. Sanitary reform, says Dr
HUTTON, caused a raising of the eyebrows: 'that is not the way we do; it is not
a custom of the people.' Epidemics therefore attack the small Eskimo settlements wholesale, and when an epidemic begins it seldom misses anyone (476);
for instance during an influenza period practically the whole population of a
station seem to catch it at the same time, the village appears deserted, and only
now and then a muffled, dismal figure moves along to the Mission-house to ask
for medicine from the House father. HUTTON reports on the other hand that the
ravages of the different diseases vary very much. Influenza comes twice a year,
in midwinter and midsummer. But within the memory of man there has never
been such a fearsome scourge in Labrador as the sickness which swept the coast
in the autumn and winter 1918—1919. When the S/S 'Harmony' arrived in
Hebron from Nain in November 1918 the natives began to show signs of having
contracted the 'Spanish influenza'; the result was that in the course of about nine
days nearly two-thirds of the Hebron people had died. One can hardly imagine
what this means. The frozen ground and their own weakness prevented the
digging of graves, they buried the bodies in the sea through a gravelike hole cut
in the ice and set fire to the houses of the deceased to get rid of the infection
(478). From Okkak the story is even worse: what had been a happy and prosperous village, was in a few days wiped out. When after the freezing of the sea
communication with the scattered homes in the bays was re-established, the
disease gradually spread to the people in camps outside, and whole families died
without being able to send for help. One little girl, the only survivor in a household, lived on in solitude with her dead parents and brothers and sisters around
her: she ate berries for food and quenched her thirst with snow. For weeks she
lived like this, and when some Eskimo boys on their travels looked in at the
windows, they thought that they were seeing a ghost. The missionaries went

out and rescued the child. In the village itself and in those lonely camps the
savage dogs became more savage still and added to the horrors of the scene. —
It was clearly quite a new form of influenza that was brought from England by
the mission's vessel and this explains the severe passage of the disease. But,
a curious fact, among the settlers the attacks were much milder than among the
pure-blooded Eskimoes.

Epidemics under such primitive conditions are a far sadder chapter than in
the civilized world, says HUTTON. CARTWRIGHT tells of one PHIPPARD who came
on an Eskimo encampment on an island in Invuctok Bay [Hamilton Inlet] where
the whole family had evidently died of smallpox. — The caribou Eskimo were
almost exterminated by one disastrous influenza epidemic (80 a).

The unfortunate effects which followed from the visits of Eskimoes to different exhibitions have already been referred to. In 1881 eight of them went to
Hamburg and then to Paris. Here they contracted smallpox and all died; there
were no survivors to take back disease and death to their kinsmen in Labrador.
In 1893 some enterprising Americans transported fifty-seven men, women and
children to the Columbian Exposition in Chicago; a few of them returned in a
destitute condition and brought with them typhoid fever and diphtheria, to
which large numbers of their fellows from Hopedale to Hebron fell victim. In
1898 thirty-three Eskimoes went for exhibition to England, other parts of
Europe, and America. Only six of them returned in 1902, sick and destitute.
They had contracted a most loathsome, venereal disease which has since spread
through all the settlements, slowly and painfully killing a large number of the
inhabitants. Venereal disease has been peculiarly fateful in these parts; it is
widespread among the Eskimo women who, according to what the missionaries
told me, are in general distinguished by extreme sensuality.

Whenever the Eskimoes have left their native coasts disease and death have
quickly destroyed them (325).

We have now quoted some known examples. We need only insert these
epidemics in their proper places in the statistics to obtain a clear idea of one
of the decisive causes of the small population increase among the Eskimoes,
even if all the information does not bear such terrifying witness as the winter
1918—1919 when their numbers at the mission stations fell from 1271 to 875
or by about 30 per cent in half-a-year, and almost wiped out some populous
settlements. The station Okkak was practically exterminated and had to be
closed. A new little society: Nutak, arose later south of Okkak.

Conclusion.

From what has been said here we ought, in my opinion, to acknowledge
that the human material as such seems physically and mentally sound on the
whole. The Eskimo, in his hard struggle for existence, lives long if he can avoid
infectious diseases. Dr HUTTON speaks of old Juliana Boase who used to live

at Okkak, one of the few survivors of the influenza epidemic of 1918—1919; he says that at the age of 82 she still travelled to and fro by boat and sledge to Tikkerarsuk, a distance of about twenty miles. There seems to be no cause to fear a decrease of the population if hygienic conditions and knowledge can be introduced systematically in Labrador as has been done in Greenland and Alaska.

So far as I know there has until the present day been only one resident qualified medical man along the whole of the north coast, namely Dr HUTTON (1907—12). Dr GRENFELL and Dr PADDON have only occasionally paid flying visits to that part of the coast where the Eskimo live. The Moravian missionaries have done wonders in the way of surgery. Certainly the effects of the missionaries' teaching appear in the somewhat improved hygiene of the Eskimo but they, of course, are powerless when it is a question of systematically arranging and supervising health conditions and getting rid of the consequences of disease. The Eskimo still lives close to nature and his physical and mental equipment is in general good; but professionally organized medical protection against epidemics and other infectious diseases is necessary. These aborigines ought also to enjoy a permanent medical service such as the rest of the population of the colony of Newfoundland already have. SUK has expressed the idea that the International Grenfell Association should take over the medical care of the north coast also, perhaps in co-operation with the allpowerful Hudson's Bay Company. He has undoubtedly hit the right note.

The present Eskimo habitat in Labrador

is seen from Fig. 222. In our days they occupy the Atlantic coast of Labrador from Hamilton Inlet to Cape Chidley, including the eastern part of Lake Melville, east of the English River. Here they share the produce of nature with the few white people and the half-breeds who have settled there. On the other hand Eskimo are the only inhabitants of the strip of Ungava Bay from Cape Chidley to the Koksoak River. Further, they alone inhabit the enormous spur towards the north-west, called Ungava Peninsula, north and north-west of a line drawn from Hudson Bay about over Lake Minto and the Larch River to the mouth of the Leaf River in Ungava Bay, this line coinciding very much with the tree line. Since olden times the Ungava Peninsula has been noted for its rich supply of caribou; in the watershed district on the peninsula the Eskimoes, both from Hudson Bay and Ungava Bay, used to assemble during the hunting period. Even at the

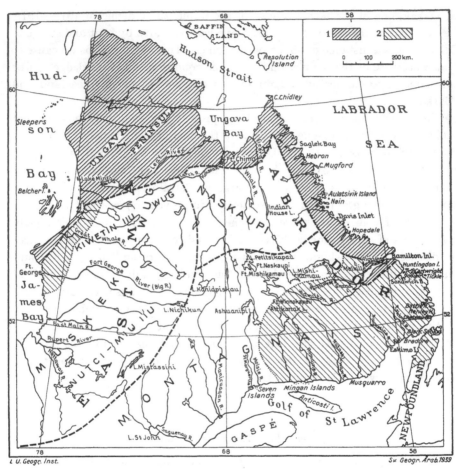

Fig. 222. The habitats of the Eskimoes and the Indians on the Labrador Peninsula (916).

1 = Eskimo habitat at the present time
2 =   »   »   in bygone days.

turn of the century, this hunting had attracted thither the Indians from the forests in the south; and as late as 1912 FLAHERTY found at Lake Minto the skeleton of an old tent of Indian type (cf. also Low). At the present time however the Eskimoes seem only occasionally and the Indians never to make their way to the inner parts of this peninsula which thus lie desolate and unused by man. On the coast of Hudson's Bay Eskimoes are found at the present time as far to the south as the George River.

There is scarcely any aboriginal people, unless the Polynesians and possibly the Tungus, who are so widely spread on our planet as the Eskimoes. The poorness of nature's resources in their habitat has driven them to divide into territorial groupings to which they themselves have given different names. In Labrador there are three such (see 965). The one living in Newfoundland-Labrador has been called S e r k i n e r m i u t, or S û h i n î m i u t, according to TURNER, meaning the people who keep to the land of the sun, i.e. on the east side of the peninsula or at the Atlantic coast. This large main group comprises the world's most southerly Eskimoes (cf. influence of the cold Labrador current). To it were also reckoned the Eskimoes who in winter lived beside Ungava Bay between Fort Chimo and Cape Chidley, because during the summer they used to wander about the archipelago round that Cape, which is also on the Atlantic.

These people of the sunny side have in their turn divided into a number of small sub-groups each of which has called themselves after the place where they mainly dwell by adding — m i u t [= people], —; see HAWKES' map (424). An example of this is *Ki'lin'ig'miut* from *Ki'lin'nik*, which indicates the district of Cape Chidley, and means the twisted, cut up, pressed together [area]. *Avanêmiut* is another example: it means northerners, and is a collective name for the groups of the north.

The old — m i u t names (see HAWKES' map) are no longer in general use within our area; a group will today take its name from the mission station near, e.g. the Nain station: *Naine'muit*, or after Nain Bay, *Nunaingumiut*. The Hopedale Eskimo call themselves *Arvekmiut*, after the Arvek' [= Whale] Fiord.

These small territorial groups have no connection in our day with any family-clans. As far as is known clan chiefs have never existed among them. It has been usual that a powerful and crafty man has taken for himself the leadership of a small group. Yet he has often been obliged to share his influence with the shaman: *angekok;* that false prophet (934) who was so clear an expression of the dogmatics of the Labrador Eskimo in former times. The struggle between the two usually concluded by the self-elected chief investing himself with the function of the sorcerer also.

The condition of the group division in olden times is less clear (cf. 196); certainly the ancestors of the present Eskimo were not wholly ignorant of ethnical distinctions, as is shown by the numerous tales of the *Tunnit;* these latter can with good reason be termed the bearers of an older Eskimo culture, the so-called Thule culture (see farther on).

The extension of the Eskimoes' habitat in bygone days.

The missionaries KOHLMEISTER and KMOCH speak of the tradition among the Eskimo at the beginning of last century according to which their ancestors from the Canadian district in the north spread out to the east and south over the range of islands on the Labrador coast. The tradition is supported apparently by present ethnographical conditions, and archaeology and agrees with what is known otherwise (cf. p. 452 sq.) following which the bearers of the recent type Eskimo culture about A.D. 1200 invaded from the west. What little has been studied of the archæology of east Labrador indicates also that the present-day Eskimoes' ancestors immigrated there at a comparatively late period. Further it is said (Hopedale) that Indians probably lived at the Atlantic coast before the immigration of the Eskimoes, for below the Eskimo cultural layers Indian relics have been reported such as sherds, and pottery it is known that the Labrador Eskimo never used.

After this invasion the people found the natural conditions at the coast most favorable for hunting. The shore is cut into by fiords, and its broad archipelago with numerous 'runs' and 'tickles' where the seals like to live, offer ideal, well-protected places for catching marine mammals and, during the ice-free period of the year, fish. In the winter again the edge of the ice fringe which forms a good way out to sea becomes a home for the seal, the walrus and the fox. On the other hand the valleys which follow the fiords inland are the home of the caribou, the bear, and fur animals. The supply of this booty has been so great on this coast that the hunters could manage without any great anxiety for the necessities of life if they were only somewhat careful. Such hunting by this predatory people has gone on for centuries on the barren but beautiful coast. The people gradually spread farther south and finally embraced an area, still on the coast, considerably more widespread than at present; cf. Fig. 222 (73, 85, 710, 873, 878).

It has been stated that the Eskimoes during pre-Columbian times probably stretched still farther south than is shown on the map; for instance, they perhaps reached Saguenay on the Gulf coast about the year 1000 (BIRKET-SMITH) and perhaps Quebec; some suppose even as far as Nova Scotia and Maine. HAWKES assumes that scattered groups of them were at one time even as far south as New York State (on the basis of certain archæological findings). Weapons resembling those of the Eskimo have been found in Ontario (WINTEMBERG) and in New York State (SKINNER); also the tools of the Beothuk Indians show a strong influence of Eskimo culture (HAWKES, JENNESS). But here we are touching upon questions which it is impossible yet to judge with certainty (cf. p. 454), and we do better to confine ourselves to the indications on the above map

It appears to me that the earliest quasi-historical information regarding the Eskimoes is obtained from the Icelandic sagas. In Markland and in Vinland (p. 43) the Norsemen met with *Skraelingar*, about the year A.D. 1000. This name they used for the natives in general, both Eskimo and Indians (920b). When Torfinn Karlsevne saw a large number of skin boats on which long staves swung in a circle in the direction of the sun and gave out the sound of flails, these must clearly have been Eskimoes rowing kajaks with double-bladed paddles, because no other people in this district are known to have used such oars. There is also no doubt that the Skraelingar Torfinn Karlsevne killed on his return journey to Greenland were Eskimoes. But on the other hand there is little doubt that the Skraelingar in Hop he had to fight against were Indians. The sagas give the impression that the Skraelingar in Markland, i.e. Labrador's Atlantic coast, were predominantly Eskimo, while those in Vinland, i.e. north Newfoundland and districts to the south, were Indians. I refer the reader to my detailed explanation of this matter (920b).

From the descriptions given by the older explorers there is often no clear picture of whether the natives they met were Indians or Eskimoes. CABOT in 1497 saw 'savage men clothed in beastes skinns and [who] ate raw flesh and spoke such speech that no man could understand them'. He brought away from these people snares for game and needles for making nets. Because of the latter HAWKES thought that CABOT had met Eskimoes; I am not convinced that he was right (pp. 54, 579). But even if CABOT happened upon Eskimo as little as JACQUES CARTIER in 1534 we need not doubt that such people were already there in the south-east of Labrador. On older maps of these areas there are fine notes of both the Eskimoes' and the Indians' life (p. 55 sq.), as on the map from 1546 of Pierre Descelière de Arques' and the splendid portolan by Nicolas Vallard de Dieppe from 1547; cf. Fig. 10. When the light of history began to dawn on the southern parts of the districts the Eskimo had already spread far to the south on the northern shore of the Gulf of St. Lawrence, where tent-rings, heaps of bone, dilapidated graves and the ruins of earthen igloos bare witness to their extension to the Mingan Archipelago north of the island Anticosti. Europeans found Eskimoes on Eskimo Island at the beginning of the 16th century where they hunted seal and walrus in the spring. One might be tempted to say that the Eskimo had there reached the limit of their natural extension; the icebergs of the Labrador current had guided the Eskimo to the Gulf just as they had guided the polar bear and the Greenland seal thither; where the icebergs ceased, there also stopped the great marine animals and with them the Eskimoes, I think.

West of the Eskimo lived at that time the Montagnais Indians from Seven

Islands and the Moisie River down to Quebec, and these Indians also occupied the country from the sea to the watershed (cf. the Jesuit Relations 1694—1750). North of the Montagnais lived the Naskaupee Indians, *'cuneskapi'*, whom FATHER LAURE met by the Ashuanipi Lake in 1731.

Thus north of the Gulf of St. Lawrence the Eskimo occupied the southeastern part of the Labrador Peninsula, both the coast and the hinterland as far as a line drawn from Mingan to the Kipekak River; place names at Lake Melville support the assumption that the Eskimo lived all round this lake and its upland till quite recently; Fig. 222 (cf. map p. XXIII in 509). They also frequented the Hamilton River (139, 325). Many people who thought that they were a distinctly coastal people are surprised to find that also in many other places they had spread inland: on maps from 1660 and 1695 they are shown as inhabiting some districts east of St Marguerite River's upper course north-east of Saguenay, where at present only Indians hunt. Clearly such Eskimo have later, family by family, been assimilated with the Indians, conditions which are known for example from Lake Mistassini and Seven Islands' district (873). This fact can be regarded as supporting BIRKET-SMITH's notion that the Eskimo at one time spread as far as Saguenay (73).

It can be assumed that the Eskimoes established themselves in the 16th and the 17th centuries, at least for a time, also upon the most northerly parts of the west coast of Newfoundland. Though it has been supposed that these latter really belonged to Labrador and only undertook occasional treks farther south; CARTWRIGHT noted that in 1771 he saw an Eskimo as far south as Fogo Island, near Twillingate in Newfoundland. There are however some curious pieces of information in the literature which quite definitely indicate that the Eskimo had moved even so far south that they had contact with the Gulf Stream off Newfoundland. With such an assumption it is easy to explain the presence of the three kajaks which were fished up in the North Sea at the end of the 17th century (615 a). Another explanation is that they were brought from Greenland by a whaler which was wrecked in the North Sea (BIRKET-SMITH). On the other hand GOSLING says that no evidence has been produced to show that the Eskimo ever occupied the eastern shore of America south of the Straits of Belle Isle, but that they were in the habit of making occasional summer excursions to the north of Newfoundland, e.g. Fogo Island; it can be confidently asserted that their range has been co-existent with that of the seal on which they mainly subsist and consequently has never been farther south than the north shore of the Gulf of St Lawrence, he says.

The gradual expulsion of the Eskimoes from the southern and south-eastern seaboards of Labrador.

Looking at the map, Fig. 222, of the extension of these people formerly and at present one can scarcely avoid a question: why have the Eskimoes deserted their former outposts? and the answer is: the savages; both the coloured in the interior and the white along the coast have competed in driving them to the north.

At their first meeting $9^1/_2$ centuries ago with the Norsemen, hate against the white man was born in the soul of the Eskimo. Icelandic sagas relate that the Norsemen raged among the Eskimo and murdered as for the pleasure of murdering. Historical sources show that CORTEREAL decoyed aboard his vessel some three-score natives, whether Indians or Eskimo is a moot point, and sent them to Lisbon, where the court jumped to the conclusion that the navigator had discovered a land of great resources, where amongst other things they could obtain brown slaves instead of black (761). Thus the white men brought misfortune and must be opposed.

A red thread running through their folklore is the everlasting struggle between the Eskimo and the Indian. This race hatred must have old beginnings: SPECK thinks that it arose during that distant period when the two peoples first met at Hudson Bay and thence spread gradually in a kind of guerilla condition over the whole of Labrador. According to FLAHERTY (265) this still prevailed in late times in the north-western part of the peninsula: the Indians made raids on the defenceless Eskimo who, attracted like themselves by the prospects of the deer, ascended the rivers from Ungava Bay. At that time they had no firearms as had the Indians, but since they also secured these weapons from the trappers they have proved themselves more than a match for their hereditary foes.

Scanty historical information also shows that for a long time a kind of intermittent state of war has prevailed between the Eskimoes south of Hamilton Inlet and the eastern Montagnais. In the interior the Indians were dangerous enemies: they knew the terrain better than the Eskimo, they were numerically superior and had better weapons. But in the archipelago the Eskimo was superior; a single kajak man would lure the Indians to follow him out to an island where a large number of his fellows lay in wait, they took the Indian's canoes away and killed their owners.

The situation of the Eskimo was made worse when the Micmac Indians from the Gaspé peninsula on the south side of the gulf also appeared in their canoes to fight them: yet the Eskimo bravely held their ground till about 1600.

In the Europeans the Eskimo found a new enemy. Almost at once a state of warfare arose between the natives and the French invaders, broken by short, calmer periods. Slowly the Eskimoes were driven out of one position after the other by the *tertius interveniens*, the white man, sometimes in alliance with the Indians. The Eskimoes both on the west and the east sides of the peninsula were compelled to withdraw to the north.

On the coast of Hudson Bay, the so-called East Main, the Eskimo district was taken over by the Indians to some extent after what was presumably a fairly stubborn resistance (DOBBS, SKINNER). In 1744 the Eskimo still occupied the area north of East Main River (52° 30'). About 1752 they appear to have been forced back to Belchers Island and Sleepers; their southern boundary seems to have lain at Hazard Gulf (56° 22') according to COATS. If this information is strictly correct the Eskimo must have somewhat extended their hunting grounds again to the south for now the boundary lies at Cape Jones (54° 50'), cf. Fig. 222.

On the south-eastern coast of the peninsula the Eskimo's withdrawal was only slowly completed. The oldest settlement there is thought to have taken place about 1704, when COURTEMANCHE began to colonize his concession which stretched from Khegaska on the Gulf coast up to Kessessasskiou or Hamilton Inlet. The most important branch of industry for the colonists was intensive fishing for which they used both French Canadians and Montagnais Indians. His work was continued after 1717 by his son-in-law BROUAGE who learnt the Eskimo language, sought connections with them and even attracted them to come south to him in the summer in order to barter. But the Eskimo also made use of the opportunity to plunder the white men. The consequence was that the station had to be closed (325) and the feud continued.

— It is curious to note how different were the white men's opinion of the aborigenes of Labrador in olden times. The Jesuit LAURE (556) for example writes in 1720—1730: Effectivement ce sont d'aimable gens [the Indians, Papinachois] pour leur inalterable gayeté. Plut à Dieu qu'ils pussient communiquer de leur temperament à leurs intraitables voisins les Esquimaux qu'on n'aprivoisiera jamais sans miracle parce qu'enfoncés dans leur rochers naturellement creuses et imprenables . . .ils ne se laissent jamais approcher d'aucune nation fut-elle basque, car on ne doute presque plus·que quelque basque pecheur naufragé sur ces côtes avec quelque Ève n'ait été leur infortune Adam. The missionary JENS HAVEN writes of his arrival in Newfoundland as late as 1764: everyone here paints the Eskimo in shocking colours. Treacherous, murderous savages, . . . no Eskimo kajak could be tolerated within gun-shot of any European boat, so great was the dread these savages had inspired (205). —

A general anarchic condition seems to have prevailed in the 17th and 18th centuries. The Indians and the white shot and plundered any small group of Eskimoes they caught sight of. On the other hand the Eskimoes used to creep up to the whites in the thick fog and terrify them with their horrible yelling so that they fled, leaving their things behind them. CORTEREAL found a sword among the Eskimo which must have come from Europeans. COURTEMANCHE at the beginning of the 16th century found a gun among them, though they scarcely understood its use. The crews of the trading ships were lured into ambush and killed; the Eskimoes used the weapons of the weak: cunning and terror. This tate of things seems to have prevailed for a long time.

For decades the Eskimoes maintained their *jus primo occupantis*. But when the French furnished the Indians with firearms the fortune of war turned. The Eskimoes were soon obliged to retire to Eskimo Island on the Gulf coast where they entrenched themselves. In 1640 a force of Indians and Frenchmen attacked and the island fell to them; thousands of warriors (?!) are said to have taken part in this affair. — On the Atlantic coast on the other hand the Eskimoes seem still to have lived in undisturbed possession of the district. However scarcely any definite mention of the people there is to be found in the literature until JOHN KNIGHT in 1606 was attacked by 'a very little people, tawnie-coloured, thin or no beards and flat-nosed'.

The information as regards the relations between the Eskimoes and their neighbours is anything but clear, but there are indications that they had become more peaceable at the beginning of the 18th century. The French had at that time (1702) begun to colonize the coast (cf. above). Again, numerous Eskimoes were to be found on the Côte du Nord (the north coast of the Gulf) right away up to Anticosti Island in the west. In 1704 the Eskimoes were, however, a great nuisance to the French stations in Bradore. Trading and mission stations were, in spite of this aggressiveness, opened up farther and farther towards the north-east; finally one factory was established at Blanc Sablon (somewhat earlier than 1733). JEFFREY relates how the French in 1753 were in the habit of buying oil and whalebone from the Eskimoes on the Atlantic Coast; 'the Eskemaux go up to Latitude 58 or further north; there leave their great Boats, pass a small neck of Land taking their canoes with them, and then go into another Water which communicates with Hudson's Streights, carry their Return of Trade into Eskemaux Bay where they live in winter; and the French made considerable Returns to Old France by the whalebone and oil produced from these People.' On a map from 1757 (LVIII) one finds on the southern shore of Hamilton Inlet toward the east of the Narrows the note 'Ance du Batiment' perhaps a trading station. Relations must then have been fairly tolerable between them and the whites. However, HAWKES quotes an old, unknown author: they [the Eskimo] fly from Europeans because they have been maltreated, fired on, and killed, and if they attack and kill Europeans it is only by way of reprisals. The whites had indeed always to be on their guard.

But soon open conflict again became the order of the day. In the track of the French the Montagnais spread to the east and with their united efforts the Eskimoes were driven still further to the north-east. At the Straits of Belle Isle there were still Eskimoes in 1760 in a kind of fortified camp on an island at the west end; but they were soon driven thence by their enemies. CARTWRIGHT in 1765 speaks of a little fort which was said to have been built in Chateau Bay as protection against the Eskimo; PACKARD describes the ruins of Grevill's Fort in Henley Harbour which is probably the same fastness; it was deserted by the French in 1753 and in 1763 was occupied by an English garrison. According to CARPENTER several hundreds of Eskimoes were living round Chateau Bay in 1765, and others lay to the west of it right away to Mingan [?!]. The fact is that in that year 1765 the British Governor PALLISER made some kind of peace and friendship treaty [!] with 400 Eskimoes in Chateau Bay; in this treaty it was distinctly agreed that thenceforth the whites should be forbidden to plunder and murder the Eskimo; it seems therefore that such action had been tolerated under the French. Relatively peaceable conditions seem to have prevailed after the English took possession of the country.

The missionary JENS HAVEN found Eskimoes in Kirpun Bay on Newfoundland as late as 1762. According to CARTWRIGHT in July 1771 there was a habitation

31

with 32 Eskimoes at Cape Charles; he speaks of other Eskimo dwellings in Chateau Bay 1786, on Huntingdon Island in 1783 and says that the Eskimoes were still often crossing the Straits of Belle Isle

Further, according to him, there were at the end of the 18th century three groups of Eskimoes south of Hamilton Inlet, called (cf. HAWKES): P u t l a v a-m i u t (Battle Harbour people), N e t c e t u m i u t (Sandwich Bay, N e t s-b u c k' t o k e, people) and A i v i t u m i u t (Rigolet-Hamilton Inlet, I v u k'-t o k e, people). It seems also that the Eskimoes lived considerably farther west on the northern shore of the Gulf at that time; JAMES MCKENZIE in 1808 speaks of Eskimo 45 leagues east of Anticosti. About 1800 Eskimoes are said to have been quite numerous on the coastal strip between Battle Harbour and Sandwich Bay.

These scattered items of information scarcely do more than indicate the course of these people's withdrawal from the Gulf coast. Yet one gets a full impression of the variations in the boundary between the Eskimoes and the whites, and that after each collective attack on the former they gave way only to begin again after some time to penetrate their old hunting grounds.

The struggle between the Indians and the Eskimo gradually spread also to the Atlantic coast. Traditions preserved and still to be heard on the spot point to several battle-grounds like Battle Harbour [?], Indian Harbour and Pompus Island in Hamilton Inlet, as well as a few other places at the outermost coast. Yet the Eskimo still sit steadily round Hamilton Inlet and seem even to have somewhat enlarged their area in the forested land beside Lake Melville. It is said that the Naskaupee Indians have endeavoured to drive them out, and according to tradition among the Eskimo there these Indians several times attacked their encampments at Lake Melville. The decisive fight took place on Eskimo Island at the outflow of Lake Melville some time at the end of the 18th century. HOLME reports something like 70 graves on the island which according to »ordinary Eskimo customs» are not deep in the ground. On the same island I found more ruins of four great-family houses and many plundered graves of stone blocks. An Eskimo who accompanied me described on the spot the battle between his people and the attacking Indians, 'hundreds of years ago'. A skull with a bullet hole showed that the attack took place after firearms reached the Indians. The large number of plundered cists containing human bones and bits of birch-bark illustrated the tales that the fight was bloody. The rearguard of the Eskimo retained the place and use it even to-day.

In the south of Hopedale the Indians and Eskimoes sometimes meet, KOHLMEISTER (530) wrote in 1811; the Hopedale Eskimoes seek to cultivate their friendship, quarrels and bloodshed seldom occur. But in Ungava the meetings now and then terminate in murder; the Eskimoes are much afraid of the Indians, who are a more nimble and active race.

I have been unable to find any exhaustive information regarding the development in the 19th century. The southern tribes soon became extinct. Intercourse with the white race proved their ruin. The European clothes and European food no doubt were the principal agents in their destruction. To which must be added also the adoption of European vices and the introduction of European diseases (325). The southern Eskimoes remained in heathendom to the last. CHAPPEL (144) tells in 1817 of a tribe of about fifty persons that visited the trading establishment near l'Anse-à-Loup. While there a woman died, her female infant was immediately stoned to death and buried with her. In the 1850's FERLAND met three or four Eskimoes at the Gulf coast between Mingan and Khegaska; they lived like the Europeans. Near Blanc Sablon, he says, an Eskimo lived. According to the missionary CARPENTER two Eskimo families were still living about 1860 in the neighbourhood of the mouths of the Eskimo River and the St Augustine River respectively. The widow from the former place married an Englishman; perhaps this was just the John Goddard whom PACKARD met, who lived on a point three miles west of Caribou Island and was said to have married an Eskimo. In 1860 PACKARD saw an Eskimo on Caribou Island off the Gulf coast. On Tub Island in 1864 he met a full-blooded Eskimo woman married to a 'William'; they had adopted a Naskaupee child. The missionary ELSNER reports in 1857 that in Hamilton Inlet there were only 31 families, ten of them Eskimoes. In 1860 McCLINTOCK (325) stated that there were said to be about 200 people living in Hamilton Inlet, but the Eskimoes who had once been so numerous were fast dying out. He was told that on an island at the mouth of the Inlet a number of skeletons of Eskimoes were strewed about the surface, showing that they had fallen victims at one time to a virulent, contagious disease (cf. p. 472). WALLACE reported in 1905 that only two full-blooded Eskimoes remained south of Hamilton Inlet (994). HAWKES as late as 1914 discovered two Eskimo in Sandwich Bay. In 1937 I saw a grown-up Eskimo girl as a servant in Cartwright, but otherwise I was unable to find any of them south of their present habitat. Their present southern boundary runs along the southern shore of Lake Melville towards the east from the English River and over Back Bay right away to Hamilton Inlet or Ivuktoke Bay, also called Eskimo Bay or Groswater Bay.

The present southern limit of the Labrador Eskimo is also a cultural-historical one. South of this limit the Eskimo had learnt to know the whites as bringers of misfortune —, murder, plundering and epidemics followed in their tracks. But at last a nobler form of the white man's civilization had found its way to the Eskimo from the north. The Moravian Mission came like helpers from above in 1752 to the northern parts of the coast and the Eskimo now met a type of white man hitherto unknown to them. The ruthless exterminating of the Eskimo south of Hamilton Inlet started by the whites and the Indians on the one hand and the loving care which the Moravian missionaries showed for the Eskimo on the other must have stood in sharpest contrast. The white people must have seemed extremely paradoxical to the dark-skinned heathens, more paradoxical than the heathens to the white man.

For reasons already given the most southerly Eskimoes in the world at present live on the sound between Lake Melville and Hamilton Inlet, The Narrows, and they still obtain their livelihood in the main from the traditional hunting; cf. Fig. 223. Their southern boundary is a line running from West Bay, somewhat above 54° N., towards the south-west to the Mealy Mountains. South of this boundary there are a few half-bloods, but they have completely given up their ancestors' way of living and assumed the so-called *livyeres'* (p. 727) habits and customs, these latter now dominating the southern part of the Atlantic coastal strip. Only occasionally can some people of compara-

Fig. 223. The salmon-fishing camp of the most southerly Eskimoes in the world at Lake Melville. On the lower picture the Mealy Mountains are seen in the background
Photo the author, 1937.

tively pure Eskimo blood be seen south of this line. To the north of it again the Eskimoes and the half-bloods, so-called *settlers*, have a small coastal strip which is only crossed by the Naskaupee Indians now and then when they come down to Davis Inlet or Nain to trade with the Hudson's Bay Company.

Within this coastal strip the phenomenon usual among primitive people is repeated, towards the boundaries their numbers are less. The Nain district is almost wholly occupied by Eskimoes and the settlers are an inconsiderable minority, while to the south the number of the latter rapidly increases. On January 1st 1939 there were 106 Eskimoes and 184 settlers in the Hopedale district, but in the Makkovik district were only 5 Eskimoes and 258 settlers; here the number of Eskimoes had diminished very early. PACKARD in 1864 tells of two Eskimoes living at Strawberry Harbour: formerly a few of them lived in the region south from that place, but they have died off within the past few years; they have gone with the game, banished like it to the arctic regions, maybe also faded away before the better armed Naskaupee Indians. When the Eskimoes disappeared the white half-breeds came in their place.

The Rev. PAUL HETTASCH told me that the boundary to the Eskimo's coherent habitat should to-day be somewhat north of Hopedale; scattered among the Eskimoes are however a few families of white and mixed stock. The Eskimoes who have their huts near this limit, have had so much contact with the mailsteamer that in many respects they have changed their economic life, are becoming Europeanized.

One more interesting fact deserves mention in this connection. While the Eskimo culture has been forced further and further north during recent centuries, a movement in the opposite direction can be noted in the beginning of our century at the northern part of the Atlantic coast; here the Eskimo have begun to drift with their homes farther to the south. The reason is said to be the lack of fuel (!) in the north (603).

Small internal migrations, on the other hand, are usual. Some years there was an abundance of seal and in others they were scarce and inadequate for the settlement's requirements. In 1840, for instance, the Eskimoes began to desert the coast north of Hebron and went to Ungava. The Nachvak people had expressly pointed out to the Mission that if they had to leave their own country they would starve, the probability of which could not be denied.

The means of livelihood.

Although the climate is hard and the country apparently so inhospitable, the supply of game both in the sea and on the land, seal and caribou, and later fish and fur-bearing animals, has made it comparatively easy for the Eskimoes in Newfoundland-Labrador to get their living. The main supplies of the sea are different seals, small white whale, dolphins, and walrus, with fish such as cod of different kinds, salmon, and sea-trout. The seals provide the Eskimo not only with their main food but also their clothing.

No less than five kinds of seal together with the white whale and the walrus are hunted. The ordinary seal — netceq — has always been his main food and it occurs near the coast at all periods of the year, in summer in the bays and in winter it is caught in the breathing-hole or out on the ice-edge, sinâ, where the winter-ice pack meets the coastal-ice belt (p. 426). The ranger, the only seal which lives in fresh water, provides a highly valued, spotted skin. The jar-seal is the most usual on the coast, its flesh is appetizing food and the skin is used for clothes, boot-legs, sacks, and formerly was used to cover the tents, sometimes also to cover the kajak. The bearded or big seal is next to the walrus in size and is common on the coast; the skin is strong and is used for boot-soles, drag lines and dog harness, and formerly also covered the umiak, the big skin boat, if there was no walrus skin at hand. The harp or Greenland seal is to be found in early spring in enormous numbers on the pack-ice; its skin is used for boots and other leather articles, also for the kajak and the tent. The hood, the next largest seal, generally visits the coast north of Nain at the beginning of May, but only in small numbers. Towards the north at the coast the catch of seal and white whale increases in importance.

Since time immemorial the caribou has been a constant object of hunting both for its skin and its flesh. The flesh of the black bear is also valued.

The trapping of fur-bearing animals on the other hand is a comparatively new source of earning among the Eskimo. It became a regular source of income only after the price of furs began to rise.

Catching cod with the 'jiggar' is really also new among the Eskimo during late summer, for cod-fishery became more extensive only after it had found a market here in the north, e.g. salted cod. Before there was a market for fish the people hunted only the seal, walrus and the black and the white bear.

These are the creatures on which the people of the coast, Eskimo and settler alike, depend for their food, and not only for their own food but also that of their dogs (319). The food of the season may fail, stormy waters or unsafe ice hinder the search for seal, soft snow makes travel for men and dogs toilsome and slow in the caribou lands, blizzards may stop all travel; even if fur-trapping brings in money to these folk, these hardy hunters can never be at their best on bought foods like flour and biscuits. The people must have meat, fresh or *nipko* [dried seal meat], for their well-being. The missionaries have therefore taught them to set aside something for the days of scarcity (478).

Hunting and fishing have always been seasonal occupations: seal in the springtime among the breaking ice, trout and codfish in the summer, seal again in the stormy waters of the autumn, and seal and occasionally walrus when the sea has frozen and the hunters can go to the edge of the ice with their harpoons and their skin canoes. In ordinary years there is always hunting of one sort or another; no sooner is one hunting season over than another begins, sometimes two sorts at a time. There is very little idleness in an Eskimo hunter's life (476).

Primitive folk's ways of living usually react very sensibly to every variation of the surrounding natural resources and to the possibilities to utilize them. In Labrador the industry of the Eskimo has had time enough to develop into correspondence to the conditions of their habitat and the possibilities to make use of the available resources. The pattern of their occupations therefore varies somewhat. This variation in the annual cycle of the Eskimo on the Atlantic coast can best be outlined by three examples: the cycle of the Eskimoes of Hopedale, of Nain, and of Ungava. As the Nain people's way of living corresponds more nearly to the habits of olden days than the others we shall begin with it. We will in the main follow DR HUTTON's descriptions and the Rev. HETTASCH's and my pilot JULIUS NATHANAEL's tales.

WORKING LIFE OF THE NAINEMIUT, THE FOLK OF NAIN.

Spring.

The *sinâ* has begun to break up and the seal hunt of the spring is finished. But the people still lie in their camps on the sound where the seal can be caught in the spring. The women fetch wood and water; they also bring home the game necessary for food, for the men do not trouble to bring it by dogs except when the distance is great. The women skin the seals, tan and soften the hide; then they make clothes and footwear of it. The men, on the other hand, after the hunt spend a great part of their lives in the camp eating, sleeping, and smoking. It is a custom preserved from the days of the purely hunting nomadism when, according to their idea, only the hunting of seal and wild reindeer was a worthy occupation for the men. Yet during the hunting season the men also are constantly on the move.

The spring has finally come in the air, the snow melts, the bay ice begins to crack. It is time to return to the mission station. But the spring flitting must be carried out while travelling is safe and easy. The air is warm by day and the snow slushy, the nights are chilly and the snow is then covered by an icy crust; so the Eskimo family starts early in the first light of the rising sun, the sledge heavily loaded. Under the warmth of the May sun the dogs will be tired if one does not start early; Fig. 224 (476).

The station is home. But do not forget this is a tribe of wandering hunters. Some families are nearly always on the move. They take trips into the inner bays to catch rock-cod for dog's food and to be nearer the firewood. Some people again have gone seal-hunting among the breaking ice, waiting for the return of the harp seal from the south. This seal disappears only with the pack ice, and after that the Eskimo is compelled to hunt the other seals. They shoot them when they have whelped on the sandy banks of some river mouths or in some bays.

The early part of June is the dreariest time of the whole year: one is shut in, no sledge, no boat, the water is packed close with floating ice, the snow is softening. One waits for the ice to float away and leave the water clear for navigation. Now, the people have returned to the village. The men mend and make nets for the autumn seal-hunt. The boys teach puppies to haul the sledges (478).

Summer.

The summer brings the halcyon days, it is possible to drift about in a boat among the islands, pitch one's tent on one of them and enjoy, as only a child

Fig 224. Sledge dog team. To the right the Rev. P a u l H e t t a s c h at Nain (205).

of nature can do, freedom from care in the magnificence and beauty of his home surroundings. Some families have already erected small huts on the islands or in the bays, Fig. 225. With that the custom of having a settled summer home is growing up. Yet many of them still live in tents which are moved from place to place so long as the summer fishing lasts; cf. Fig. 226.

Not all the families that flit in the springtime go seal hunting. Some Eskimo travel along the coasts and collect the eggs and down of the water birds. These eggs are saved until they have a specially 'high' taste, and play their part at the winter festivals as a great delicacy. Puffins and guillemot are caught in the summer with nets and preserved in fat for the winter, it is said.

Sea-trout fishing.

Some families take their nets to the haunts of the sea-trout. In the early summer, when the ice has broken up and disappeared, the trout come down to the sea from their winter home in the fresh-water ponds of the far interior, and after a short time in the salt water they become fat, so the fishing mainly begins in July. Fishing for sea-trout has prevailed among the Labrador Eskimoes since ancient times, but it is believed they did not then know the use

Fig 225. An Eskimo summer dwelling at Port Manvers and its 'mount guard' (lower picture). Photo the author, Aug. 1939

of nets; they erected a stone barricade across the mouths of the rivers in which the fat trout went in with the tide and when the pool thus obtained dried up with the ebb, they speared or took up the fish with their hands. This method of catching is now forbidden and, as the Eskimo is very law-abiding, he now uses nets spread out in the shallow water where the big rivers enter the sea. The catch of trout continues as long as they remain in the bays, until the middle of August or a little later. As far as money goes trout-fishing is more profitable than seal-hunting; it is always fairly certain and salted trout always fetches a good price at the store. It is the women who split and salt the

trout. The catch can be large, many barrels or so (476).   The Eskimo also
dry the fat trout for winter food and then call it *pepsi*.

In olden days the sea-trout was taken by spearing under the ice (530).

Cod-fishing.

In July, when the cod fish throng the deep channels among the islands,
the Eskimo goes afishing. A good season not only pays the debts which the

Fig. 226. Eskimo camp at Ryan's Bay   Photo the FORBES-GRENFELL EXPEDITION,
Aug. 1931.

family has piled up at the store during the winter and spring, but gives him
new tools, guns etc. Therefore the news of cod — *oggak* — in the bay will
send many people post-haste to the fishing; very few stay at home, though
there are of course a few thriftless people among them who linger in the
village long after the beginning of the work. In the months of August and
September this occupation keeps them busy and the whole of Labrador smells
of fish. Yet the cod shoal never stays at the same place and he must find
out where it goes and follow it in his boat. The Eskimo is a wonderfully

strong oarsman; he can row for hours without resting; he knows too how to make his boat travel quickly by short, sharp strokes of the oars. He has all the grit and simple perseverance of the trained fisher and hunter (475).

He will at last find the right place, and here the fishing begins with a *'jiggar'*: a bright piece of lead shaped like a little fish and armed with two barbed hooks. The patient jerk-jerk-jerk of the arm goes on for hours and the cod-fish tries to swallow the jiggar before him and is caught. Thus the fisherman jerks his jiggar up and down, and if he has success pulls up fish in goodly numbers. But on bad days he may 'jig' for hours, moving his boat from place to place in search of a shoal of fish, without a catch. In stormy weather he has to remain at home (475).

Often the shoals of cod come into the bay near the mission station as in 1939, and if the fish is biting well, the bay will soon be dotted with boats, in every boat one or two men are jigging. They stay on the water for hours. But, even in the bays the sea can be high enough to swamp a boat, and in water at only two or three degrees above freezing point a swim is an utter impossibility. Therefore, when the spray comes whipping along with the rising wind the fishermen take warning and head for the shore. And though there are no finer boatmen in the world than the Labrador Eskimo, they have thought many times that their time had come, HUTTON stated.

Day after day men and boys 'jig' from morning to night. Patiently they jerk the bright leaden lure up and down within a couple of meters of the sea bottom. There are times when the fish is so plentiful that they are on the hook before it is well sunk. The jigging is done only by the men and boys, the catch is given the women and girls for splitting and salting. They take much care in the drying of the fish on the racks; they scurry when a sudden shower of rain comes and threatens the half-dry fish.

The quantity of codfish in the sea is here astonishingly great, they must teem in myriads along the coast; year after year not only Eskimoes but hundreds of schooner crews gather the fish by tons and year after year the fish is there again quite as plentiful as ever.

During the summer months the mission settlements are almost deserted as nearly all the able-bodied inhabitants are away at their fishing camps. Some families live scattered along the shores of the bays and runs. Some have a few little huts there (and earlier *igloos*), whither they return year after year. But most of the people live in tents which they pitch on some little island. At heart they are wanderers, Dr HUTTON wrote, restless beings following the call of their hunting and making a temporary home where their work is. The Eskimo is very industrious at this time; my experience is that it is no easy matter to get a guide when the codfish has been seen in the bay, everybody is busy.

At the end of September the main rush of the codfish is over. When the autumn days come the codfish is moving away to the deeper water and jigging is giving little result. Many of the people then make their way home with their great bundles of fish and mend the seal nets.

When the summer fishing comes to an end, not a few of the Eskimoes remain away at their camps till near Christmas, when they can return with their dog teams over the sea-ice which is always strong enough some weeks earlier. The Eskimo now spends his days on the water looking for seals. He goes out to the various hunting-places. Hour after hour he sits like a man of stone waiting for the wary seal. The kajak has been an excellent craft for this work, easy to turn any way and riding the waves like a cork, but steady because the hunter sat low.

When they were away at their seal, trout and cod fishing places the village was almost deserted; the savage mother-dogs nursing their litters of puppies under the doorsteps are the only signs of life. When the sea freezes the people begin to make their way home to the village.

Caribou-hunting in bygone time.

In former times before the Eskimo began cod-fishing they used in July to move with their whole families to the heads of the fiords, there they left the old ones, the women and children while the hunters went into the land to hunt *tuktu*, caribou, in August and September. Four of the Eskimoes went up the country from Saglek on July 4th 1811 to hunt caribou, KOHLMEISTER (530) writes. Farther inland in Tessiujak fiord there is a place name which still bears witness to this: Ittukorviak (= 'the place where the old ones are left'). But at that time there were feuds between the Eskimo and the Indians and the former never ventured far beyond the heads of the bays, so the result of the hunt was probably limited.

This custom is no longer known and only a few individuals make their way in summer to the interior to hunt caribou when they are tired of a fish diet. In our days the caribou is hunted chiefly at the end of April and the beginning of May, — curiously enough just at the time when the flesh and skin of the animal begin to be less good. The reason must be that the possibilities of communications are just then at their best, for all the brushwood in the river valleys lies under snow. Yet some hunters pursue the caribou also during the fox-hunting period in late autumn and winter; then *tunnu*, the suet of the caribou's back, is a favourite tit-bit of the Eskimo.

Autumn.

Seal-hunting.

Formerly as at present, the hunting was chiefly concentrated around the seal, as already stated. The Eskimo must have seal to keep well. Still to-day the seal is of integral importance for the people in Nain; a winter without seals means that the Eskimo must give up some of his hungry dogs. Without dogs he can neither go to the ice-edge nor fetch firewood; he can have no boots, no clothing, no meat, no blubber. The seal seems to have learnt to go northward rather than stay in the neighbourhood of the Eskimo villages, and nowadays they do no more than pass by in the autumn and the spring. Added to that seal-hunting is his greatest sport. Never shall I forget a boy who nudged me in the back and whispered: *puijei* (= a seal), his face grown tense and eager. It is not easy to discover the animal. He looks just like one of the rocks close by. The mimicry is perfect in the autumn, an imitation of the black boulders with their coating of ice (476).

To better understand the seal-hunting it is may be of interest to give some information here about the biological habits of the principal game, the harp (cf. 45 a).

Late in the autumn the harp comes from Melville Sound, and from even more northerly waters during November to February, going southward. About the first of March they bring fort their young on the ice-floes drifting south off the coast as far as the Magdalen Islands in the Gulf of St. Lawrence. For this they herd together in enormous groups on the floating ice. The young are almost all born on the same day. Absolutely unable to escape, from 150,000—500,000 animals gathered in these herds are killed by the Newfoundlanders. The young, the »white coats», have an exquisite fur (364). The mother at last forces the remaining pups to take to the water, a mysterious instinct at once teaches them to go north, and by the end of May these 'beating seals' have mostly passed along the Labrador coast. The next winter the 'beater seals' return south again as 'bed-lamers', the second winter as 'young harps', in the fourth year having the dignity of 'old harps'.

The harps which have escaped slaughter by the Newfoundlanders are still not safe in the north. In May and June huge frame nets enclosing a great space are put out from a capstan on the land, all along the shores, and when the harp enters them the net is raised on to the land. The Eskimo catch the harp in ordinary seal nets. Also the harps are shot as they noiselessly rise into the fissures between the pans to take breath. The dead seal floats for a short time at this period of the year, and the body must quickly be hauled up on the ice.

The last weeks before the definite freezing of the sea are a busy time in the villages: seal-hunting is beginning.

The Eskimo move out to the islands where the seals collect in the channels with a strong current which do not freeze so soon. The hunter has noticed the places where the seal has a habit of diving and there his net is drawn along the bottom across the channel. In the beginning of the century, there was no lack of nets. From the middle of October the season may be said to have begun for seal-hunting. Then herds of Greenland seal appear, first the young ones and then the old.

The Eskimo will avoid competition in hunting. Therefore he often goes to the seemingly most unpromising places, near the restless Atlantic. Many of the places are pictures of utter desolation, like the Salomon Island which rises steeply from the sea, and is surrounded by mere jagged reefs sticking out of the foam in the shelter of which the seal loves to swim. However, my friend Julius's minute sealing-cabin nestles there among the rocks. Where the coast is broken by swarms of tiny islands there it is especially good to put out the ground nets. We must agree with Dr HUTTON, that such places, to us Europeans no more than a picture of grandeur, looks different to Eskimo eyes.

In older times harp-hunting used to begin on the first autumn ice in the fiords, Dr HUTTON reports: Strong winds had cooled the surface water for some weeks, the west wind died down somewhat the cold increased, the sky was clear, and finally one morning the water was no more to be seen because of the frost mist. Everyone knew that the fiord was freezing, and that now the seal was moving to the channels and the bays, and the hunters started out. First the animals went to certain places in the bays; when these were blocked by the ice from the rivers and froze with the increasing cold, the seal moved on to the channels kept open by the strong currents and there herds of the Greenland seal would remain. There the hunter crept out on the ice only twenty-four hours old, harpooned or shot the animal and dragged it up on the ice. Under favourable conditions each hunter could average ten to twenty seals. And twenty seals amply provided for a small family for the whole winter. This life suited the Eskimo well.

A curious method of hunting the Greenland seal still prevails in cold winters. The animal remains in some places along the shore and the narrows where the battling currents give the ice no chance to set even in the bitter cold of January.

Then when the channels finally begin to freeze the seal climbs up on the ice and begins to creep towards the sea. They sometimes mistake the woods for the open water, get lost in the woods, and there are surprised by the tree-fellers and killed. On the ice the Greenland seal moves so quickly that the hunter must run in pursuing it. Sometimes a whole herd of seal will come up in the last open hole: in Hopedale some years ago about twenty large Greenland seal were taken with an axe out of a single ice-hole.

In modern times, however, nets must be the order of the hunt before the sea freezes, nets stretched along the sea-bed in some favourite channel or inlet. They will lie there over night while the hunter waits in a smoky hut till morning when the nets will be hauled in. The bay seals especially are captured in nets anchored to the bottom; they will soon be drowned as they cannot rise to breathe. Sometimes they too travel like the harp long distances over land. In winter they move out to islands in the open where the sea does not freeze soon. It is a cold job, ropes and nets are frozen and the heavy seals are stiff and dead. Sometimes the hunters have to pile up a sort of stockade of seals taken from the nets.

For a fortnight the hunters are busy with their nets. But the sea freezes, and the nets freeze with it and they have the awkward job of hacking them out. In the middle of November the Okkak Bay begins to freeze at the edges. After a fortnight the bays freeze and when the 'sikko' (= ice) covers the sea, the seals are away to their winter haunts at the edge of the ocean ice (476).

— The use of seal nets was probably introduced among the Eskimo by the white people. CARTWRIGHT mentions them as early as the 1770's (134): in the south-eastern part of the peninsula the seal nets must be out by November 20th. The length of the sealing seems to have depended upon the temperature conditions; when the anchor ice, or 'lolly' (p. 270), began to form the nets must be taken in, he reports. —

There are of course many variations in the normal habits connected with the supply of seal or the weather conditions. For instance if it is very cold and masses of snow fall early the time is not favourable for seal-hunting. If the store of food is running low the hunters must first devote themselves to hunting caribou in the interior. Yet of late years seal-hunting has been neglected in the autumn because of the lack of nets. When the Moravian Mission managed the trade with the Eskimoes its people used to lend them nets, but since the Hudson's Bay Company took over the trade, although the Brethren in England have collected money to procure seal nets for the Eskimo, it has not been nearly sufficient, and they now seem to be so poor that they cannot afford to buy nets, for they cost several hundred dollars. The seal nets used in the autumn are enormous, 3.5 by 50 fathoms, with a fish-chest. It is heavy, costly equipment and was therefore often owned by a group of Eskimoes who used it together. Formerly it was said that a catch of two to three hundred seals per net could be expected in the autumn season. The co-operative principle was even in olden times not unknown among these people. If one had a seal net he often lent it to one of the younger men and shared with him the resulting catch. Dogs, too, were lent to a neighbour, the owner getting a couple of logs or other firewood in payment (475).

In November when the sea is freezing and the snow begins to drift upon the land the hunters live rather quietly in their tents or huts mending their nets (476). A large number of the people even to-day pass the late autumn and early winter in the bays as already mentioned and beside the channels employed in this kind of hunting and catching seal until the water has completely frozen some time before Christmas. Then they get their sledges ready and a few days before Christmas the hunters return with their families to their winter homes in the mission village.

Some of the booty is carried home on dog-sledges and divided among the families of the hunters, being used at once for food and clothes. Another part is left upon the ice as the hunters' personal property to lure the foxes, and is laid out in traps with good results.

A few of the hunters at this time have visited the inner parts of the fiords or the lakes to catch foxes. But even they generally make their way to Nain before Christmas in order to celebrate the holy-days. By the middle of December the village looks fairly busy (476).

Winter.

Home for Christmas is the great idea! Now begins a large influx of visitors. It is a cheerful sight to see the scores on scores of brown faces shining with good food, their bright eyes twinkling with pleasure, Dr. HUTTON relates. The excitement grows and grows, culminating on Christmas Eve.

Work at the station.

But life in Labrador does not lend itself to fastidiousness. The Eskimoes work hard in general.

Only on stormy days when they can do nothing out-of-doors do the men sit at home smoking and talking, making and mending nets, dog's harness, dog-whips, sledges etc. Earlier some of them used to carve ivory, but now this is practically a lost art among the Nainemiuts; walrus tusks are scarce since this animal has been scared away to the north; also time is more precious nowadays and the market not good, so the young men of to-day do not bother to learn to carve in ivory (476). Instead the father teaches the boys all a hunter's tricks and dreams of a day when the boy has grown up into a handy and clever hunter and can be the stay and companion of his father's old age. The boys under their fathers proper supervision are occupied with the puppies, teaching them to draw the *koma'tik* (=the sledge); cf. Fig. 224 With an indescribable sweep of the arm even the small boys learn to send the thirty feet of walrus hide lash hissing through the air, and with a sharp flick catch a disobedient dog. He drives about with the sledge and dogs and fetches firewood or helps at the chopping. They are as full of fun and mischief as it is possible for a boy to be (475). On stormy days the children also make baskets.

The Eskimo housewife spends the day working in the dull light of the hut at the seal that her husband brings home. She cuts and stitches with neatness. She has to scrape the skins, cut out, chew the leather to soften it, stitch and mend. Eskimo teeth are made for chewing, and the chewing of the boot-leather is women's work from one end of life to the other, HUTTON states. She makes boots and clothing for the family. The art is passing but the skill is still there and only needs rousing. The sure eye is still seen in the fact that the bootmaker with her eyes only can judge the size required by her customer, she has no need to use a measure. It is a surprise to see the teeth of the women often worn almost to the gums. This is due to the fact that when making long boots they are obliged to chew thoroughly the edges of the sealskin to prepare it for the needle. It is also the wife's horrid duty to chew the boots into comfort for the husband's feet if they have become hard and stiff with wear and salt water (761).

The mother has until most recent years cut and sewn all the clothing for the whole family. The clothing of the bigger ones descends to the smaller ones in turn, so that from one cause or another the peculiarities in the cut of the father's clothes reappear in the rest of the family wardrobe; it is the inherent tendency to imitate. — Besides this the housewife has to scrub the floors, floor-washing is an established custom in most of the houses — and the wood-chopping for the stove also belongs to her business (476). The old widows become 'blubber women': they are employed to cut up the seal blubber and boil it: also to clean the eiderdown.

The growing daughter is the most useful member of an Eskimo family: she minds the babies, she chews the leather and scrubs the floor: she goes out fishing; she also fetches the water, or in winter she fetches large lumps of ice in a bag and thaws them for drinking water.

Caribou and fox-hunting.

But shortly after the holiday and the celebration the hunters again disappear from the stations. Most of them go off on their dog-sledges over the fiords and up the valleys: they go partly to hunt the fox in the valleys themselves and partly to hunt the caribou in the surrounding tracts (1018).

These hunts may be repeated several times in the same winter, and in between them the hunters stay with their families at the stations and gorge upon the spoil they have brought home.

The real fox-hunting season does not begin till after the New Year. Some of the families who have not stayed there all through the festive period, then return to the bays where they hunted seal in the autumn. Then they proceed to set out their traps. With the same object in view some families move to the farthest islands where they have small huts sometimes also used when netting the seals in autumn and to which the foxes have been drawn by the seal carcases left there. As the days become longer, especially in February, a few hunters — not the worst — go off to the plateau in the interior where they kill caribou to be used as bait for the foxes. These hunters stay in the interior till the dogs' food begins to run low and then must be fetched from the autumn stores. Soon the white fox comes; it generally follows the caribou herd to eat up the remains of the meals of the wolves which always pursue the caribou. The fox-hunting season finishes on March 15th and April 15th for the coloured and white fox respectively. The white fox, however, also comes down from the north with the drift-ice which gradually freezes on to the ice-edge along the coast; thence he wanders over to the farther islands and there he is trapped. But on the ice-edge also the fox is caught. When the Eskimo discovers the tracks on the *sinâ* he knows the fox has come down from Baffin Land to feast on the remains of the seals left on the ice-edge; he follows them with his dogs, lays out his traps, or even shoots this highly valued fur-bearing animal; in 1921 no less than 1100 white and black fox-skins were sold, the wearers having been caught in this fashion. For the rest it may be said that the Eskimo takes anything of nature's store which can be of use to his way of living.

In our days trapping is of great importance to the Eskimo. A night's trapping can make a poor man rich at a single bound. Moreover the trapping suits him. The best trapping places are pieces of broken land dotted with stunted trees and hidden behind the hills. Here he sets his fox-traps.

During the long winter the men are seldom idle, they are mostly absent from their homes. There is always the strenuous work of log-chopping to satisfy the big need for fire-wood.

Seal-hunting.

As the sun rises higher in the sky and the hard frozen snow cover is formed by the wind in the later part of the winter, the Eskimo packs his family, his kajak, and his necessary household utensils on to a 'komatik'; the dogs are harnessed and off they tear at a furious pace to the distant islands, there to pitch camp.

Now the early spring seal-hunting begins! From the camp the trappers make their way to the edge of the ocean ice, the *sinâ*, where the seals sport in the chilly water or clamber up on the ice to rest. Hunting the seal and the fast dwindling walrus can begin as early as February. But it is in May, when these animals and the polar bear are making their way north again, that they can be most successfully hunted. And this has been the Eskimo's favourite occupation. From Nain the hunters often drive thirty miles over the vast barrier of islands to the ice-edge of the open sea to catch seals. The keener men stay among the lumps and hummocks and hollows of the sinâ for some days: sleep all night, hunt all day. The stolid hunter does not know what danger is, climbing on to a floating pan of ice he paddles away with his hands to fetch his seal. The clear calm days are real days for the sinâ. The hunter tries to find the seals' 'blow-holes' where it comes to breathe. With an inherited, unlimited patience he stands at the hole ready for the animal's next visit,· and with a shot from his gun or by a flash of the harpoon sunk in its fat neck its fate is sealed. A skilful and strong hunter takes it up simply by the tail. He hauls his catch on to the ice, cuts a hole in the neck and takes his drink of blood (761).

Another kind of seal-hunting is the *otok*-hunting. An otok is a seal that basks upon the ice in the warm spring sun, and lies for hours as motionless as a log. No sooner does the Eskimo see an otok than he arms himself with a protecting shield of white cloth and crawls along as cautiously as possible so that the otok never notices the slowly-moving body before it is near enough to let drive with the deadly rifle. The old method was to creep up to an otok-netsek with a harpoon without any protection by lying flat down and waving a foot to represent a seal; then the hunter used also to cover himself with a seal's skin and imitated the animal's movements.

The gray seal, too, the largest of all Labrador seals, is shot as it plays along the ice edge, but occasionally is caught in sunken nets. Yet by far the best fortune is to catch sight of a clumsy walrus — *aivek* — resting on the ice, and to rush boldly upon the formidable, apathetic beast and shoot it while it is too dazed to move. The walrus belongs to the outermost belt of the coast, and when the new ice forms at the sinâ, it creeps up on it. Thither the hunters also make their way, but only the boldest. For if a ground swell comes suddenly and the covering of new ice begins to bend and curve then one needs every bit of skill and speed to save oneself; often in such a case the game has to be left behind. It was the Okkak people who gave birth to the boldest and most skilful walrus-hunters. The walrus is an enormous lump, fifteen feet long and fourteen feet round the middle, weighing some 2500 pounds. An old Eskimo custom was to cut it up at the sinâ, because it was too heavy a load; every one who has seen the capture has to have a share. The capture of a walrus meant for the Eskimo: meat, clothing, footwear, light, house, boat, nets, and oilskin jumper for the rainy season. The ivory of the tusks was of great value.

To-day opinions differ as to the walrus and its importance to the Eskimo. Those best acquainted with local conditions seem to believe that this animal is about as plentiful, or perhaps more correctly as scarce, now as it was in the time of the older generations. From what the Rev. PAUL HETTASCH told me, the walrus is not now so important, because it is scarce and difficult to catch. This is also true of the white whale.

When on sinâ hunts the Eskimoes are in the greatest good humour. One remembers what JOHN DAVIS wrote of them in the 16th century (210): They are never out of the water, but live in the nature of fishes. And this was described at the end of the 19th century in the following words: Evolution and environment have produced a type of human being which has actually some points of resemblance to the animals upon which it principally subsists (325).

As the spring passes and the ice breaks the Eskimo chase the seal among the floating ice pans; boats are needed and speed is necessary, the seal has lost its winter coat of fat and sinks as soon as it is dead. He must act immediately when a seal's head pops up.

The old style hunting in kayaks with their ingenious walrus-tusk harpoons, one of the most marvellous weapons of primitive man, is nowadays still in use only in the northern parts of the peninsula. In our days the hunter prefers guns to harpoons. The picturesque is gone, the rifle makes the hunt so much easier. But it has at the same time been disadvantageous: many of the killed seals sink immediately, especially in the spring, and the detonation chases the animals away, and they go to the far north, where they can fish and gambol unmolested (476).

The danger of hunting on the sinâ can even take the hunter by surprise. It sometimes happens that, engrossed in his watch for seals, he fails to notice that the ice on which he is waiting has broken away from the main ice-field and is drifting out to sea before the wind. The off-shore wind freshens as the day wears on, a wide strip of water opens between him and the shore, the block is swept out to sea in the grip of the gale, the current too is strong and finally the land is no more than a low, blue strip. The sledge and dogs are with him, he has seal-meat and will not starve, the snow upon the ice will serve to quench his thirst. He is moving farther and farther in waters where at that time of the year there is no hope at all of any ship. The hunter has only to be reconciled to his fate: to be lost by the breaking or foundering of his ice-pan. But it also happens that the wind changes and the floating island turns onshore. It happened thus for those Nain men who were adrift for fourteen days on a pan of ice, but landed near Hopedale, 100 miles to the south, safe and sound. GRENFELL describes still longer journeys on an ice pan. Seldom does it happen on this coast that an Eskimo loses his life in this way (478).

Early spring.

At Easter the Eskimo visits his village again. It is good travelling time, the ice is sound and the days long. But he will stay there only for a short time.

Caribou-hunting.

When the sealing at the sinâ and trapping in the woods come to an end, the »*tuktu*», the caribou, occupies their thoughts. Custom had fixed Easter Tuesday as the day at Okkak for the beginning of the reindeer hunt, the great event of the hunter's year; it is late but no Eskimo would cut the religious services to go a-hunting. Year by year the same start (476). In the other villages one was more liberal. When the news comes that the caribou have been seen great excitement prevails. There is stir and bustle all day long preparing for the journey up the frozen bays and rivers westward to the moss-covered plateau where the animals are pasturing. The hunters start out together but soon separate. One or two stay in solitary snow huts whence they go in search of the deer. Sometimes, if the weather is stormy and there are many snowdrifts, they remain there for weeks (1018).

— The Eskimo hunts where he chooses. In contrast to the Indian he has never had any family-controlled hunting-ground; the sea is the most important and that is open to all. This is equally true of the caribou hunting-ground in the interior, he thinks. He has therefore never a thought of singling out any special hunting area, for to him the wild game belongs to society as a whole. His ethics prompt him to give way rather than to quarrel on account of it. On the other hand when caught the game is shared out according to complicated rules. In consequence the abodes of the hunters often move, one summer they are here, another there. —

At the beginning of this century the caribou was still being generally hunted near Nachvak and Saglek, Nain and Davis Inlet, where small, more or less stationary herds of about 70 used to appear. But then the hunter never went far inland — never beyond a certain distance — their limit was that of half the quantity of dogs' food they could carry with them, they had to retain enough for the return journey in case they killed no game. The best dogs' food is seal flesh or caplin; a coarse paste of oatmeal is kept as emergency food (761). The large cache of dried or frozen caribou meat which the old folks talk about belongs to history; the frozen meat was eaten raw, but the dried meat was soaked and then boiled.

The small herds mentioned above seem to appear no longer in the neighbourhood of the coast, and the hunter must now seek out solitary or small parties of caribou.

The hunters follow the tracks and circle the herd so as to come upon it from the lee side; they know by the freshness of the tracks where approximately the herd is to be found. The dogs shall not come too close to the deer, they become wolves again, keen to do a little hunting for themselves. When they see the gray spots on the hill-side digging with their forefeet at the moss the Eskimo turns the sledges upside down and makes the team of wolfish dogs lie down; they cannot drag the upturned sledge. The hunters approach the herd, in May the cows have their little calves beside them. When he shoots the cow, the calf will not go away, it is killed too. Finally they can proudly return to the home village with the sledges piled with meat and skins.

The Eskimo seems always to be on the track of one sort of animal or another. He shoots many black bears too.

A stray white bear, *nanok*, sometimes comes along the coast with the ice sailing southward on driving ice-pans, competing with the Eskimo for the seal. It becomes the object of a furious hunt. But most of the polar bears have now retreated to desolate spots in the north where there is no smell of man. Sometimes a polar bear has been caught without guns or spear. The Eskimoes fell to their oars and got the boat between the bear and the shore so as to head him off. Then they chased him to and fro until he began to tire, and then they assailed him with their oars, hammering prodigeously at his head until they had him stunned and helpless and so could tow his carcase ashore.

Seal-hunting.

When the spring has advanced so far that the ice-edge begins to break up, some Eskimoes begin to manoeuvre their way back to their Islands on the last pieces of ice.

But when the ice at the sinâ begins to break up the Greenland seal comes back to the bays, and after a time again goes out to sea. Many Eskimo catches them during this visit with the net and gorges on the flesh; however, the seal is then very thin and may sink if it is shot.

When the drift-ice comes to an end in the last days of June, or perhaps not until the middle of July, this harp seal withdraws to the north. Now the family hurries back to the mission station, the fixed pole in their yearly round, and usually they have with them plenty of game. Then the Eskimo has to be satisfied with hooded seal, bearded seal, ringed seal, and ranger. They shoot them when they give birth to their young in the sandy river mouths or any-where in the bays.

The spring seal hunt brings the Eskimo's hunting year to a close.

A retrospect. The metamorphosis.

From all this we must conclude that the annual cycle as a whole and all its divisions are extremely well balanced in relation to the produce of surrounding districts and nature's rhythm. If he follows it, the Eskimo can manage quite well by himself without any food supply from foreign countries. Dr HUTTON for instance reminds us that the waves of the World War 1914—18 reached the shores of Labrador only in faint ripples. Fishing and seal-hunting went on as usual, the people set their fox traps in the ordinary way. But the market was disturbed with the rest of things and there was not only the difficulty of getting the goods to market but also of selling them. Food and fishing became troublesome for the white people, but the Eskimo lived well. More than ever Labrador became a little world of its own.

But sometimes the seal-hunting is a failure and hard times come to the little community. Then even the Eskimo can see the shadow of famine rise. In these conditions the hunters' generosity appears, HUTTON says; his catch he hands over to the old people of the village and goes to hunt caribou. The neigh-bouring villages give friendly help, sending blubber and meat during the bitter days, and some skin for boots and clothing. This is why the Eskimo hunting people never starve to death like their neighbours, the Naskaupee Indians, they are always neighbourly enough to find food for one another and for the worst days. When there is sickness in the camp it may happen that a dog must be killed to make a meal for the rest. Especially in the springtime, when the caribou have gone and the seals have not yet come the dog food is often very scanty, sometimes old bed-skins are chopped up for dog food and moistened with a little rank oil. The Eskimo say that over-feeding makes the dogs savage!

Since olden times the ways of living have been almost the same as that just pictured. REICHEL describes the ways of the Eskimo living at this Moravian Mission station in the middle of last century (787 a), and this is given here as a short review of the main lines of the annual cycle:

The New Year finds the Eskimo flocks assembled in their winter huts at the mission stations; at this time the attendance at school and church is the best, and the spiritual work goes on regularly. The chief occupation is the capture of the ptarmigan and fox. For the latter traps are used, which are often placed at a great distance from home and must be visited daily for fear of the bait and catch being carried off by wolves. At this season both men and women are employed by the Mission to cleave wood and clear away snow, for which they receive payment. In February, many of them go in sledges to the ice-edge along the sea-shore, where they catch seals in their kayaks, if they meet with open water, carrying with them in general, besides their gun, a telescope to discover the seal at a distance. Towards Easter the boats are repaired or new ones built. During Passion week all make a point of assembling at the stations. After Easter they usually go inland to hunt reindeer, especially from the northern stations, where the deer abound more than in the south. At the end of June eggs are gathered on the islands and the fishing season commences; this lasts till September, when the haddocks are taken. Salmon and salmon-trout are salted for winter consumption. In October the Eskimo repair to the mission stations for nets to catch seals, and remain in the bays in general till Christmas. This is the chief season for taking the seal: at times they are cut off from the sea by the sudden formation of ice in the bays, and then they can be shot in greater numbers.

Cf. also KOCH (528).

The Eskimoes to be met in Nain to-day are thus no 'human fossil' as rumour described them before I met them in person. Certainly their outward appearance has been tidied up, their manners and customs thoroughly improved and their tools and equipment modernised: if one compares them with their forefathers of a century ago and their racial cousins in Baffin Land (68 a, 223), Hudson Strait (742) or East Main (183) at the present day, it is no exaggeration to say that they have undergone a p o s i t i v e m e t a-m o r p h o s i s (more of this later on). Yet below the new creation it is not difficult to perceive the old threads in these hunting people's way of living fairly unchanged, for in the economic life the most important of the original manners and customs have been preserved to a large extent as before. Their very disposition has still, after the civilizing process, the same, sunny, open stamp which has always won the sympathy of the peaceable stranger.

At every kind of hunting, except that of the caribou, the hunter is usually accompanied by his family: the hunting nomadism continues. Yet the

Eskimoes of Nain are not hunting nomads in the strictest sense; they have traditionally at least one fixed winter abode and around this circles t h e i r h a l f n o m a d i c l i f e at the present time.

WORKING LIFE OF THE ARVEKMIUT, THE FOLK OF HOPEDALE.

When the spring seal-hunting is at an end the hunter vegetates for a time in his hut at the mission station in Hopedale. But when the shoals of caplin make their appearance he devotes a good deal of his time to catching it; some of it is eaten by the people, but some is covered with a little train oil and given to the dogs. Salmon is not much caught, but when it is it totals only about 5000 lbs.

Then follows a dead period if the cod do not appear at the usual time. Some devote themselves to putting their nets in order and the house is brightened up.

Normally the cod-fishing begins about the 15th July and lasts well into September. Usually September and the first part of October are chiefly employed in felling trees and cutting firewood, and this can continue till the hunting of fur-bearing animals begins.

About the 15th of October, when the fur has grown and become beautiful, the Arvekmiut begin trapping, especially the fox. The hunting-ground is not far away, at the most some 50 miles from the station; mostly the Eskimo lays his traps out on the islands where the seal can also be caught.

Most of the Hopedale Eskimo begin to hunt seal when the fiords and sea begin to freeze, for then comes the Greenland seal from the north, the greatest prize. Those who have nets can catch seal even earlier in the sounds and the fiords, but their poorer brethren must shoot or harpoon the seal in the open holes or on the ice. At this time the hunter's base is at Hopedale, but he extends his hunt for miles out to sea, seeking the breathing holes and the crevices where the seal lies; on many of the islands wooden shelters have been set up and these give the hunter some little protection from the elements. Thishunt often lasts only a week or a few days depending on when the seal comes and how cold it is, for when the sea freezes all over it is impossible to pursue them.

After the seal-hunting of late autumn the Eskimo devotes himself to trapping. But if the ice conditions are favourable he goes off now and then to the sinâ, 10—20—40 miles from the coast, and there seeks the breathing holes. This form of hunting is precarious, for if the drift-ice comes down from the north or all the holes are frozen over that puts an end to it; a ground swell, too, can force the hunter to return home. Sometimes a few walruses have been caught far out on the sinâ.

After the 15th of March it may be said that sinâ-hunting is at an end: the Eskimo then takes to felling and chopping fire wood once more.

But again the snow begins to melt and again the Eskimo makes his way out over the ice to seek the breathing-holes and catch seal.

Later, when the winter-ice has broken up and the great herds of Greenland seal wander along the coast on their way up north, the spring seal hunt begins at once with net and harpoon, for the seal disappears with the drift-ice. With this hunting finishes the cycle of the year.

These facts I received from the Reverend GRUBB, who for decades has had a good deal to do with these Eskimo. From what he related it is clear that the main characteristics in the working life of the Eskimoes have been preserved in this district also.

There are, of course, some differences in their seal-hunting methods conditioned by the different geographical characters of the two districts. In Hopedale the islanded area is comparatively limited and open, with low islands and small fiords, where the height of the tide can be nearly levelled out. The Nain archipelago is extensive, the islands larger, the fiords narrow and deep, and the tidal currents attain a considerable strength and size. In Hopedale the semi-nomadic way of living has almost ceased because of the short distance to the trapping-places, which makes it possible to use the station as a base.

Seal-hunting is also in principle the main earnings of the disparaged Hopedale Eskimo, and their continued existence as Eskimoes stands or falls with the exercise of this form of livelihood. But in this respect it is said that their position has become critical since the Moravian Mission handed over the trade to the Hudson's Bay Company. The Mission used to be responsible for the seal nets and at the same time saw to it that the seal products were sold at a reasonable price. The trade of the post of to-day is said not to be interested in seal products, and it attaches importance to the fur trade only. Very few of the Hopedale Eskimo own seal-nets and trout-nets, so the catch is decreasing and the people are becoming poorer; of the population on 1st January, 1939: 106 Eskimoes and 184 settlers, there were only one family of Eskimoes and one settler family not living on poor relief. The Hopedale Eskimo therefore views the future with anxiety; all his equipment is wearing out, the houses decayed, his clothes are ragged, and when baking is to be done his wife must make her way over to a more fortunate neighbour who owns an oven. Nothing can be repaired, it was said, because they do not earn enough for their needs; cash was paid only for fur-skins. Things were especially bad in the winter of 1938—1939 when the catch of both seal and fur-bearing animals was very poor. The people were exhausted and suffered great privations, and the resulting diseases began to rage, e.g. scurvy, beri-beri, carbuncles and tuberculosis gained ground. In the spring of 1939 many of them had no bread from March to July, only some few could procure flour from Davis Inlet and that at the unusual price of $11.50 per barrel.

— If I am rightly informed, what has been said above about the Eskimo way of living is true also of the settlers in the district of Hopedale. It is clear that nature herself has compelled also them to adopt this way; the changes for good or evil in the economic conditions affect them both alike. —

We now move to the farthest north of the peninsula and survey

THE ESKIMOES ON THE EASTERN SHORE OF UNGAVA BAY.

In sketching the annual routine of the life of the Ungava Eskimoes TURNER begins his report with the breaking-up of the ice in spring. This period may vary ten or twenty days. When the season has sufficiently advanced, all the belongings of the family, boat, kayak and other personal property were put on sledges and transported down to a place where they were stored on shore until the outside ice of the bay was gone. While waiting the men hunted various kinds of land game and fowl.

When the bay was free from ice the hunters crept along shore to the objective point. Sometimes the party divided. The men sought seals, hunting in kayaks, the women and children searched the islets and coves for something edible, every accessible bird's nest was robbed of its contents.

By the 25th June the people thought of returning, as the seals were becoming scarcer and the birds had again laid eggs. The Eskimoes arranged to drive white whales when the season came about the 12th July. The whales were driven into an enclosure of which the sides were formed by nets. There the whales were speared by the natives, dragged ashore and skinned.

Then began the setting of the nets for the salmon, and this fishing continued from 25th July to 1st September.

Later the natives must be up the river to spear caribou crossing the river (cf. p. 683). This hunting lasted until the deer had begun to rut.

The season was now so far advanced that the river might freeze; the hunters had better begin to descend the river to a place near the sea where the winter could be passed. When the snow had fallen, winter quarters might be taken up. Traps must be set for foxes and other fur-bearing animals. Ptarmigan arrived in large flocks and were eagerly hunted, and the people subsisted on their flesh and the meat of caribou killed in the fall until the ice was firm enough to allow the sledges to be used to transport to the winter camp. Hunting excursions were made to various localities for stray bands of caribou. Also with their dog-teams they hauled wood to the river banks to be floated down in rafts when the river opened in the spring.

The life of these Eskimoes was not so very unlike that of the Naskaupi Indians (p. 681 sqqq.), which, as will be shown later on, was wholly supported by the wild reindeer. But in the account just given it is evident that their working life includes the elements of the coastal culture too. They were in the beginning of this century in the habit of moving out to Killinek in the spring to hunt there and about Tutjat (Button Islands), where there is plenty of game, both seal, walrus, and polar bear. It is a very difficult hunting district to move about in, this island area being full of dangers because of the violent tidal currents and storms. Therefore the men must be excellent sailors. In this connection their spring moves were revealing, for they were as amazing as they were bold. Up to recent times they would pack themselves, bag and baggage, on to an ice-floe in the district north-east of Fort Chimo, where they made the summer haul, this very original vehicle was then allowed to be carried slowly along by the current the whole distance of 180 miles down to Killinek (at Port Burwell), while they passed the time hunting foxes and catching seal, all the time surrounded by their families, their dogs, sledges and household utensils. When they arrived at Killinek they abandoned their frail bark and made their way up on to the land (578). They stayed in the district till the beginning of the next winter, when they returned to the Ungava district by dog-sledge. It is to be regretted that more detailed information regarding these Eskimoes' material culture and the changes it has undergone is not available.

— FLAHERTY (65, 66) has given some reports of the Eskimo living further to the west. —

We shall now consider the little group of Eskimoes living farthest to the south from whom can be obtained an idea of the variations within what is to a great extent an exclusive Eskimo culture:

Working life of the Rigolet Eskimoes, the most southerly in the world.

The Karwalla or Rigolet Eskimoes (208) are clearly the survivors of those of Hamilton Inlet who were once so numerous (cf. p. 466). On a map from 1757 annexed to the work of DE LA HARPE (405) the note 'Habitation des sauvages' is placed at about the contraction of the sea that leads to Lake Melville; also outside the present Rigolet on the southern shore of Hamilton Inlet can be read 'Ance du Batiment' [presumably a trading station: see farther on and p. 481]. In the olden days the wanderings of their ancestors stretched not only far to the north and south along the coast, but also some distance inside Lake Melville, even up the valley of the Hamilton River. The struggle with the Indians drove them back towards the east as we have seen (p. 482). But in their present area they steadily remain.

This group though small is very active, and keeps to the district around the sound between Hamilton Inlet and Lake Melville. It is composed of eleven half-nomadic families, which form what may be called the rearguard of the Eskimoes. They belong to the Church of England in contrast to the other Eskimoes of this coast who are Moravians: they have therefore scarcely ever enjoyed the advantages of the missionaries' direct care and guidance like their fellow tribesmen in the north.

The families were in 1937:

Mark Palliser	Winter station	English River.
James »	» »	» »
Hugh »	» »	» »
Joseph Palliser of Mark	» »	Snookes Cove.
» » » John	» »	Webers »
Mark Mucko	» »	Peter Lucy's Brook.
Peter Adams	» »	» » »
William Ikey	» »	Back Bay.
Charlie »	» »	» »
Wilfred Shiwak	» »	» »
William »	» »	Rigolet.

It is said that in 1842 the total number of these Eskimo families was eight (208); among them at least there can be no talk of a tendency to die out. To me they seemed to be fairly pure of race.

Although these eleven families have lived for a long time apart from their fellows in the north, they have preserved the same semi-nomadic habits and customs, and their life is divided into the same seasonal sections. They, too, choose their camping places with consideration of the natural resources of the district at different times of the year.

As an example can here be given the migratory life of their 'alderman' MARK PALLISER, called »King Palliser». He spends the winter with his family at English River on the south-eastern shore of Lake Melville where he owns a hut; there he sets out traps for the fur-bearing animals, especially the fox. In the spring he moves out to Big or Henrietta Island and stays there as long as the salmon-fishing lasts, that is, the middle or latter part of July (cf. Fig. 223). Next he moves with his family to Hamilton Inlet, where he spends the cod-fishing time at Turner's Bight 18 miles east of Rigolet. Finally in late autumn he moves back to his winter hut on Lake Melville.

Of course many long excursions are made in different directions from these seasonal dwelling places. For example, the Eskimoes of the Narrows go out to the ice-edge with their dog-sledges to hunt the Greenland seal on its way northward. Later they net seal, and that is still being done in Hamilton Inlet at the end of May, at times even until the end of June. From all that one hears of them it appears that these Eskimo too, catch seal as their favourite sport; for them, too, the seal is the most useful game as regards food both for themselves and for their only domestic animals, dogs; it provides them further with clothing and objects which they can sell.

It would be difficult to think out any other arrangement of living which fitted in better with the resources of the districts and the possibilities for their utilization. A proof of this is the fact that the natural conditions have compelled also the livyeres of the area to adopt the same life and customs. In a later connection it will be seen that the white trappers have also found it wise and practical to imitate the Eskimo ways in Labrador (p. 726).

These Eskimo are regarded as being fairly well off. The salmon-fishing brings them a considerable income: for example, I saw Peter Adams on July 12th, 1937 take up in one catch 70 salmon with an average weight of 12 pounds. Because of this prosperity some outsiders have joined them; for example a gifted young man (whose white mother belongs to an old Labrador family and whose father is a Montagnais Indian) settled down with these Eskimo and later married an Innuk he met in Baffin Land who appears to be unusually pure of race.

* * *

Primitive people are children of the moment, they have many fancies and love variations and changing their habits and customs somewhat. For this reason many deviations from the above main plan can be observed also among the Eskimoes of Labrador at the present time. Here, however, there is not sufficient reason to go into such details.

In any case the economic life of the different groups of Eskimo on the Atlantic coast follow the same cycle in principle. Some seasonal stages are added or omitted according to the natural character of the district, and some habits are changed somewhat. Nomadism is most developed in the extensive Nain archipelago, and — strangely enough — the Rigolet Eskimoes are still half-nomadic. The hunts of the Hopedale Eskimo are not very widespread, for, like the former Okkak Eskimo, their winter village can also be used as a base for the winter hunts. The Eskimoes at Hebron, who are on the very edge of the Ocean, feel least of all the need of a migratory life.

The part of the skerry fence the Eskimo knows is nowadays relatively small. He circles in a restricted area year after year and in this the habits soon become set in a fixed mould, a local variant. The knowledge of the country is therefore limited to the few miles he tramps to fox traps and the longer journeys that he makes in the spring in the tracks of the reindeer. It was otherwise in bygone days (cf. p. 512).

The coastal culture of the Eskimoes in Newfoundland-Labrador is becoming more affected by the inland culture.

The description given above of the annual socio-economic cycle among the Eskimo of the Atlantic coast reflects chiefly the basic elements of that culture which goes under the name of the A r c t i c   c o a s t a l   c u l t u r e. It seems probable that a purely coastal culture formed the basis of the livelihood of their forefathers when, once upon a time, they wandered into the Atlantic coast of Labrador. This has not only been the foundation of the subarctic Eskimoes' material culture, but has also been its consolidating element. Capricious mass migrations among the land animals can disturb or even completely alter the customs of the hunting nomads if they have been founded exclusively upon hunting, and compel them to seek a new habitat. On the other hand the supply of sea animals does not seem to be subject to any such sudden and thorough fluctuations as we know among the land fauna, so that a life based chiefly on capture in the animal world of the sea is seldom exposed to such capital risks as, for example, that of the Caribou hunters (cf. the Naskaupee later on). Clearly this is why the culture of the Eskimo at the Atlantic coastal strip is so stable compared with that of the Indians in the interior. A contributory cause is the fact that, like the Eskimoes of the Pacific who were excluded by the Indians from the attractive hunting in the woods and on the mountains of the interior, the Naskaupee and the Montagnais Indians have done their best to exclude the Labrador Eskimoes from the chief haunts of the caribou and the fur-bearing animals in the interior (cf. however p. 441 sqqq.). This is another reason why the Eskimoes have become more or less exclusively dependent upon the resources of the coastal belt.

Among the Labrador Eskimoes chieftainship has been reduced to a minimum, sociopolitical organization is lacking, and hunting is the all-engrossing need of the people (606).

From the older descriptions scarcely any definite impression is obtained that the half-nomadism of the type we know to-day stamped the economic life of the forefathers of the present Eskimoes of Labrador. It seems more probable that they then, like the wandering tribes of the Canadian Arctic, moved along the coast over very large areas and stopped here or there for a time when the prospects of catching seals were good even as late as when they first came into contact with the white men. These people have probably never formed any more permanent territorial clan groups. At the primitive stages of culture nature herself in time brings about the best possible equilibrium between the customs and habits of people and the natural resources of the habitat and their utilization. As the natural conditions vary very much in the different parts of the coast, and as in each part equilibrium can be attained only as the yearly cycle of their nomadism is constantly adapted to them, this equilibrium could not be fixed in former times because the groups constantly displaced each other. The territorial grouping based upon fixed winter dwellings which now exists and upon which we have based our description, can be assumed to have developed after the arrival of the Moravian missionaries on the coast, some time after the middle of 18th century. The missionaries have certainly greatly influenced its rise, and they have even planned it with such thorough knowledge and such far-sighted judgement as the Eskimo themselves could scarcely have shown. The habits of the natives then became more specialized on the above lines.

— In passing it may be right to correct a rather general misunderstanding. It is natural that the missionaries invited the Eskimoes to collect around the stations because the good shepherds considered it wise to keep their newly won lambs under their eyes lest the wicked wolf in the guise of an *angekok* (shaman) should go off with them again. Superficial observers have been heard to say that by this reform the Eskimoes' means of livelihood deteriorated. This is quite incorrect. The missionaries themselves have done their utmost to get the Eskimoes to retain their inherited, sound ways of living and prevent their proletarisation in closely packed societies. They have therefore encouraged the natives to spread themselves during the hunting and fishing seasons out over the area where they hunt in agreement with the principles of nomadism. The very complete literature of the work of the Moravian Missions in Labrador (see bibliography) gives full proof of this. The Eskimo's way of living has hardly been injured by the village life. We have seen above that the huts in Nain are only lived in during the rest periods, so that nomadism continues and perhaps even more systematically than before. Even if the winter dwellings are now the same from year to year, this in itself does not seem to have been at all injurious. My personal impression is rather that this reform has been of the greatest advantage for these small groups, their preservation, and civilization. It seems me much more probable that the Eskimo themselves discovered the advantages of living near the missions and that is why the winter village groups moved to the mission stations from the different fiords where they used to live. There they were within reach of care and attention if their health demanded it; the children had the chance of going to school, the old people of trading and all kinds of entertainment, especially music. My personal impression is, therefore, that the custom of living in villages during a part of the winter became stabilised rather spontaneously through the Eskimo themselves. Naturally this stabilisation was assisted by the fact that the missionaries had secured rights not only over the reservation areas but also to decide who should be allowed to settle within those areas (in this way they kept out traders of the bootlegging type). This resulted in the Eskimo reservations.

At the present time one has the impression, however, that in the last few decades the material culture has been evolving. It has been affected by the a r c t i c   i n l a n d   c u l t u r e, the culture of the caribou-hunters; see e.g. the Ungava people. The Eskimoes have of course always hunted the caribou when they found them on the coast; their folk-tales relate how the spirits of the Eskimo migrated to the interior to hunt foxes, and how in consultation with Superkugssuak the angekok could send caribou to the Eskimo when he hunted there. In older times, too, they used to trap some fur-bearing animals. But at least already about 1900 hunting the caribou began to be practised on a big scale in the fiord valleys and the moors in companies, assuming such proportions that on one hunting excursion made by the Eskimoes from Okkak and Nain 700 caribou were shot, according to McGREGOR.   The income from trapping fur-bearing animals became more and more important in the household budget.   It is therefore clear that some shades of the inland culture are more and more sharply showing themselves in the local arctic coast culture. This development has been possible because of the peaceful relations with their neighbours the Indians, as formerly the Eskimoes did not dare to make their way into the interior.   The tendency of future development is to me quite clear, for, as the Naskaupee Indians disappear from the country and the hunting districts become evacuated, the energetic Eskimoes will take possession of them (compare above the Western Ungava Eskimoes).   It seems, therefore, that a way of living will develop such as we shall find among the trappers in Dove Brook (cf. p. 721) and such as can be said to form the ideal for the hunters on the Atlantic Labrador coast.

The gradually increasing addition of inland culture to the Eskimoes' economic life is only a link in the series of new developments and progress which can be observed throughout the whole of their culture, both the material and the non-material.   This complex of changes is perhaps the most interesting anthropo-geographical phenomenon of the sphere we are studying.   It may be said that no more or less than a c u l t u r a l   m e t a m o r p h o s i s is taking place in the Labrador Eskimoes mostly to the advantage of both the individual and the race.

More concerning the cultural metamorphosis.

The frozen sea still locks up the land for eight months of the year and this keeps the Labrador coast within its own icy loneliness.   It does not seem that the country is more accessible by any change in its nature or in its climate! But the summer travellers have brought Labrador more clearly on to the track.   A change has come over the people of Labrador during the last fifty years, Dr HUTTON stated in the beginning of our century.   Civilization in its march has touched the coast with its fringe.   The explorers see a world to conquer.

Changes regarding equipment and its use.

The Eskimo was for a long time slow to accept new ideas and slow to change utensils; like primitive tribes in general he liked things in the same groove from generation to generation.   The very lack of school education was maybe one of the principal causes of their narrow-minded conservative-ness. The influences that combine to educate the mind in other countries are to a great extent denied him in the loneliness of Labrador. Labrador was all the world to the Eskimo. But finally he became accustomed to the modern ways of life and to new inventions through the white pioneers. And what do we see today? the Eskimo playing the gramophone, for instance. And even the 'movies' have been seen in Labrador. The wireless has also come, there are several stations. The radio enables the Eskimo to set his watch by the striking of Big Ben. The noise of the airplane, *tingmi'suak* (= the great bird), has been heard by the Eskimo, and some of them have taken trips in the air. All these marvels are accepted by this childlike folk without question and without fear. They no longer associate such things with the supernatural; their wise guides, the Moravian missionaries, have steered them safely through the troubled years of change, HUTTON writes.

In order to understand this we shall study some aspects of their dwellings, clothing and food, though it is not my intention to give here a systematic ethnographical description of the very captivating Labrador Eskimo; there exist already some fairly good ones by CARTWRIGHT (134), HAWKES (424), HUTTON (472—478), LOW (588), MCLEAN (610), TURNER (965), etc. We shall perceive a slow change for the better, the chief cause of which seems to be the Eskimo's own psychological nature. They are good observers and, like their ancestors, are distinguished by their receptivity and capacity for quick apprehension. For this reason it is not at all surprising that the Eskimoes of Labrador have with avidity seized upon and assimilated all kinds of new acquisitions with their inherited culture, if they in any way found them advantageous for their way of living and the efficiency of their occupations, resolutely discarding old customs and equipment.

DWELLINGS.

For the old Eskimo, as for most primitive peoples, the home has been only a shelter against cold, rain and wind; it is somewhere to sleep securely and to keep the captured game safe from prowling dogs. The dwelling therefore remained very primitive until recent times, and even nowadays it is almost only a mean little shack built of rough tree trunks or planks and floored with packing-case boards. But even this change must be considered an enormous progress from a sanitary point of view, as their medical man, HUTTON, soon found. However, it took some time before the northern Eskimoes became acclimatised to the warmth of the wooden huts.

A century ago the Eskimo of this coast still dwelt in hideous stone huts, i g l o o s, dug in the sodden slopes of a hill on their winter places; Fig. 235. The interior was dark and noisome, the floor, the walls begrimed with soot and grease, the air stuffy. A small allowance of light filtered dimly in through the membrane of a seal's bowel stretched across a hole in the roof. One joined two or three houses together: living room, bedroom and storehouse. In some cases two or more such earth huts — w i n t e r   h o u s e s — were grouped together into something like a village society. At that time the Eskimoes used to wander along the coast as they were inclined, sometimes a family joined one of the already existing groups of dwellings, sometimes they would burst away and, as they became interested in another group, join them. — In Ramah in the winter of 1906 sixty-nine people were gathered in the *igloo-suaks* heated with stone lamps (995).

In the turf and stone igloo the only ventilation was the occasional breath of air that wafted along the tunnel-like porch. Dr HUTTON points out: I cannot imagine anything more dismally unhealthy. Imagine yourself on the floor of a badly lit and worse ventilated Eskimo turf igloo in Labrador. The seal-oil lamp is burning; it is no more than a half-moon shaped trough, hollowed from a piece of soapstone and half-filled with thick brown seal-oil. A flat wick of moss leans on the edge of the trough, dipping into the oil and burning with a steady white flame. The part farthest from the entrance is the sleeping-place, occupied by a platform of moss and earth spread with skins. Upon this platform-bed the hunter snores, when he has returned tired from his hunting trip, while his wife turns his boots to dry and patiently rubs them supple, ready for the next excursion. Wherever you look is dirt and vermin and nauseous food. You have maybe to eat this fishy-flavoured meat handed to you by fingers innocent of washing except for the dipping in the sea when hunting the seal. The floor is strewn with the litter of work and scraps of sealskin, fishbones, bits of putrid meat flung down at meal-times. Some folks have to share an igloo with some other family and this leads to endless quarrels and jealousies (478).

Such was the primitive winter dwelling of the Eskimo on this coast. The picture had remained unchanged for many centuries, the gloomy homes were still dark igloos of turf and stones; cf. in 472 picture facing p 40. What few pieces of wood they needed for building they gathered from the drift-wood on the beach. There were no stoves in these houses; the cooking was done either over seal-oil lamps or over fires of brushwood out-of-doors.

After the Moravian Mission arrived successive improvements in the dwellings took place. In the beginning of our century the people still had huts of wood built into the earth, looking from a distance like heaps of turf or sods, with a tin pipe sticking out of the top and long low tunnels leading up to the side; cf. in 472 picture facing p. 308. It required no little labour to erect such a dwelling and consequently a family stayed in it as long as possible so that it became fantastically dirty and foul; besides this it had not even such ventilation as is possibly with the draughty walls of an ordinary wooden hut. A new acquisition was the little iron stove. But inside all was still smoke-blackened and shiny with grease, the evil, rancid, fishy smell of stale seal-oil and putrid flesh remained, and the only suggestion of air wafted sluggishly through the low porch. The result of all this was, of course, injury to the people's respiratory organs, and the only remedy they knew was recourse to the *angekok* (p. 548 sq.). The heat from the stove combined with the smell made the atmosphere still almost unbearable. In some of these igloos, in winter, with their long snow tunnels to keep out the cold and at the same time effectively the fresh air away from the door, I have had to gasp for air, said HUTTON, how can folks be healthy in homes like that, in this sort of air? he asks. Even the craziest shacks must be better than these dwellings.

More wholesome were the temporary dwellings: tents and snowhouses.

The frame of the tent was a bunch of poles asymmetrically stuck out through a hole in the top; a seal skin cover was thrown over them and kept in place by big stones laid upon its edge; cf. Fig. 226. The rocky ground make the use of tent-pegs impossible, and in stormy autumn nights the tent may blow over. Therefore one tries to find a rocky wall or something similar as shelter against the north-west wind. In bygone days the tents were of the same type but covered with reindeer-skins stitched together with sinew and stretched on poles with the hairy side outward. But nowadays only the lighter calico tents are in use, and they afford just as good protection from the weather. They are also easily mended and can be packed up into a neat little bundle. Various types exist; cf. Figs 226 and 223. In the tent the air is always fresh. But it is not always good for the tent dwellers on a stormy and rainy night; the tent is not proof against torrential downpour, the inhabitants are like drowned rats, but show the same placid smile (476). The Eskimo in the north use rough masonry lean-to huts as well as tents.

In the north, where wood is lacking, villages of snow are built. But on the Labrador coast the Eskimoes consider the turf or wooden house to be more secure and lasting. Here the real snow-houses are used only when the people are moving about the country. It is indeed not expensive to make a provisional Eskimo home in winter. A snow-house for a short journey can be built in a couple of hours. It is, however, generally on towards December before the snow is beaten so hard by the wind that the building of snow-houses can begin.

The snow-houses used on sledge journeys in Newfoundland-Labrador are but poor imitations of the real thing with its ice-window and its protecting wall and porch. Quickly the Eskimo cuts slightly curved blocks from the frozen drift and inside the circle piles them one against the other in spiral fashion from left to right and stuffs the cracks with powdery snow. The house becomes so strong that one can sit placidly inside while a man crawls over the top. The window, if put in at all, is a sheet of clear freshwater ice. The air in the house scarcely rises above freezing-point. The snow-house is always protected by a wall of snow a few feet away, which keeps away the destroying wind. In the better snow igloos the door is reached by a tunnel running uphill (476). In May the snow-house begins to melt and threatens to tumble in upon the occupants, the tent must be used.

— It had no doubt a favourable influence on their health when these Eskimoes lived all the year round in tents or snow-houses and knew nothing of the permanent winter houses of grass-sod etc., which are veritable hotbeds of tuberculosis (80 a). —

Little roughly hewn wooden huts in the midst of a tangle of forest or shrubbery have in the end sprung up like mushrooms. Such dwellings have grouped themselves into a little hamlet round the central buildings complex of the mission stations, church, school, shop and, later, boarding-school cf. Figs 236 and 237. Nowadays the hunter wants a wooden house, he does not want to go back to the dirty, unhealthy earth mounds in which infantile mortality must have been terribly high. He seeks out the best of the stunted trees in the woods, chops them down and carries them home. With his wife's help he saws them into beams and rafts and planks. He lays a foundation of stones from the beach and erects his beams and joists upon it, he works intently and seriously and before the next winter the house is ready.

The Eskimo square wooden hut is at present a very humble dwelling, but it has immense advantages in comparison with the old igloo of turf and stones and a few rough beams, HUTTON writes. The white man has often to stoop because the doorway is too low for him. One must be careful in passing the dog tenants of the porch, if one of them is struck with the foot it raises its head to snarl and shows wolfish teeth. Through the doorway you enter a square room. The air is not so bad, although steaming heavy and warm, and a tremendous fishy smell comes from the cooking, the floor is not very dirty. In the winter you may stumble over a couple of plump seals brought in to thaw, behind the stove on the wall is stretched an oily sealskin to dry. One corner harbours a little table on which stands a lamp, it is also strewn with cups and spoons and knives and fishing-tackle. Various queer looking objects lie at the edges of the floor: snow-shoes, dogs' harness, whip, bladders, slabs of dried meat, bundles of straw for basketmaking, boots and clothing, etc., and a keg of water stands near the wall. A big, rough home-made bedstead fills a corner. It is made of rough boards and spread with deer-skins and a patchwork counterpane. The children of the household are playing on the floor. But the characteristic Eskimo smell cannot be abolished. Every home, dirty or clean, has the same odour.

Many of the houses are architectural curiosities; cf. Figs 238 and 239. Some of them show signs of much patching; on the grey and weatherworn boards you will see light pieces added; the front looks curious with its windows covered with boards, a necessary protection against the wild draught dogs who may try

a.

b.

Fig. 227. Eskimoes in Hebron. Photo the author and I HUSTICH, 1937.

to break in and steal; it has indeed happened that the dogs have torn up the roof to break their way in and steal when they have been left at home at cod-fishing time. In time the house is made larger and it can completely lose its individuality with such a process, the walls being pushed back to meet the needs of the growing family. Then one can find behind the square room, for instance, a big, oblong shack with a sloping roof, placed crosswise, so that the little, square part of the house looks into it from one side. In the middle of the room is a table surrounded by the customary wooden box that serves both for seats and a storage place. At two ends are the sleeping-places, roughly partitioned off

Fig. 228. The family Boas Obed on M/S '*Maraval*'. Photo the author, 1937.

A peep behind the partition discloses an array of bunks where the children sleep, sometimes one above the other like berths on board a ship, sometimes side by side. Some houses are a real credit to their owners: there are sofas and a harmonium, linoleum on the floor, etc. In some houses you might find even some books. the Bible, a hymn book, the Pilgrim's Progress, etc. About the beginning of this present century CHRISTIAN SMITH introduced an Eskimo newspaper: *Aglait Illunainortut* [the Paper for Everybody] (476, 478). PRICHARD 1910 tells of a very industrious man who possessed an extremely comfortable house with lace curtains at the windows He had ordered fine furniture from St. John's for his sitting-room, which had armchairs, rocking-chairs, a piano, etc. The owner was besides always very good to the poor and to the widows.

Warmth is one of the most serious things to be considered in the wooden house. The modern Eskimo has lost to a great extent the coating of fat that his

forefathers possessed, and the stove must now take its place as his natural over-
coat. One cannot have more than one stove, the firewood is far away.

Like primitive folk in general the Eskimo love to build. Here and there
one can see piles of sawn planks near the mounds of firewood. But they can
construct larger buildings too. During some stormy days when it was not
possible to go fishing in the boats the Eskimos built a church with their own
voluntary labour at Uviluktok near Hopedale. Even the decorations of the
interior were carried out by the Eskimoes. With willing hands and happy
minds they rebuilt a new church
at Nain, when their old one had
been destroyed by fire (478).

Thus the dwellings are better
to-day than some decades ago
and the aim of every man to have
his own house has been almost
realized all along the coast.

As appears from the above
account, thorough changes for
the better have come about with
time. Of course, further changes
are necessary and we may hope
that they also will come. But
the possibility of realizing them
is now a question of money. If

Fig. 229. My pilot Julius Nathanael from Nain
on board the Strathcona. Photo the author,
1939.

the products of the Eskimoes' work are justly paid for, there is no doubt that
their dwellings will become increasingly healthy and comfortable. Under
present conditions they can scarcely improve more than they have done.

CLOTHING.

Primitive people having adapted their clothing to their surroundings
during a long period it is quite usual that it remains unchanged in shape and
form century after century. The winter clothing of the arctic and sub-arctic
natives especially is in such close agreement with the climatic exigencies that
there is little possibility for variation and improvement, and the outfit of the
fur-clad people in winter is mostly reminiscent of the dressing of ancient times
and still unsurpassed in its suitability.

From top to toe they were in bygone days clad in sealskins. Curtis (196) tells in the 1770's that between the dresses of the different sexes there was no variety, except that the women wore monstrous large boots, and their upper garment

Fig. 230. Eskimoes at Cut-throat. Photo C.-G Wenner, 1939.

was ornamented with a tail, Fig. 240. In the boots they occasionally placed their children, but the youngest was always carried at their back, in the hood of their jacket. However, it will soon be a century or even more since one was able to say that 'the Eskimo is a singularly composite being, — a link between savage and seals, putting the seals' bodies into their own and incasing them-

selves in the skins of the seals' (325). Since the Labrador Eskimoes on the
Atlantic coast have been in close contact with the whites, their clothing has
gradually changed, for they like to imitate the white man's fashions. At

Fig. 231. Eskimo at Cut-throat. Photo C.-G. WENNER, 1939.

present they like to dress in European frocks, or jacket and waistcoat; cf.
Figs. 219 and 233. Instead of the men wearing hooded smocks — cf. Figs
228, 232, 241 and 234 — and the women long-tailed native *sillapak* — cf.
Fig. 240 — you will find the men in fishermen's jerseys and sea-boots, the
women in skirts and jackets — Fig. 243 — and hats of many fashions, HUTTON

stated already at the beginning of our century. PRICHARD, however, wrote at the same time that in Killinek the people were just emerging from the wild state. They wore skin garments and the babies were carried quite naked in the hoods of their mothers *attigeks*.

Otherwise on some of the southerly Eskimoes there is practically nothing left of the original Eskimo costume but the boots.

Meanwhile the dressing of the majority has still some traits which give them the specific national stamp, and also correspond well with the conditions of their life; cf. Figs. 227, 228, 216, 220, 217, 229, 218, 230, 231, 232, 241 and 244.

In the summer they like to dress in blanket cloth, bought at the store, and cut and sewn after the old Eskimo pattern by the wives and daughters of the household. The men wear a blanket smock or *'dickey'* (corruption from the Eskimo word *attije*), which is a hooded frock. Ordinary trousers and sealskin boots, *kamiks*, complete his costume. The women wear a *sillapâk*, which is made of white calico, with a short tail in front and a long tail behind, it is 'the outside thing of all'. The wealthy ladies generally wear lovely fur round the hood. In olden days the women used to wear trousers; Fig. 240. However, probably because they were afraid of being laughed at, they adopted the skirt. Kamiks they still wear. A girl is however not allowed to have the big hood until she is old enough to get married; so the little girl who sets out to act as nursemaid borrows her mother's; she would be helpless without a hood, to carry the child in her arms would never satisfy it. The hood has been the Eskimo's cradle longer than memory can tell.

An interesting habit is the use of different colors in the clothing. On their sillapâks and kamiks married people use blue bands, unmarried red. Good custom demands that a young girl ties her plaited hair with a pink ribbon, a married woman uses blue, a widow white. Instead of the mysterious hand-bag of our ladies it is true Eskimo fashion to use the leg of the woman's boot as a pocket. From thence she brings out the most varied articles: handkerchief, pipe, hymn-book, biscuits, etc.

Dr HUTTON is no doubt right when he considers that the clothing of the modern Labrador Eskimoes corresponds well with the demands of health and comfort in the natural conditions of Labrador. Moreover, it is an interesting fact that they all, men and women alike, have clean outfits for Sunday wear, they would never go to the Service in a dirty state. The men wear snowy *sillapâks*, the cotton covering for the hooded blanket *attigeh*. The women are fastidious about their hair, which they wash and comb (476).

FOOD.

Many curious stories can be heard concerning the Eskimoes' diet. CURTIS (196) in the 1770's wrote: When they are pressed with hunger, and have nothing to satisfy it, they make their noses bleed, and suck the blood to support themselves! It is true, they have many peculiar dainties, but their main food is amongst the healthiest one knows.

In the subarctic conditions the well-being of the people depends essentially on the kind of food available. The mostly good health conditions amongst our Eskimoes witness to the fact that their diet is appropriate.

During his long stay amongst the Eskimoes at Okkak Dr. HUTTON studied their diet very closely. The list of their foods is long, he states. seal-meat raw, dried, boiled, fried, made into a stew with flour, seal blood; the flesh of reindeer, foxes, bears, hares, sea birds of all sorts, eggs of gulls, sea-pigeons and ptarmigan, trout and cod and salmon, boiled skin of white whale and the walrus; raw reindeer lips and ears. Dried trout stomachs, seal-flippers and half-hatched gull's eggs are real tit-bits for the Eskimo. An epicurean dish is seal's flipper which has 'hung' some time so as to be ten-der; nothwitstanding this the party at such feasts can fall seriously ill. The Eskimo says only strong (= more primitive) people can eat rotten seal flippers! Vegetable food they ignore. Mushrooms they do not eat. But different herbs are used: the shoots are plucked and mixed with the meat. In hard ti-mes they also eat reindeerlichen and tripe-de-roche' (p. 343). Roots and 'tubers are collected by the women. Berries are a great boon, the people gather barrels full, especially crowberry and *akbık* or

Fig. 232. Eskimoes at Port Manvers, Photo the author, 1939.

bake apples (cloudberry) About 20 kinds of berries are eaten, mixed with fish or meat. The cloudberry is the most important, next the bilberry.

Fish play no small role in their diet; they are split, washed and dried well in the open. Cod, trout, salmon and caplin with sea-trout are all equally liked. They dry them without salt on poles and make *pepsi*.

The Eskimo in still fond of raw meat and rancid oil, a partridge eaten raw and warm is a real delicacy to him. Ptarmigan and white hare he likes very much, and a dinner of fox meat followed by a luscious supper of the precious blubber makes him happy. The Eskimo's principal dainty is freshly killed seal. The children's eyes glisten and their mouths water when the seal carcass flops in-to the room. The people take the raw, red meat from the half-warm carcass with delight. It is the food they love most; in the Eskimo's opinion no European delicacy can compete with the greasy, raw flesh and blubber. The Eskimo waste nothing of the seal. The meat is the nicest thing he knows, the skin is made into clothing or boots or bedding, the sinews make thread for sewing, the shoulder-blade makes a handy scraper for skins, formerly he used to split the bowel and stitch it for window-panes. Whatever is left can be used as food for the dogs.

But some pieces of the best of the meat are set apart for drying without salt: the dried meat — *nipko* — is a great delicacy. It shrivels and blackens on a pole until it looks anything but appetizing. The leathery nipko is a real Eskimo dainty, and it is also very sustaining food (475).

The most gluttonous Indian would turn green with envy to see the quantities of meat the Eskimo can stow away within his inner self at a single sitting; but on the other hand he can live, and work hard too, on a single scant meal a day, just as his dogs do (994). The Eskimo feast high while there is plenty of meat and drink melted snow or tea with it. They seems to have no idea than to live from hand to mouth. However, the reindeer meat too is kept for the time when the ice is cracking and seals are hard to find.

Fig 233. The Eskimo Julius Nathanael and his wife at Nain. Photo the author, 1939.

The people live thus literally from the wild animal produce of the country. There are no sheep in their Labrador, no cows, no milk and honey except the kind in tins. They live well and happy, and their health condition will continue to be marvellous as long as they avoid the food that the white man uses, Dr. HUTTON pointed out. All along the coast the Eskimo still keep to the staple diet of raw meat that earned for them in olden days the epithet Eskimo (est'ski'mau = eater of raw flesh). Here we find the reason why the Eskimoes are practically free from scurvy: it is the vitamins in the black and nasty raw seal meat that keep them healthy.

Unfortunately, the use of the white man's food is said to be on the increase. In 1910 PRICHARD stated that in the southern settlements both store-food and store-clothes had gained in popularity, partly because the skin garments were regarded as old-fashioned, but partly, no doubt, because store-garments were obtained so much more easily. But already then it had become clear that the white man's food is not only unsuitable for the Eskimo, but actually sows in them the seeds of weakness and disposition to disease. It is a well-proven fact that European clothes are not adapted to the climate.

HUNTING AND FISHING GEAR.

The transition affecting the boats has slowly continued. For about the year 1860 the following interesting statistical notes are given by REICHEL (787 a):

										Total
Sailing	boats at Hopedale	9,	Nain	10,	Okkak	12,	Hebron	2,		33
Fishing	» » »	40,	»	20,	»	14,	»	10,		84
Skin (*Umiaks*) »	» »	—,	»	4,	»	4,	»	6,		14
Kayaks	» »	49,	»	58,	»	61,	»	46,		214
Tents	»	27,	»	23,	»	31,	»	38,		119
Houses	»	35,	»	32,	»	36,	»	25,		128
Families	»	46,	»	55,	»	75,	»	68,		244
Individuals	»	248,	»	275,	»	327,	»	313,		1163

About 1860 the large skin boat, the women's boat, *umiak*, was rarely to be seen. It had been replaced for fishing and sailing by a wooden boat about the size of 8 tons (787 a). In the period 1861—76 the number of wooden boats increased from 117 to 237 but the umiaks decreased from 14 to 4, the kajaks from 214 to 154. Now, among the means of transport they have the lustily puffing motor boat — cf. Figs 245—247, 244— which is replacing the sailing-boat which displaced in its time the skin boat and to a very great extent also the kajak; cf. Fig. 242. During my journeys I saw only two kajaks in use in Labrador, and no umiak. It is, of course, natural that the motor boat is an invaluable help in the methodical seal-hunting when the sea is open and the old vessels are unable to compete with it. The Eskimo themselves say that the motor boat has wrought a change of the first order because it has increased the capture from the sea 10—20 times. But on the other hand the noise of the motor frightens the seals avay.

The transitions in connection with the dwellings have already been treated. The heavy, clumsy *tupek* of the Labradoreans, the skin tent, is said to have never given complete protection from rain or wind. It has therefore been ex-changed for the tent of a fairly water-tight, cotton material (cf. p. 517). This tent is light to carry, it gives more effective protection from the rain, and can be warmed when necessary with an iron stove, just like the trappers' tents. Another reason for this exchange is that the cotton material is not particularly tempting to the appetite of the dogs. For the temporary homes within the fiords — cf. Fig. 248 — as also here and there on the islands — cf. Fig. 249 — small, wooden huts have begun to be set up, especially on the places which are used permanently in catching seal and trapping foxes. On the other hand, during their winter journeys for caribou-hunting, the snow huts, the *igloo-wiuk*, are still the only dwellings used.

Among the hunting implements the gun has long since displaced the bow; the latest information I found regarding the Eskimoes' use of the bow was when Bell speaks of the Eskimo near Little Whale River at Hudson Bay having killed several caribou with bow and arrow in 1877. Prichard, being an expert hunter, considered, after his stay in Labrador, that it was doubtful whether the introduction of the rifle had been an advantage for the Eskimo. In the old days when the sealer harpooned his quarry he rarely lost it after he had struck it. But with the rifle many more are killed and a good many are lost, for the carcase often sinks immediately. It is obvious, therefore, that the supply is decreasing and all the game is becoming frightened. As an example he takes the walrus, which at the beginning of this century was very seldom seen on the coast except in the waters of the far north, and the seals off the Labdor coast are far more wide-awake than those of the Irish and Scottish coasts. The harpoon is, however, still being used as well as the gun in seal-hunting, according to what was told me in Nain in 1939. For this reason I was somewhat surprised by Reichel's information that already in the 1860's they had given up the practice of harpooning seals or taking sea-birds by means of darts.

Some Eskimoes have traps for cod-fishing, instead of hook and line.

The earlier kindling of the Eskimoes, which consisted of some pieces of pyrites and *Eriophorum*-wool, has completely given way to matches.

* * *

We do not need any more examples to draw the conclusion that among the Newfoundland-Labrador Eskimoes the whole complex of dwelling, clothing, food and equipment has been irretrievably scattered since the people came into contact with the Europeanizing culture, and with the modern means of communication there is scarcely any limit to its extension. People living in a hostile milieu will always take over the products of the factory when they serve their purpose more adequately than the tools of their own making. Yet, the Eskimoes are sufficiently conservative not to give way immediately to what is new. Only after a most careful testing has the older gear been exchanged for that of the new which was found more suitable for their hard battle with nature for the daily food. On the other hand, some of their old tools, which they found better than the new, have been retained.

The specific knife of the Eskimo women, *ulo* (633), for example, is still exclusively used among the women: there is, presumably, no European invention which could supplant it in its varied uses. Ease of manipulation as well as utility are always considered before an article is accepted and incorporated in the local cultural pattern.

One change, however, I find very difficult to explain as resulting in greater efficiency: the light and handy kajak seems to have been indispensable when the hunters were waiting for seals at the edge of the ice. If the ice cracked away from the shore and floated out to sea, the kajak could save them. But a heavy wooden boat covered with ice can scarcely be transported at the sinâ.

NOTWITHSTANDING THE CULTURAL METAMORPHOSIS THE HUNTER IS, IN ALL ESSENTIALS, STILL AN ESKIMO.

In spite of all these changes the Eskimoes' working life still follows its traditional rhythm determined by nature. The technical introductions do not affect the precepts or mores of olden days. The Eskimo still travels out to the trapping areas in the wooded valleys of the fiords where traps are set for the fur-bearing animals, and for the caribou-hunts on the mountains, or out to the sinâ on his dog-sledge drawn by a splendid span of dogs; cf. Figs 224 and 250. This his ancestors did already thousands of years ago and he and his children will continue to do it. On the forestless camping-places he still builds the winter snow-hut as has been the custom since time immemorial, and in this he rests between the hunts, till a new camping-place is sought out, and there, in a few hours, he sets up a new snow-hut. On the other hand, in valleys where there is plenty of timber and fuel, many Eskimoes have already built blockhuts which, in spite of their inadequacy considered as dwellings, greatly surpass both the snow-house and the sealskin tent. When the sea rolls free he places himself, with wife, children and dogs and all his baggage in his motor boat and makes this journey to the sealing and fishing places in a few hours, when before it would take days in a paddled umiak.

The changes which have occurred were necessary for the maintenance of their welfare in the struggle with competitors from the outer world. The specific Eskimo material culture connected with catching animals had already reached the limits of method and technique within their own possible development, it had — like the Eskimo language — become petrified over the whole of the very extensive area they inhabited. Then the Moravian Mission landed on the coast of Labrador, bringing new equipment and methods.

Fig. 234. Julius Nathanael on the look-out for caribou at Webbs Hill. Photo C.-G.
WENNER, 1939.

The European culture was unyielding: it demanded assimilation or the extermination of the old.

The good shepherds of the Eskimo realized this at an early stage, when the
stranger element began to force its way in, and they helped most considerably
their protégés to break away from the charmed circle of inherited tradition and prejudice which usually prevents the savage giving up what is good
for what is still better. For to change his customs and habits at a primitive

stage is indeed one of the most difficult things. Yet, if he does not manage to break away he is doomed.

As already shown, it is a long time since the changes began among the Eskimoes. Captain A. CROFTON in 1791 gave an description of the Eskimo in southern Labrador which I give here for what it is worth (325): These came from the north in a large shallop [umiak] belonging to a tribe of 'Eskimaux Indians' and consisted of six men, five women, and seven children. They were to remain the winter with the English fishermen at Bradore and be employed in the seal-fishery. They brought with them some oil and whalebone to barter for English provisions, which they were very partial to, preferring European clothing to the sealskin dresses they formerly appeared in! They were so much civilized as to abhor raw meat [?], having cooking utensils with them. They had likewise laid aside the bow and arrow for musquets. The traders' agent said they were strictly honest and well-behaved. Of the Moravians they seemed not to have any knowledge.

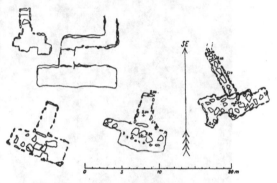

We know nothing of the fate of these people, but we know that there are no longer any Eskimo south of Hamilton Inlet.

Fig. 235. Plans of varied types of Eskimo igloos at Hopedale dug out by JUNIUS BIRD JR. and measured by E. H. KRANCK and the author (cf. also 920).

I think we may now conclude that there is no reason for pessimism concerning the future subsistence of the Eskimo of Newfoundland-Labrador, providing the hygienic conditions continue to be improved and he can go on earning his livelihood as hitherto. Meanwhile, one important circumstance must be taken into consideration and we will do that here.

THE ECONOMIC FOUNDATION OF THEIR LIFE HAS ALSO UNDERGONE CHANGES.

The whole economic picture among the Eskimoes has undergone a complete transformation since trading-posts were introduced.

Since time immemorial the seal-hunting provided them not only with food, clothing, and boots, but with lamp-oil, window panes, thread, to mention only a few of the commonest things. In bygone times it also gave them products which they could barter for other things useful in the Eskimo household: seal

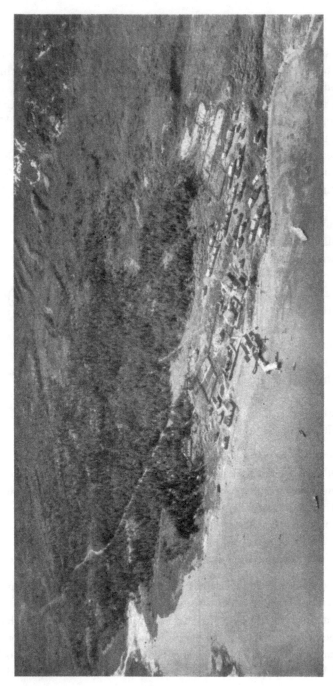

Fig. 236. Nuin. Above the wharf are seen the Mission House and the Church toward the right and the Hudson's Bay Company's house at the left. At the edge of the park is the school. Toward the right are the Eskimo houses. Note the gardens. At the right edge some raised shores are visible. Cf. Fig. 199. Photo the FORBES-GRENFELL EXPEDITION, 1931.

train oil, sealskin and its products, such as boots, mitts, etc., slightly salted, dried fish, slightly salted or smoked sea-trout at the northern stations, and a smaller quantity of salted salmon at the southern stations. In certain years the fox skins were also important. Now, however, instead of barter *en gros*, the Eskimo is in contact with systematic detail trade. This brings the people a whole lot of modern and useful articles which before they had had to do without, just like so many isolated Eskimo tribes in the north to-day. These

Fig. 237. Hopedale. In the foreground Eskimo houses and some husky dogs on summer vacation. In the background the Church of the Moravian Mission and its business offices. Below the Church is the park. Cf. Fig. 328. Copyright WILMONT T. DE BELL, 1937.

new commodities enabled the Eskimo to develop and enrich his own and his family's life in many essentials.

The Moravian Mission, realizing these advantages, felt it a duty to find a market for Eskimo produce, and to keep a stock of things necessary for living and for hunting to barter to the people. The young men coming as recruits to the Mission station's staff were obliged to spend their first years behind the counter; in that way they had practice in the Eskimo language and learnt much of Eskimo character and ways of thought. It was no easy business. In the store the Eskimo always wanted more and more: I cannot think that you love me, he says to the storekeeper, as a last attempt at exaction (478).

The first aim of the missionaries' trade was to prevent the Eskimoes from falling into the hands of unscrupulous traders. Nearly everywhere when the pioneer of the whites, the trader, has forced his way in among primitive peoples, the same traffic has been carried on, bringing ruin in its train: alcoholic liquor in exchange for furs or other products offered by the people. In Labrador, too, it is said that a drink of whisky for a fox skin was regarded as a good bargain

Fig. 238. Eskimo architecture at Nain. Photo the author, 1939.

for both seller and buyer. The missionaries, however, have succeeded in putting a stop to the traffic in spirits — at least in principle: on their initiative the regulation was introduced in Chapter 130 of Consolidated Statutes, Sec. 51: no intoxicating liquors shall be sold, given, or delivered to any Esquimaux or Indian under a penalty of 200 dollars.

When the missionaries had got so far they could begin to realize the second important point in their programme: the growing generation of Eskimoes were to be educated to a competitive standard in which they could in all their ways of life meet the new period which must come to Labrador as to other places. And here I should like to repeat a statement by HUTTON, worth remembering when discussing this question: the missionaries worked on the imitative faculties of these people and trust that reforms will become habits; the Eskimoes do the thing you want from mere obedience so long as your eyes are on them, but, leave them to their own devices, and they slip back into their

Fig. 239. Eskimo house at Nain. Photo the author, 1939.

old ways at once: 'it is not a custom of the people' (476). The Eskimoes had to learn the demands made by modern trade upon both buyer and seller if they were to avoid economic slavery. Now the missionaries, of course, tried to make their own trading concern self-supporting. In this way they could also most easily teach their charges to see how profit and loss must be levelled out in the long run. They taught them that to balance the expenses the sources of industry should be as varied as possible and every branch of productivity developed to effectivity. The mission store had therefore not become only a kind of emporium, where the Eskimo could sell his products. It also encouraged the Eskimo to increase and diversify their products and paid fair prices for them. The missionaries lent them traps and seal nets; they supplied them with ammunition. They avoided balancing the trade by debiting the products the Eskimo bought at too high prices, but helped them to develop and stabilize their economic situation. Let us for example consider the importance of the capture of fur-bearing animals. From what one hears in Labrador one has the impression that this industry was worked up to importance only by the intervention of the storekeepers of the mission. The Eskimoes, still during rather recent times, dared not make any long journeys into the interior of the country, so they were obliged to be satisfied with catching foxes near their homes on the seashore and in the fiord valleys. But they are so energetic that they have already been able to show surprisingly good results from even these

Fig. 240. Eskimo women in the old clothing gathering sticks. (205.)

restricted areas of capture; I was told for instance that they deliver many more coloured foxes to Hudson's Bay Company's factories than the Naskaupee Indians, who live in what ·is a special fox country. This has not so little importance in the Eskimo family's budget. In 1939, the value of the fat, skin, footwear, etc. which an industrious Eskimo family in Nain could produce was valued at 300 to 700 dollars in the year. To this must be added the occasional income from precious skins, but this is extremly uncertain, sometimes it will give a family only 100 dollars more in the year, in another year perhaps thousands.

Yet the amount of products which the Eskimo can present for trading is not very large, and it was therefore rather difficult to balance his accounts. Trade was also more difficult because the giving of credit was unavoidable for purely humane reasons. The consequence was the same for the trade there as among other primitives: time and again a more or less considerable writing-off had to be made of the client's credit account. The mission has been much criticized for this. One cannot blame the rather cunning, but childish natives that they exploit the credit just as recklessly as many civilized business men, and that in this way they have made difficulties for the mission. Another circumstance which hampered the mission's trade was that both in theory and practice it was absolutely free from compulsion. The client received in exchange for his products flour and other ordinary things at reasonable prices, but if he was not satisfied with them he could have his money and seek another market, even if a man was heavily in debt to the mission store. It also happened that the natives deserted their benefactors and took the bait laid out for them by unscrupulous, strange traders in the form of rum, tobacco, or trumpery articles. However, during the winter that same individual and his family became chargeable upon the mission funds.

Yet all promised well from the standpoint of the mission, but then came the World War — catastrophe! That meant a break in the chain of development of the mission's practical educational work. As business men, the missionaries were perhaps not experts, but they had most honestly tried to undertake the general education of the people in ordinary life.

To the economic difficulties political complications were added. Originally the best relations had prevailed between the local representatives of the Moravian Mission and the Hudson's Bay Company, but soon a keen rivalry grew up (HAWKES). The result was that in 1925 the mission was obliged to lease to the Hudson's Bay Company — on very generous conditions from the company — the trade at the mission stations. The fate of the Labrador Eskimoes now lies in the hands of this trading company.

When primitive people are under discussion, it is quite usual to state that any change in their old habits and customs must lead to decline and ruin. Yet, from what has been related here, it ought to be clear that the Eskimoes in Newfoundland-Labrador continue to subsist on the resources of their own districts. And they also have earlier managed quite well with them, as is shown by the fact that they never need to experience any years of real famine such as sometimes ravage their neighbours the Indians of the interior and the whites on the coast. Some of my informers, officials of the Hudson's Bay Company, certainly did not consider that economic conditions among the Eskimo were very prosperous, but the more northerly of them lived in any case free of debt; only exceptionally have the Eskimo there required some assistance in the shape of flour, sugar, tea, etc. from the authorities, it was said. — There are, however, signs that the Eskimoes' earlier wealth is sinking, and in some parts varying numbers of them are becoming a burden on the public assistance authorities, like the settlers, e.g. in Hopedale (p. 506).

But whether the relative prosperity which must have prevailed quite generally before 1925 and still prevails in certain districts will continue, depends practically wholly on the organisation of trade. Since money economy with its extensive credit system has superseded an economy based on barter, and the consumption habits of the Eskimo have become so different that they do not even wish to go without the white man's food, it is quite easy for a man having a trade monopoly to create a kind of economic serfdom among his clients by arranging prices. It is said that cash is an almost unknown commodity on the coast, and the price for the Eskimoes' products is given him in trade equivalents only!

The local population think that the prices for their products can only be justly regulated if free competition in trade is warranted. Here I shall content myself with stating this opinion. On the other hand, it is wise to remember that as production is very limited among the Eskimoes, it is a question of who will take the risks of competition.

A review of the changes in the material culture.

The marvels of modern times, which have caused other countries to move forward with giant strides, have left the Labrador coast practically untouched. For a long time the years passed with a sameness that necessarily marks a world which nature has hemmed in with a broad barrier of ice. Moreover, there is in every primitive people opposition to rapid cultural changes, and a disinclination to depart from the accepted way of living is customary. The Eskimoes of Labrador were also slow to change, and so the customs of bygone years continued to prevail for a long time (cf. 476).

The material culture now represented among the Labrador Eskimoes is no longer of the primitive type which met FROBISHER when in 1577 he sailed along that coast, and not even that which met the first Moravian missionaries when they landed in 1752. The hunting nomadism was certainly then in principle the same as now: from certain groups of igloos excursions were made in different directions and the families accompanied the hunters. The whole summer they are generally rambling up and down the coast, CURTIS wrote in the 1770's (196). They used the umiak for their summer migrations, that flat-bottomed, square-ended boat of drift-wood and seal hide in which the women moved the children and dogs and equipment from place to place, while the men travelled in lordly fashion in their kajaks. If food ran low, the families had to separate, the men went in their kajak to track large game, while the women and children in the umiak were given the task of searching the shores of the islands for anything edible. The results of their hunting were relatively small because of the gear of those days, and the igloo groups had always to be on the move. The yearly cycle can be supposed to have been in the main the same as now, but their occupations were stamped by considerable irregularity. The people were wild in the real sense of the word. Other countries advanced in material culture, but the Labrador of the Eskimoes long stood still.

Then the missionaries came. Small fixed centres were formed. Decisive changes took place in morals and education. Even the housing became different. Changes also occurred in the life connected with hunting and trapping. A review above of the whole complex of changes leaves the impression that the new accomplishments have not only been advantageous to the Eskimoes in making life easier for them, but have also here as elsewhere become the foundation for increased efficiency in economic life, which is, of course, the demand of the day in Labrador as elsewhere in the world, if one is not to be handicapped in retaining his position among his competitors. But here it may be objected: how is it with the Eskimo themselves? Do they still remain a nation of hunters?

There is some misunderstanding among the occasional visitors: the nomadic hunter has gone, the Eskimo of the present day is a villager, they say.

This incorrect notion of their ways of living goes far back in time and has been frequently repeated. HIND wrote about 1860 that the ob-

Fig. 241. Eskimoes at Ryan's Bay. Photo the FORBES-GRENFELL EXPEDITION, Aug. 1931.

ject of the Moravian Brethren was to collect the Eskimoes in villages around them and, besides instructing them in the truths of Christianity, seek to teach them those simple mechanical arts which may contribute to their comfort and wean them from the wandering life to which they are so prone. This last is an important misunderstanding, because as early as 1860 REICHEL, who was a most prominent Moravian missionary, wrote (787 a): it is evident that the Eskimo must for ever remain a huntsman and fisherman. In this way alone can he obtain a livelihood in his sterile country.

We let Dr HUTTON, with his expert knowledge of these people, answer this question (476):

In the village, in the ships, in the presence of foreigners their native ways are not much in evidence. The casual summer visitor sees the Eskimos in European clothes, curing codfish, they see them drink tea and eat biscuit that they have bought from fishing schooner folk from Newfoundland in exchange for skins and sealskin boots. They think the Eskimo has gone and may exist only in the icy solitudes of the Arctic. But that is a mistake. You must see them at their seal hunting and you will find that the Labrador Eskimo still is an Eskimo to the core. And he is perhaps a more successful hunter than his ancestors the witch-finders, he does read and write, he is able to inform himself of all the modern instruments for chase. Peoples who have lived among them for a good many years, who speak their language, who have travelled with them and base their opinion upon what they have seen themselves, report that the Labrador Eskimo is still a kind-hearted, openhanded, raw-meat-eating Eskimo. He is no less a Christian because he sucks blood from a freshly killed seal. At these latitudes the aboriginal hunters are almost like pure animals: to eat or to be eaten. And you will therefore see the hunter slit the seal's throat and suck its warm blood. That heartens them, that keeps the cold out. The hunters are Eskimoes yet.

Even to-day (1939) we find that in general most of the people live in small groups in more or less distant bays and sounds and only come to the stations for the festivals or for Sundays in the summer. They still like to depend on hunting for their income, and are still the semi-nomadic hunters of bygone days. For their mental ability this is very important (303 a).

Fig. 242. Eskimoes in their kajaks, Ryan's Bay. Photo the FORBES-GRENFELL EXPEDITION, Aug. 1931.

It is most interesting to learn that it is in reality the Moravian Mission that has kept the Eskimoes true to their tradition in their ways of livelihood. They soon came to the conclusion that the hunter's life is the ideal life for an Eskimo. The strenuous fight of the Moravians to protect their charges has been waged not only against the ignorance, superstition, evil practices and primitive religion of the savage but also against the sins of the so-called civilized people. When the tide of immigration swept up the coast it brought with it most of the evils of the outer world. The missionaries recognized the danger and endeavoured to keep the Eskimo to the food and the mode of life which had sufficed them since olden times, and which they knew would give them a better chance of ultimate survival. But civilization is insidious: the Eskimo learnt from the invading whites to like coffee, flour, tobacco and strong drinks, not good things for a constitution accustomed to blubber and raw seal meat, which they had loved to gorge upon in earlier

Fig. 243. Modern Eskimo dandies at Nain. The expedition ship in the background. Photo
C.-G. WENNER, July 1939.

days, writes PRICHARD. Yet seal meat still ranks in the Eskimo mind above
all these things as food and a delicacy.

One has said, and no doubt rightly, that were it not for the farsighted and
heroic work of the Brethren there would not be a single Eskimo on the Atlantic coast of Labrador to-day; the composition of the population towards the
south from Hamilton Inlet may prove it (761; cf. 614). In this isolated wilderness they have in silence carried out a task of human love of inestimable value
for the native themselves by carefully and slowly bringing them into contact

with modern civilization, while at the same time persuading them to preserve
the valuable characteristics in their own material culture. Christian culture
can rejoice over this victory. The civil administration has again in this way
received a firm foundation on which it can build if it so desires.

Changes in the sphere of non=material culture.

Having illustrated the metamorphosis in the Labrador Eskimoes' material
culture with some isolated examples, it may be interesting to read some

Fig. 244. Eskimoes at Hebron unloading firewood (577).

statements which throw light upon the radical changes in the sphere of non-
material culture. The romantic glow of a golden age which is usually thought
to envelop the life of the natives will pale considerably.

In order to obtain a background for these changes, let us remind ourselves
of some traits in the spiritual and social life of the Labrador Eskimo in olden
times. For this a prime collection of examples is to be found in the Moravian
diaries from Labrador. The old seafarers and traders down to CARTWRIGHT
(see Bibliography) give many hints about the Eskimo on the Atlantic coast.
TURNER (963, 965), who knew the Ungava district so well, has left a description of the northern Labrador Eskimo in the 1880's, and in this there is
also some information about their beliefs as to the hidden powers controlling
man's fate and to which we ought to devote some attention.

The spirit world of the pagan Labrador Eskimoes.

The pagan Labrador Eskimoes considered that all phenomena and events in life were dependent upon the actions of certain definite spirits. Each of the spirits could determine things regarding his special creation or thing. According to what TURNER and HUTTON found there were spirits of different ranks: spirits of a lower order under the rule of a great spirit. The most important spirit of the sea, for instance, was held to be an old woman, whose home was at the bottom of the sea, but who used to come up to breathe on the shores;

Fig. 245. The starting of the fishing-party from Hebron, July 1937. Copyright WILMONT T. DE BELL.

all living things in the sea were under her control. Thus the demonology was very extensive.

The idea of a good spirit did not enter the old Eskimo's minds; the spirits of their pagan life were all ill-disposed and apt to sulk, they would hinder their hunting and cast an evil spell over them. Therefore the spirits, the powers of mighty ill-will, beings of malice and cruelty, must be appeased. Some chosen men met to approach the dwelling of the principal ghost, the *Torngak*, in the gloomy mountains. They took with them offerings of clothes, food-stuffs, etc. in order that his anger might be quenched at least for a time (476). The great spirit of the sea too had to be appeased; if not, she could make the seals avoid the hunting-places of the Eskimos. She could feed the codfish and make them lie fat and sluggish while the fisherman plied his hook and line in vain. Broken knives must be sacrificed, bits of meat, etc. and cast into the water for the old woman that she might be in a good humour (476).

It is worthy of remark that some of the personalities in the spirit world had changed roles among the Labrador Eskimo. *Tornarsuak*, the great spirit who had taken the form of a giant polar bear, and was supposed to live in a cave on Cape Chidley, and the other great divinity, *Superkugssuak*, the old woman who was said to live at the bottom of the sea and control its inhabit-

Fig. 246. The Eskimoes' modern fishing motor-boat, Cut-throat. Photo C.-G. WENNER, July 1939.

ants, and before whose demons the pagan Eskimoes trembled, had opposite functions in Labrador: on the Atlantic coast Tornarsuak ruled over the sea animals.

Under special circumstances a certain spirit was materialized in the form of a doll which one carried about hidden under one's clothes, and the doll's assistance was called upon when difficulties arose. If the materialized spirit was unable to help the hunter, it had to be punished. If this did not help, the last resort was to transfer it to a fellow creature, but this must be done without the knowledge of the receiver. Not until he had rid himself of the unwilling spirit did the hunter succeed in carrying out his task. Against the bad temper of the spirits it was good to wear charms, as for example against the spirit of rheumatism; tattooing was a protection against the evil powers.

Every living creature in the world was accompanied by his special spirit. This spirit was united with the body, but when the body was lost by death, the spirit released itself, for it was indestructible. Like most children of nature, the pagan Eskimo were thus firm believers in a life after death. Their idea

Fig. 247.   Eskimo fishing-party on the outermost coast off Cut-throat.
Photo C.-G. WENNER, 1939.

was like that of the Happy Hunting Grounds, with the difference that the best hunting ground to their ways of thinking is the sea.

This is made clear by the method of burial. Moss-grown heaps of stones mark the grave which was just an oblong rough masonry of boulder stones; the people know nothing of digging because the soil is not deep, scanty, and often frozen. The place was chosen on a lonely cape or height, as I found on Fermoy Island for instance, so that the hunter could overlook the sea. Under the stones lie the hunter's remains. Heavy flat stones were laid on the top of the pile to keep away dogs, foxes, and wolves; in the cold, pure air the bones are preserved from decay for a long time. At the head of the tomb in many graves under a separate little heap of stones lie the hunters tools etc.: arrows, harpoon, fish-hooks, stone knives and stone-headed axe, his lamp of soapstone, cooking-pot; this equipment

tells of a people who lived in the Stone Age long after the discovery by white
man. In some graves, in which the bones have turned to dust, the woman's
knife, *ulo*, and the skin scrapers tell that a woman was buried there, and a little
child's bulging grave is marked by the childish toys piled up in it. The people
seem to have believed that the future life would be like the present, and they
wanted to have their tools in it. In the Eskimo graves on Labrador we thus find
the same trait as for example in the graves from the Iron Age of the Varanger
Lapps in the most northerly part of Fennoscandia; in these, too, gear and tools
were placed, but destroyed first, they must be 'killed' for the dead can, of course,
only use things that are also dead. In a child's grave the Eskimo sometimes
considered it was useful to place a dog's head: children are rather lost when
they come into unknown parts, so they thought the dog's spirit could show the
child the way to the dwellings of the dead.

The clever hunter was believed to be happy in his grave. His great adventure
had begun, he hunted every night. Many had seen his footprints in the snow. —
Often have the Indians profited from this superstition, says TURNER; they plundered the graves. I have looked into many an old Eskimo grave; they have all
been plundered except for the bones, but in one grave near Hebron I found a
little copper amulet. When the Eskimoes later noticed that the articles had
disappeared from the graves, they were absolutely convinced that the spirit
of the departed had really taken the objects and definitely started out on the
long journey to the dwellings of the dead. It was not necessary to fear that the
spirit would begin to haunt them if it became angry.

The existence of these hunters of the nether-world appeared in local beliefs
until quite lately. The Eskimoes would not live in a house where a death had
occurred, believing that the spirit of the departed would haunt the place (995).

Every animal was also supposed to have its special spirit. The spirit of
the animal was like that of man, but belonged to a lower order. Nor could the
animal's spirit be destroyed; if an animal was killed, its spirit abandoned it
but continued to live as a spirit and in time became re-incarnated. For this
reason it was not necessary for the hunter to limit his activities, the number of
animals could never be decreased in however great numbers their bodies
were slain. The animals' spirits also were after their kind under the rule of a
greater spirit. The spirit of each caribou, for example, obeyed a powerful
spirit who took the form of an enormous white bear, and it depended on this
spirit's goodwill whether people would be able to kill caribou in the future
when they required its flesh.

When a person wished to approach a spirit with a request, it was necessary
to observe a great deal of ritual, otherwise the spirit would be incensed. For
this reason there was a functionary, the *angekok*, the shaman, who was the
intermediary between mankind and the spirits (see below); only the angekok

could get the great spirit to send the caribou to the humans when they needed food, but on his side the spirit made many demands for expressions of reverence from the people, and he had often to be pacified. If an accident happened,

Fig. 248. Eskimo trapper dwelling in Fraser River Valley. Photo C.-G. WENNER, 1939.

e.g. if the dogs had found a chance of gnawing at a caribou's thigh bone, the great spirit of the caribous was offended and might perhaps in the future refuse to send more of the animals to the hunter. It was still more unfortunate if the dogs had got at the caribou flesh. As compensation for this it was necessary to cut off a bit of the tail of the guilty dog or clip its ears, for blood had to flow: — in this special way the guardian spirit of the caribou spirits could be propitiated.

A special class of spirits ruled the great objects of nature: the sea, the land, the clouds, the wind, etc. Every bay, cape, island, etc. in Newfoundland-Labrador has had its special guardian spirit. This class of spirits obeyed the will of the great spirit called *Torngak* (cf. above). Torngak was no less than the spirit of the dead (965). He would worry and trouble people so that they should tire of their existence on earth and their spirits desire to move over to Torngak and live with him. For this reason he attacked people of all ages; hunger, disease, and death he sent out to catch the desired spirits of mankind and bring them to him.

The ordinary person, as already indicated, was not worthy nor able to have direct contact with the spirits, and their relations had to be entrusted to the angekok.

Anyone could obtain for himself the qualities of an angekok. An Eskimo who felt urged to it could, by fasting and abstention from contact with other people, obtain for himself the supernatural powers of the shaman. These abstentions made it possible for him to learn the great secrets of Torngak. He therefore lived for some time alone in the wilderness. When Torngak got to know this, he pitied the lonely one and took him up into his fellowship. Now he entrusted him with the great secrets of how to maintain life, to drive out the evil one who caused death. In time the disciple himself became a conjurer of spirits, the angekok. When he had entered upon his vocation he returned from his retreat to his tribe and told people what he had seen and done and how he was now prepared to fulfil his mission among them. When someone came and wished to have his services, the angekok let them understand that he was entitled to compensation for them; and he was not at all uncomfortable in saying right out that the greater the compensation the greater would be the effect of his arts.

The hunter was very often obliged to ask advice from the angekok. He was the oracle of the weather and hunting. A great conjurer was believed to possess amazing powers. It was he, for example, who could compel a large number of game to find themselves just in the district where his village group were planning to hunt. To achieve this it was sometimes necessary that the angekok hung a magic doll or a stylized image of some famous hunter or angekok on a staff at the place.

The client had to obey the angekok blindly. The Labrador angekok's rule was one of terror, TURNER considers. As time passed, he says, the shaman therefore became more and more exacting and impudent. If one got on bad

terms with the angekok, the only possibility to avoid his often really idiotic
ideas and bluff or revenge was by fleeing from the camp and abandoning
the hunter group in which one had grown up.

This can be illustrated by a quotation from HUTTON: The dreaded sorcerer
stalked in, a weird and filthy fellow, bedaubed and betasselled. A silence fell
upon the company, the people sat around in the dim and smoking light of a seal-
oil lamp, talking in undertones. As the sorcerer spoke, a silence fell. He trumped
his seal-hide drum and began to chant in a nasal voice. Maybe he had a grudge
against some unhappy fellow: There is the fellow, he sings, who is making the
storm. He forgot to give me meat when he caught a seal, and the spirit of the
storm is angry. You will be hungry. The strong hunters trembled (475).

Fig. 249. Eskimo summer shack in the skerry fence near Makkovik
(536 a).

The angekok was also the doctor of the camp, and it was believed he could
restore the sick to health. In doing this, he did some very curious things. For
example, a patient with fever was laid on the ground and the magician worked
himself up into a condition of mad rage and began to drive out the evil spirit
who was supposed to be causing the illness. CARTWRIGHT speaks of the shaman's séances. The magician threw himself in holy desperation upon the
patient and by kneading began to chase the evil spirit from the place where it
was to another spot, and so the kneading continued. The sick person got
blows of all kinds, shoves and jerks, while the shaman gave voice to the
most horrid yells and hideous cries. After a while he declared that the victory
was won; the operator stated that he now held in his hand the spirit who had
caused the illness. If, however, it escaped from him again, and again got into

the patient, which was shown by fresh pains, the angekok began his cure again and went on with it till the patient got well or died. If the patient died, well then, the cause of death must be something which was not under the control of the shaman: a sudden appearance of a colour shade in the flaming Northern Lights, a sudden snow-fall or some other alleged tool in the hands of the evil spirit which had succeeded in escaping from the magician was concocted *ad hoc*. To pacify the spirit the shaman sometimes recommended that the tail of a living dog should be cut off, and it is said that many dogs ran about the camp mutilated at the order of the shaman. TURNER (965) states that a greater rascal could scarcely be found in the Ungava region that the shaman Sapa, who was as impudent as he was famous, and was still alive at that time.

From what has been said we may conclude that when the white man came into contact with the Labrador Eskimoes, they met no free sons of the wilderness, but terrified wretches of a world full of unfriendly spirits and trembling under the terrorization of the angekok. It seems that the angekok in Labrador could drive them to do whatever he wished, like a crowd of resistless animals. Sometimes in descriptions of the medicine man's feats we find an outbreak of pure madness, and from this we see to what an abyss of terror and suffering superstition had opened the way among these Eskimoes. It was the Moravian missionaries who, in the face of danger and privation led them away from this misery with loving hands and tactfulness.

Some notes regarding customs and habits in pagan days.

TURNER (965) relates a good deal of the customs and habits among the nearest neighbours on the west, the Innuit at Ungava Bay. The social conditions which appear in this description must be considered as existing among the Innuit of the East at an older period of time and, indeed, when the missionaries first came among them.

Married relations bore the impress of might being right. The man who had the power took as his wife whoever he wished and had as many wives as he could. It is said that girls were often married before they reached puberty. They were not very particular about kinship, though MACGREGOR mentions that brother and sister marriage did not take place. But according to TURNER it once happened that a son took his mother to wife; as, however, the other hunters urged him to give up the connection, he agreed and took instead two other women.

Thus the inclination to polygamy was general, and the reason was purely one of convenience: the more wives a hunter had the better he was served in the camp. When a missionary once reproached an Eskimo for having increased the number of his wives, he answered — not the least disconcerted — that he needed them all to man his boat.

The woman was in every respect subject to her husband; if she was disobedient or neglectful she was thrashed until she improved; the custom of the people is still for the wife to follow, and she follows, in public the husband must be the 'boss'. On the other hand a model husband has always been affectionate, kind and faithful, a smart hunter withal, well able to keep his little household in proper Eskimo plenty, says Hutton. — It has been said that polyandry prevailed among the Labrador Eskimo because of the lack of women, but I have been unable to find any support for this statement.

Through the influence of the missionaries marriage relations among the northern Eskimoes have changed and the wife's position has been improved in certain respects. In 1905 there was only one man who lived in pagan style and had two wives, and that was in Killinek, according to MacGregor. As far as I know polygamy has now completely disappeared from among our Eskimoes. The old savage mentality has also been very much tamed.

Sudden outbreaks of rage were the expression of the strongly asocial predilection which distinguished the Eskimo societies; there was lust and greed and hatred, bitter feuds and no forgiveness, sometimes even murder (475). If things came to such a pass that there was murder or manslaughter it had to be avenged; and with this aim in view the other men watched for an opportunity to surprise and kill the murderer, usually by stoning him; this was the established custom, and those who took part in it made no secret about it. According to their code of honour it was the duty of the murdered person's nearest relative to avenge him. The crippled and the infirm were treated without pity, Turner says; if they became dependent on others they were simply got rid of by suffocation or left behind to die alone when the camp was moved to another place. The dead were treated most unceremoniously.

Cartwright, in 1792, mentioned that the Eskimoes were fond of a game of ball. All those of the west were at the time of his visit much given to pleasure and amusement, says Turner; games of many kinds had been preserved among them from olden times. The love of gambling and card-playing are still common among the natives; sometimes they play every night in the huts; both sexes loved games of chance to such a degree that they could even risk their own lives. It happened sometimes that a woman gambled on her only dress and nearly all her other rags rather than be left out of the game. Feastings were frequent in the huts. In 1777, according to Crantz, the Eskimoes near Nain had a temporary winter club house (public gaming-house or pleasure-house; cf. the old promiscuous tent) sixteen feet high and seventy feet square, built wholly of snow; in this they collected on festive occasions, to the horror of the missionaries. Nowadays gambling and card-playing are said to be much less prevalent.

These happy savages, however, were a danger for the white man who wished to exploit the resources of the great wilderness. When reading older descriptions, one has the impression that the Labrador Eskimoes, a century and a half ago, could be reckoned as North America's most dangerous and asocial savages, who were the cause of very great difficulties for the British Colonial Government. After DAVIS some other seekers of the North-west passage: HALL, HUDSON, BUTTON, GIBBONS, KNIGHT, and others tell about the Eskimoes of northernmost Labrador or in Hudson Strait. They had all the same story: friendliness first, then attacks. GOSLING exclaimed: the Eskimo seem to have been the Ishmaels of North America, — their hand was against every man, and every man's hand was against them. The Eskimoes were then fierce and treacherous, if strangers came to them the process was a short one: KNIGHT, in 1606, sent a boat ashore and seven men were immediately murdered; the Moravians who landed south of Hopedale in 1752, disappeared without leaving any trace, clearly murdered in ambush (cf. 205). By a show of friendliness they would entice men from the fishing vessels ashore, and later attack and massacre them with merciless ferocity. Their habit was to rise from their haunches, utter in unison a mournful cry and fire off their guns. It was this cry which terrified and frightened away the fishermen in the 17th and 18th centuries.

The Eskimoes had lived at war with the Indians for generations, and cruelty and treachery had become part of their very being. It might be said that these 'savage heathens' later had followed the example that Christendom had given them. The white invaders had so to say only educated the Eskimo to self-defence, and in this was included the overawing of their enemies by cruelty. Both the Indians and the whites had, at least since the 17th century, forced them to hold the view that every moment they must be on the watch, yet the causes of this must be still older. The Icelandic sagas tell how, as early as about 1000 A.D., TORFINN KARLSEVNE slew many natives on his Vinland journey and kidnapped two Eskimo boys (920 b). In 1501 GASPARRO CORTEREAL kidnapped 57 Eskimoes to test their suitability as slaves in his homeland; it was a kidnapping in the direct manner analogous to that of collecting for a zoological garden; — in this humane way began the study of the natives of North America ILWRATH wrote (606)! SETTLE (848) writes in 1577: At the sight of our men the people fled, we tooke one, and the other escaped. Seekers of the northwest passage tell about the Eskimoes of northernmost Labrador or in Hudson Strait.

— Their hand was against every man, and every man's hand was against them. In the 17th century it is said the French started a systematic extermination of the Eskimo in their most southerly district by supplying the Indians with firearms and egging them on against the Eskimoes, who can not indeed have been very difficult according to what has here been described (p. 480). The Eskimoes were, however, bold and courageous: 'When the Eskimoes come on board, the sailors go right on to the forecastle with guns, bayonnets and cut-lasses', is a description given at the end of the 18th century; if the Eskimoes discovered the French flag on a vessel they seem really to have run amok (205). Repeated bloody fights took place between the trappers and the traders on the one side and the Eskimoes on the other.

When one knows how rapidly a rumour spreads and how long the memory of unusual happenings can live among primitive people, it is not to be won-dered that the Eskimo suspected all white invaders and expressed it in many kinds of horrors with a view to preventing their coming again. It seemed to them quite natural to murder the white encroacher at once rather than be murdered by him.

— It was impossible for us, PARRY (44) says in the report of his second voyage, not to receive [1821] a very unfavorable impression of the general behaviour and moral character of the natives of the entrance part of Hudson Strait, who seem to have acquired by an annual intercourse with our ships for nearly a hundred years, many of the vices which unhappily attend at a first intercourse with the civilized world, without having imbibed any of the virtues or refinements which adorn and render it happy. The same thing seems to have struck LYON in 1824.

At that time, it seems, reconciliation between peoples was not thought very highly of; yet voices demanding a better understanding of the Eskimoes were not lacking. The unknown author of the »Memoir concerning Labrador, 1715—1716» (325) made the following interesting statement: The Eskimoes are consi-dered extremely savage and intractable, ferocious and cruel; they flee at the sight of Europeans, and kill them whenever they are able; but I believe they fly from Europeans because they have been maltreated, fired on and killed, and if they attack and kill Europeans it is only by way of reprisals. — — — Messieurs JOLLIET and CONSTANTIN [cfr the Carthography annexed p. 899] who had visited them, had received a thousand tokens of friendship. M. COURTEMANCHE (p. 480) said, that they are good, civil, mild, gay and warm-hearted men and women and that they danced to do him honour. They are very chaste, dislike war... are more timid than savage or cruel... there will be no difficulty in civilizing them if proper means are taken: to forbid the savage Indians to make war on them, to forbid the French fishermen to fire on them or to offer them any insult, or to sell them any intoxicating liquors...

However, the Eskimo band on south-east Labrador numbered about eight hundred and they had firearms of various sorts in their possession; it was sup-posed that some Europeans had taken up their abode with them. They were much dreaded by the white for their robberies, which were often accompanied by murder. COURTEMANCHE also had much trouble with them.

Peace or a sword was the alternative which the British government was forced to lay before the Eskimo. Then the Moravian Brethren stepped into the breach and saved them from themselves.

It was fortunate for all parties that Newfoundland and Labrador at that time had an unusually wise governor, Sir HUGH PALLISER. He gave the Moravians an opportunity to try the first alternative before the musket and cutlass were used by the naval forces. It was not an easy task; the white voyagers to the Labrador coast, as elsewhere, had certainly treated the natives abominably, and their cruelty had fostered cruelty and treachery amongst the Eskimoes. PALLISER was right. It needed only that the missionary JENS HAVEN met the Eskimoes, talked to them in their own language, before the 'vilain Huskies' cried: Our friend is come (205). The herd of barbarous savages was already in 1773 found to be in a fair way to become useful subjects (196). The savages were slowly tamed, war and murderous reprisals passed away for ever from the coast (761).

The educational work of the Moravian missionaries has certainly caused the psychological attitude of the Labrador Eskimo towards the great questions of human life gradually to undergo a cardinal change. It is therefore most interesting to see how their religious views of their local world have altered in a comparatively short time. They have absorbed the articles of the Christian creed and its social ethics surprisingly quickly, and as far as is known, they have ceased to bow the knee to the heathen divinities. Their receptivity in the sphere of the non-material culture has been as great as it was in that of the material cultural. Even the language has been modified, partly because new conceptions have been introduced by the foreigners, partly because the missionaries and the traders could not learn to pronounce the difficult guttural idiom of the natives, so that the latter have gradually adapted their pronunciation to suit the newcomers.

The struggle between Christianity and superstition.

About 1860 there was something like 200 pagan Eskimoes in Newfoundland-Labrador (787 a) out of an estimated total of 1500, of whom 1163 belonged to the Moravian mission. At the beginning of the present century it was said that there were still some thirty pagans in Eclipse Bay far away in the North on the coast, and in 1905 there was one pagan even in Hebron (604). Nowadays there are no more pagans.

The missionaries have had a hard and varying struggle against the angekoks and the spirit world of the Eskimoes; I refer the reader to DAVEY's instructive work: 'The Fall of Torngak.' The struggle between belief and superstition is generally a long one among primitive people. The Labrador

Eskimo have also long preserved flashes from the superstition of olden times.

TURNER, in the 1880's, considered that the conversion of the nomadic Ungava Eskimoes was only apparent, for as soon as they got away from the watchful missionaries and their influence they placed themselves again under the direction of their shamans; it was in any case, they thought, the angekoks who since time immemorial had caused the spirits directing the life and movements of the seal and the caribou to let these herds move towards the hunters when they went out to kill them.

It was said among the Eskimoes at the beginning of this century (577), that while it was not possible to know exactly how ideas were connected with reality, it was best to be careful, not to forget the old spirits for the sake of the new divinities. It might happen for example, that the spirit that lives in the moon is of course comparatively well disposed according to the old belief, but at an eclipse of the moon it was still considered to be safest in any case to make the spirit gracious by cutting off a little of a dog's ear for her; this must in any case be done when the Northern Lights are flaming blood-red, for this spirit is then in such a humour that it could send misfortune if not propitiated in time. People earlier also protected themselves with amulets or tattooing against the spirits. But tattooing has gradually died out: HUTTON saw one of the last tattooed people at the beginning of the century at Nachvak.

PERRETT stated at the beginning of this century that the natives of Killinek certainly still believed in Torngak. The only one who would tell him something about this spirit, said: Torngak still plagues me, though I have renounced him. He wants me to go back again, but.I won't go. The other Eskimoes don't tell me they are still worshipping him, but I feel it inside me when they are in connection with him (478). In 1910 the missionary WALDMANN in Killinek (989) said that the Eskimoes suspected that he was having secret telepathic connection with their hereditary enemies the Indians. Their fear of *allat* (= Indians) was then still very great. They believed that allat could pluck away the fish out of the net while the Eskimoes slept.

However, not many traces of superstition still linger, HUTTON reports at the beginning of this century. But going in and out among them he found a few, especially connected with the hunt (476). A young man became very frightened, he relates, when he had caught a fox with a peculiar mark: I shall die soon, he said. They are now also afraid of the presence of death; sometimes they dismantle the whole hut where a person has died and build it up from the same material elsewhere. And HUTTON concludes: The fears and fancies of the old times have scarcely passed away. For instance, in the night, the people are fast asleep, but they like to keep a light burning; like children they are timid in the darkness; they superstitiously dread some vague and malignant power. Most interesting is that some decades ago they still feared to be photographed; they believed their spirit would be caught in the box (= camera), HUTTON noticed when he first came to Labrador.

Since the missionaries succeedéd in getting the people to retain their nomadic way of living the contact between them is no longer the unbroken one of the older days. The nomads now have much freer possibilities of testing things than their forefathers had when the missionaries began their work and their charges lived a long time at the station. It would be astonishing, if, when the nomads are left alone again with nature and their traditions they should not give way to the temptation of seeking re-insurance for their welfare from the old spirit world too. This has proved to be the case, and even in our days the angekok has not yet completely disappeared from the consciousness of the people. Here and there the old ideas peep out again and reflexes from the past can temporarily fill the Eskimoes with enthusiasm. For example they are sure that a mad person is quite simply possessed by an evil spirit. Then they can be caught by mass suggestion and, relying upon the angekok's old mishandling, they try with blows and slaps and other bodily means to drive out the devil; this happened still in 1939. Not long ago, WALLACE noted, in 1906, the wife of one of the Eskimoes in Ramah was taken seriously ill and became delirious. Her husband and his neighbours, deciding that she was possessed of an evil spirit, tied her down and left her until finally she died, uncared for and alone, from cold and lack of nourishment (995). It seems that anyone, at any time, can be driven into such excesses, however bitterly he may have regretted it after an earlier action of this kind. From this and other facts one has the impression that the Labrador Eskimoes are distinguished by an extremely over-excitable mental life which reacts very easily to small impulses in an abnormal way.

In normal cases, however, the angekok's bluff has lost its authority; scarcely anyone now believes in his instructions about sacrifice. His old trick of recommending the cutting off of a dog's tail or a dog's ear for the purpose of curing a patient has fortunately gone out of use; I looked very carefully at the dogs but never saw one with a mutilated tail. From what I saw and heard in Labrador it may be safely said that among the Eskimoes of the Atlantic coast the Stone Age animism has practically ceased to exist and they have become mild, sincere Christians. 'Wild' Labrador Eskimo are scarcely to be found anywhere except perhaps on Belcher Island in Hudson Bay among those who trade at the Hudson's Bay Company's station Great Whale River.

The slow transformation of the Eskimoes both in their mental processes and customs is most impressive. Any other missionary undertaking would do well to imitate the proceedings of the Moravians in Labrador. Even with regard to the transformation of their mental ability it may be said that a miracle has been wrought. The missionaries have systematically educated the

Fig. 250. Husky dog at Nain. Photo C.-G. WENNER, 1939.

Eskimo both spiritually and in worldly matters and thus prepared them to receive the blessings of western European civilization.

THE BLESSINGS OF EDUCATION.

In the olden days education meant amongt the Eskimoes that the parents were training their children for the practical demands in the hunter people's life. In all their games the children were chiefly training hand and eye. Above

all the Eskimo boy likes to feel himself a hunter. This fashion still prevails.
Dr HUTTON saw the boys use the crossbow to kill birds for dinner, and boys
of thirteen or fourteen went up the valleys with real guns hunting hare and
ptarmigan. They enjoy an imaginary seal hunt with harpoons.

The puppy dogs are always handed over to the boys, it is training for both
boy and dog, for the boy uses all the tricks he has seen his father use in driving
the big sledge. It is said that only one who has tried to drive a team of Eskimo
dogs can know what a stock of patience and perseverance the child must have
to teach the puppy to keep its traces tight and to know and obey the words of
command. These adventures are commonplace to them. One says that a well-
trained dog team driver is strong enough to strike a dog with a single blow of
the whip from the full length of the lash of thirty feet of hide. These powerful,
quarrelsome and half-savage animals — Figs 250 and 224 — must be kept under
absolute control by the formidable Eskimo whip. With his short-handled whip
the driver can not only touch any dog he pleases, but can flick a fly, drive in a
nail, or kill a willow-grouse with a turn of his wrist (761). To be a good whipper
is the highest ambition of the young men of the rising generation.

The mastery of a boat seems to be another of the Eskimo's inborn gifts. He
is often on the water by himself before he is three years old, lugging away at the
pair of oars. Soon he learns to respect the waves, the wind and the currents
between the islands.

Thus the children educate themselves at this point. The boys are petted and
pampered far more than the girls. There is no such thing as punishment in an
Eskimo household. The children hear few harsh words. The little boy still
grows up like his father once grew up; he has the same pose, the same way of
holding his hands, in everything and especially in little mannerisms he is his
father over again (476). But nowadays he acquires other faculties too.

Times have changed. Nowadays life demands more of Labrador children.
The mission therefore teaches the children in a school at the mission
stations during the winter months. The children begin to attend after
their sixth birthday. Usually the smallest are taught by an Eskimo, later the
schoolmistress or the missionary himself takes them in hand. Many of the
teachers were once Eskimo children and have learnt from the missionaries to
present things in a way that the Eskimo child can grasp; at the present time
professionally educated schoolmistresses come from England and the Nether-
lands to teach the children (476). The success of the schools has been astonishing. McLEAN (610) was, surprised already in the 1840's to find even all
the Eskimo half-breeds far from the mission stations able to read and write;
the task of teaching devolved upon the mother or a neighbour mother, because
there were no schools or schoolmasters. Here we see the far-reaching and
good effect of the work of the Moravian missionaries, he says.

There is no time for the schoolmistress to mope and feel lonely, Dr HUTTON remarked during his stay at Okkak, she had to dispense with all ceremony and follow the customs of the land and its folk. The first thing was to learn the language of her pupils, and this is not at all easy; the Eskimo says: Kablunak [white man] has a different mouth from the Innuit. I once asked the Rev. PAUL HETTASCH if he considered that he could talk Eskimo perfectly. Not at all, he answered, I have spoken it for thirty-seven years, but I am not perfect. My daughter on the other hand who was born here, should I think, speak it like an Eskimo. HUTTON says that. the language has, however, one immense advantage: it is beautifully grammatical, governed by plain straightforward rules, with only a few exceptions. The remarks made by an Eskimo are said to be always applied literally (476). In practice the words are almost ludicrously long — twelve, fifteen, seventeen syllables, and built up bit by bit according to a pattern that can be learnt and understood, like chemical formulae. It takes some time for a foreigner to be able to talk freely with the people; the learning involves a prodigious feat of memory too, there are so many words for the same thing under different circumstances and the many-syllable words are built up by putting in all kinds of tags and bits between the unchanging roots of a verb and its grammatical expressive endings: e.g.: *tikkipok* = he comes; *tikki 'niara' suarkôr' pok* = he will probably try to come, etc. On the other hand there are short words in this language, as in that of other primitive peoples, which express some picturesque idea, e.g. *ôtok* = the seal which is basking on the ice in the spring sunshine (476).

All in all this language is a very storehouse for the philologist, not least the phonetic and the picture words evolved by the missionaries (750). Since the Eskimo has been in contact with the whites the idiom has changed somewhat (cf. p. 554), so that there are now really two languages: the original with its difficult, guttural sounds is spoken by the Eskimoes among themselves, but with the missionaries and others they use a distinct language and choose with discrimination words easily understood or borrowed words (478). A Labrador-Eskimo dictionary was compiled by a German Moravian missionary in the '60's (525 a, 95 a; cf. 750 and 972).

It is no easy task to teach the children their A.B.C. and it is most difficult for one who does not possess the instincts of the Eskimo. The range of ideas in the minds of the Eskimo is inevitably narrow. As an instance PRICHARD tells of one of the difficulties the Moravian Brethren met with in their translation of the Bible: In Labrador there are no domestic animals save the dog, and when they had to describe the Patriarchs with their herds of horses, cattle and sheep, they were driven to present them under the guise of seals and fish, interpreting the riches of the earth by the riches of the sea. Many comic points arise during the reading lessons, e.g. when the children have to spell out the words beneath pictures hung up on the wall: the cow gives very good milk! The dear souls would never see a cow, never drink fresh milk (478)!

Teaching Eskimo children of six to read words of twelve syllables or so is not easy for any instructor. To make things easier the missionary PERRETT composed a reading-book in Eskimo. This »*A—B—pat*» with the sub-title: *Okautsit illiniaraksat sorrutsinut* [= Learnable words for children], is a very thin book of twenty pages (478). The success of the schools has however been marvellous: already in 1884 KOCH (528) reported that at the whole coast there were no Eskimoes who could not read and reckon, and in 1905 Governor McGREGOR stated in his report that even at Hebron a l l c h i l d r e n o v e r s e v e n y e a r s c o u l d r e a d. We must fully agree with Dr HUTTON's conclusion, that by their quiet labour in remote Labrador the missionaries have worked out the saving of a nation. Now the Eskimoes should be able to control their business themselves.

The missionaries have also taken charge of an Eskimo newspaper: *Aglait Illunainortut* [The paper for Everybody], which deserves the title. It is by no means a daily paper, rather it takes the form of an annual bulletin; it tells of the doings of other lands and helps to stir the Eskimo's loyalty as British subjects. The people like their paper. As far as I know there are today at least twenty books printed in the Labrador Eskimo language.

Though true to all ordinary Eskimo habits and living, men and women bear witness both in word and way that their '*sagiarnek*' [conversion] is a real one. By their simple faith and their willingness to follow the teachings of the Gospel this artless race has risen from superstition and heathen practices to a life of Christian kindliness (478). The Eskimoes obviously delight in the church services and are very regular in their attendance. The organist is mostly a native and the native 'chapel-servant' frequently leads the service and can invariably preach and pray with fluency (cfr 761, p. 154). I can myself confirm that the devotion during the service is impressive, sometimes you will see tears flow down the brown cheeks of the eager listeners.

The Mission's teaching and training has thus completely transformed the earlier savages. Every one who is fond of these humble people and remembers what this kindly race at the mission stations was long years ago (cf. p. 552 sqq.) must sincerely admire the devoted Missionaries when he sees the blessing which has followed their efforts to save the Eskimo's body and soul. Instead of the mysterious dances and promiscuous orgies of pagan days a balanced harmony prevails in the small societies. The whole race has been lifted to a higher plane of living, Dr HUTTON, that fine observer, says.

It has been said that PERRETT and Hopedale are inseparably associated in the minds of all who follow the work of the Moravian mission in Labrador. So also HETTASCH and Nain are inseparable for all the different visitors to the coast: schooner men, steamship crews, tourists and explorers. The Eskimoes learnt to respect them because they had a learning which the natives had not, and because of what they brought into their lives. This is because the missionaries in all their preaching and teaching have never forgotten that the Eskimo must for some time remain an Eskimo if he is to earn his livelihood as a hunter; t h e i r p o l i c y h a s b e e n to m a k e t h e E s k i m o a b e t t e r E s k i m o. These, the Rev. PERRETT's words, give us the principles followed by the missionaries which made possible the transformation of the Eskimo who seventy years ago lived almost untouched by the outside world: In the old days when the stations were more shut in, the Eskimoes were more like children. We are trying to make them more independent now, more self-reliant, more manly. We are trying to do what we can, and we feel sure that this is not without blessing (478). The natural isolation of Labrador has also helped them to stand between the people and the vices that civilization might bring if it were grafted on their nature by careless minds. We hope that good shepherds will follow them and carry on their work until a future day when the Eskimoes are able to look after themselves.

The result of the missionaries' educational work among these primitive people, as sensitive as it was sensible, presents the solution of a problem much wider in extent than this special case. The wildness of the Eskimoes was legendary in the 18th century. Journeys along their coast were as dangerous as those in the areas of notorious pirates. The British Government many times were embarrassed in the work of bringing about peaceful relations. KOHL says: Die armen Eskimaux wurden als Teufelsanbeter gefürchtet und verabscheut. Die englischen Matrosen zwangen zuweilen diese Leute ihre Stiefeln auszuziehen um nachzusehen ob sie nicht gespaltene Hornfüsse hätten wie der Böse. It is no wonder that very few foreigners seem to have won the confidence of the Eskimo. What British statemanship could not do was, however, done by a handful of people who were conscious that the childlike mind of the savage is just as easily developed in the service of good as in the service of evil. The missionaries could never use compulsion; it is mainly by means of good example that they have succeeded gently and slowly in transforming

36

the savages into a free, glad, good-hearted and unusually sociable people whose life is now distinguished by great stability. One who has met the Eskimo and seen their devotion during the Church Services in Hebron and Nain would scarcely dare to believe in their past savagery, if reliable historical sources did not guarantee it. A n e w e t h n o l o g i c a l d e r i v a-t i v e i s t h u s a r i s i n g i n N e w f o u n d l a n d-L a b r a d o r, as has already occurred among the Eskimoes in Greenland and perhaps in Alaska.

We have thus happened in Labrador upon a rather rare, clever solution of the very difficult problem facing the administrative authorities all over the world when a primitive people passes over into modern civilisation. The splendid educational work of the Moravian Mission among the Labrador Eskimo can well be compared in plan and execution with the cultural achievements of the Danes among the Eskimoes in Greenland, especially when it is remembered that the work of the former was necessarily chiefly carried out with private, philanthropic means.

There has probably been no one who understood the Eskimoes better than KNUD RASMUSSEN (79) and it may therefore be important in this matter to hear his voice (779):

We see that along different paths it has been possible to inculcate new habits, new customs, in fact an entirely new culture in the Eskimo race. But what about all the ancient and peculiar, the primitive human traits that have had to perish before the conversion? Is the price not too high for what has been obtained?

It may be; but sentimental considerations and regrets have no place in the realities of life, nor do they prevent the forward march of progress or the differences in the human races from being levelled out, while at the same time the white man's culture is setting its stamp upon the dwellers of all the world.

There is nothing for us to do but submit to circumstances and find consolation in the fact that the splendid romanticism and the enviable lightness of heart in the care-free settlements have given place to a reality which contains this conciliatory factor. That was necessary if the struggle for existence was to be made easier for fellow-beings less advantageously situated. The uncertainty in an Eskimo community, the fear of nature and of the famine periods, and the crudeness of habit born of such anxieties, have been replaced by greater security. Primitive man's narrowminded family egoism, the neglect of orphan children, pitiless indifference towards the aged and worn-out; and assassination, blood vengeance, and merciless persecution, bred by envy, have disappeared in the face of the justice of the law and a humane, public care which surely represents to the very highest degree the elevation of human ideals.

And if we mourn the fact that the ingenious material culture of the Eskimoes will soon be demonstrable only by men of science in ethnographical museums, that the beautiful, poetic legends are sinking into oblivion when the old story-tellers become mute and are replaced by literature, or if we feel sad because we shall never again experience the inspired ecstasy of a shaman, or the fierce drum-dance, or the unsophisticated, fine, and appealing exclamations of the choruses, we must beware of exhibiting a one-sided under-estimation of what is taking their place; for it would be futile conservatism not to acknowledge that the Eskimo mind in Greenland has already found new expression through native painters, writers, composers, and poets, as children of their time.

In the ruins of the old hunting culture there is vitality enough to feed blossoms that have both beauty and the power to grow.

Knud Rasmussen was inspired principally by the fundamental changes which have been carried out amongst the Eskimoes in Alaska and Greenland by the state administration. But why should not both the outword prosperity and the mental and spiritual advancement of this people be secured in Labrador also by a prescient government. It must be to the advantage of a state to assist producers that can exploit the resources of a country with a minimum of expense to the exchequer.

The Newfoundland-Labrador Eskimo of to-day.

As soon as the ship has dropped anchor the brown-faced and wind-blackened Eskimoes come in their crowded boats and swarm on board. As HUTTON stated: you are at once aware that Labrador is a land of smiles; their brown, contented faces are all smiles, Fig. 227. One finds them a lovable folk from the first moment; a happy people. The white people who live among them also say that the Eskimo are on the whole a light-hearted community, ready with smiles. When they are working together, unloading cargo for instance, men and women laugh and joke the whole time; cf. Figs. 218, 229, 233. With my companions of the 1937 expedition we called them at once 'the smiling people'.

Very soon in this crowd on the boat one finds many fine specimens of the Eskimo race, lively, good-humored, affable, and one feels strangely drawn to the bright-faced folk. One can become fond of them, one soon forgets their rough and patched clothes, their air of poverty, and the smell of harsh fish-oil and blubber odour they spread around them.

Another crowd of Eskimo has gathered on the wharf to see the strangers disembark. When you greet them with the traditional: *aksuse*, be strong! — the Eskimo greeting through the ages — they grunt politely in reply.

When one comes up to the village and meets an Eskimo waddling along with a train of starving dogs after him, one is undoubtedly sorry that no conversation is possible, only an exchange of smiles. But each feels they are friends, and it is well to confirm it with a handshake. You have surely a lovely impression of this good-natured, easy-going folk, which faces life with a smile; they leave in the mind a memory of kindliness, says Dr HUTTON. It may seem a curious paradox that while the spirits formerly worshipped by the Eskimo seem without exception to have been ill-disposed towards mankind, and his only domestic animal, the husky dog, is an extremely unfriendly and cunning animal, the Eskimo in this district are the sunniest and most friendly people to be met on the earth. The explanation must certainly lie to no little extent in the success which has crowned the educational work among them of the missionaries. Those Eskimo whom the first travellers and missionary pioneers came upon were anything but mild and friendly towards the strangers, as we have seen. A few hundred years ago the ancestors of these same Eskimo were considered an extremely truculent race, cruel and wild.

My personal contact with the people has been mostly occasional; but from my conversations with the missionaries, the doctors and the Eskimo themselves, as well as from my studies of the literature (cf. 126, 400, 402, 424, 472—478, 480 a, 485, 491, 530, 534, 577, 578, 603, 604, 703, 710, 764, 860, 963, 965, 989, 995, 1012 a, 1017, 1071, 1076 a, 1082, 1104, etc. and especially HUTTON's fine and penetrating works) I think I now have a rather true picture of them and I will sum it up in few words.

The happy, delighted grin upon the beaming faces of the Eskimoes bears witness that a certain degree of good humour is their natural state of mind. They mostly look glad and happy. There must naturally be a little reserve with strangers, and it needs a lot of coaxing, Dr HUTTON stated, before their shyness will wear off, then it takes but a little to bring them to a bubbling-over stage. When you ask them about something they respond with alacrity. And when they talk in their own language together, one has the impression that pathos and humour is in their talk. The women mostly chatter loud and shrill with many bursts of laughter. The men too like to chat and joke and to exchange gossip, and the climax is followed by boisterous laughter. With the missionary they speak slowly and seem to weigh their words; they are said to have a rare gift of graphic and fluent speech in their own language, and they are delighted when they have an opportunity to talk things over. — The Indians are quite different.

The white Labrador people I have spoken with about the Eskimo character vie with one another in praising their honesty and reliability. Most of them are said to be capable of much gentleness and affection and are absolutely worthy of the trust placed in them, the residents say. They are said to have a dog-like devotion for a man who has won their confidence; silent and faithful they sat for hours watching and waiting in the missionary's room, while I sat writing at his table. If you show them a little friendliness you will surely have the support of their goodwill. — In this respect also they are very different from the suspicious Indians. — The manager of one of Hudson's Bay Company's factories, who had lived among them for decades, expressed the opinion that in his district the Eskimoes in character are 'more white than the white-faces in Labrador'. Another curious characteristic was, he stated, that the Eskimoes who were his clients were mostly afraid to run into debt; — the contrary is indeed not unusual among both primitive and cultured people: If one increases one's possessions by availing oneself of credit, one of course becomes richer.

HUTTON during his five years' stay at Okkak heard very little of family feuds or longlived quarrels. The old blood feuds have gone; perhaps the fact that the drink evil has been abolished helps them to be more reasonable. They are quickly roused but as easily appeased, like children, quick to be angry and quick to forgive. An Eskimo may be in a terrible temper, but a few minutes afterwards he is friendly and smiling again, bearing no malice. They are accustomed to take the rough and the smooth together, says one.

The children are charming with brown, rosy-tinted cheeks, glossy black hair and limped brown eyes. Shyly they look at you, — Figs. 220, 217, 221. But childlike curiosity soon comes uppermost and the little one approaches you. All the love of the Eskimo's nature centres on the children. I remember the rapture with which my friend and pilot Julius — Fig. 229 — daily spoke of »my little Sousi», a girl of two; Fig. 219. When afterwards Julius landed happily with a cheque I had just written out for him for 80 dollars in his hand and caught sight of Sousi, a pocket edition of himself, he let slip the cheque, which was caught away by the wind, in order to take the child in his arms. One of the things an Eskimo loves most seems to be to make little children happy. It seems that there are scarcely any naughty children among the Eskimo, in spite of the fact that all observers agree that punishment seems to be unknown among them. Games never seem to lead to quarrels. — Already at the age of ten or twelve years both boys and girls are eagerly trying to help in the household duties.

They are said to be very courageous when hunting or travelling. There is however some kind of inherited fatalism in the Eskimoes. He has often to meet dangers, but the hunter meets them and hardship with his sublime smile, HUTTON states. While the settler toilsomely trails over land the Eskimo travels over freshly frozen sea on ice that bends beneath the sledge; quite safe they say! Their native gift of pathfinding is miraculous, HUTTON says. The travellers have many times had their hearts in their mouths when the sledge touched the edge of a sheer precipice. In the blinding snowstorm the wonderful path-finder is in his element, it is impossible for him to be faint-hearted. At an early age the boys are allowed to accompany the hunters on their excursions and during these times they obtain a detailed knowledge of the geographical character of their district. It is their pride to find out the best paths. They remember their observations very exactly, and are seldom deceived later when they must serve as guides even in a snowstorm. During the dark period it is the pole-star which guides them on their travels; their paths are co-ordinated round it.

But, though incomparable as a dog team pilot, it is better to be careful and not omit supervision when he pilots on the sea. The coast requires an almost superhuman instinct from its pilots and this seems to be a heaven-sent gift in the Eskimo. He takes his duties very seriously too, no doubt, but sometimes he seems to find it difficult to understand that for instance a steamer cannot float where he has gone in his motorboat. I say it is a wonder of wonders that we were not swamped with 'Strathcona' in the Nain archipelago under the guidance of my Eskimo pilot.

However happy the Eskimo is, like other children of nature, he does not at all like to be ordered about; he can then be sour, irritable and crotchety, and he is perhaps tempted to run away. If a white man has played a trick on an Eskimo, it is said that the former can never win back the trust of the latter. On the other hand if he is met with goodwill and sympathy he shows the primitive man's inherent devotion and discreetness. These qualities combined with reasonable cleanliness make him an incomparable helper and comrade, at the same time peaceable, happy and gentle. Only jealousy can unbalance the Eskimo, it is said.

In the hard living conditions of Labrador it is surprising that the Eskimo do not worry much about the future, they lay up no considerable store for the days to come; such is their nature. They do not, however, lack forethought and a frugal mind, like the Indians; no, they dry codfish, seal-meat, and deer-flesh for the lean times of autumn and winter. They also gather berries in bags for use in times of want. Yet there is something one admires in their care-free habits; they live from hand to mouth and, come what may, they always seem to be happy, HUTTON found.

A very lovable side of the Eskimo's character is the well-known willingness to share the last bit of food with a hungry neighbour. If one starves, all shall starve. All the widows are remembered and cared for. The hunter invites the neighbours to a meal of fresh meat; a fine joint is set aside as a present for the missionary, he sends to the sick and the crippled a knuckle-bone, and outside the hut the dogs will busily demolish their share. A guest is always welcome to share the dinner of frozen seal-meat or of hard dried fish and baked dough-cake, with cold water to drink, or weak sweet tea. The natural hospitality of the people simply boils over when a friend comes to see them; it will be a delightful picnic party, with food and drink. There is much to talk over: health and hunting, weather, news of friends and relatives. In civilized lands

Fig. 251. Mrs K. M. Keddies', Cartwright, collection of old Eskimo sculptures in walrus tusk. Photo the author, 1937.

one will scarcely find a community who so willingly and liberally help one another as the Labrador Eskimo, says HUTTON. This generosity, openhanded to a fault, celebrating a wonderful day and then often living from hand to mouth, is characteristic of the true hunter all over the world.

The good hunter is respected not only for his ability to hunt, but because he soon gathers together such things as are wealth to an Eskimo: a fine team of dogs, a roomy house, a good seal-net. He becomes a prominent man of the village.

Two special characteristics surprised me and in my eyes appeared to be distinctive for these children of nature: Their artistic gifts and gift for mechanics.

The Eskimo is a remarkable artist and in this respect he reminds one some-what of the paleolithic hunters (cf. 780, 861, 220, 213, 444, 488). The quaint little figures in ivory of walrus tusk of men and sledges and all the animals that an Eskimo knows, were once real masterpieces and bore the true impress of the hunter's instinct; cf. Fig. 251. This is also true of their free sculpture in steatite: the motion in these can in some cases almost be compared with that of the artistic creations of the paleolithic people in the south of France. Their artistic skill has reached virtuosity in relief maps, in which they also give material expression to their geographical observations by cutting out the district in relief in wood or walrus tusk. PRICHARD (761) describes the Eskimo Filipus' map of the hinterland of Nain which proved in the event both useful and correct.

The spontaneous gifts for painting of the Eskimoes can be appreciated on the walls of the International Grenfell Association's tourist shop in St. Anthony.

All the Eskimoes can sing. At the Church service they sing in time and tune and harmony. The band leads the singing and the choir sing beautifully. It has been argued that the Eskimoes have never possessed any music of their own, the old Eskimoes had only the sense of rhythm and the thumping of a drum, one says. Yet the Eskimoes are nowadays one of the most musical of peoples, and this fact can scarcely be explained without the supposition that the Labrador Eskimoes possessed some native music when living in a pagan state, like many other primitive folk, for instance, the Lapplanders. It is not long since the Eskimo obtained musical instruments: violins, cornets and trombones; and nowadays they find expression for their gifts in modern instrumental music. In Nain the Eskimoes have formed a string orchestra, which in 1905, according to MacGregor, consisted of 7 violins and a 'cello. They are especially fond of music, said Prichard in 1910, and he writes of the excellent brass band which played at the consecration of the church, and added much to the general sense of rejoicing. They all delighted to own a gramo-phone, and those who could afford such a luxury purchased it. They also sang heartily, and there were many whose beauty of singing was impressive. They still have (1939) a brass band at Nain. A number of the people could play the harmonium, and at the church service one of them plays the organ. But it sometimes happens that, when the last tender chords of a hymn are dying away and the thoughts of the congregation are still hovering in higher realms, the organist suddenly forgets himself and begins a fiery jazz which he happened to hear the evening before in the radio, probably for the first time in his life.

The Eskimo's feeling for natural scenery finds expression in their place-names.

The Eskimoes of Labrador have always been regarded as a very ingenious race, and generally superior to their neighbours, the white people and the Indians. Their mechanical talent is amazing, say all observers. It is, for example, quite common for a white Labrador man, unable to start his motor, to call an Eskimo to help him and, strange as it may seem, it is generally not long before he has the engine running again. It is said that an Eskimo can make up small mechanical miracles from machines that have been thrown away, and from scraps of hoop-iron they soon make skates for themselves.

A peg to hang things on is the Eskimo's will to imitate, HUTTON says, he never saw any one to equal them, and in contrast to many other primitive folk the Eskimo is as a rule very industrious. On the other hand, in general the Eskimo does nothing in a hurry; with great dignity and deliberation he inspects the task he has to fulfil. But after he has once started he works with a will. There are naturally also indolent men, but most of them are active as an ermine. They work to the last and slip away like tired children falling asleep, HUTTON stated.

My observers told me some interesting traits in this respect of the modern Labrador Eskimoes. They considered among other things that the farther north the Eskimo live, the less energetic they are in capturing game; e.g. in Hebron the Eskimo have shown themselves to be quite the most skilful and the most hardy hunters, superior in everything belonging to this branch of their occupation. But at the same time these men are the laziest of the whole coast, I was told. One can guess at different reasons for this. For these primitive people things are the same as elsewhere in the world, that as life becomes easier the farther north he lives on the Labrador coast the less reason he has to exert himself in the northern waters. Even if capture on land is less profitable there, yet seal and other sea mammals are more common there than in the south, and therefore easier to catch. These facts agree with a piece of information which MACGREGOR in 1905 gives in quite another connection: that in Okkak it was reckoned that a man caught 50 seal per year, in Hebron 150, and in Port Burwell 250—300. Placing oneself in the position of this primitive people it does not seem difficult to understand that circumstances are just what one Hudson's Bay Company manager described. It is indeed the inborn inclination of every child of nature which here is expressed in the northern Eskimo: To lie on his back and to sleep or joke, when his stomach has been over-filled — is the supreme enjoyment of life. With little trouble he has a real *dolce far niente*. These facts combined with others seem to show that the impulses which distinguish the psyche of the hunter continues to live at least in the north among the Eskimoes. Those in the farthest north are also said to exhibit the rather royal indifference of primitive man to the worship of money in European civilisation. The Eskimoes in Hebron speak little or no English; they still wear much of the national dress. At the most southerly stations, where communications are livelier, the traffic greater, and intercourse more varied, it seems that money economy in combination with ready-made clothes has already obtained a firm grip on them.

The tie of the Eskimo blood is said to be generally strong. If he must live and work in a foreign milieu for some time he will easily become homesick, he feels strongly the call of his native land. He then wishes to return to the chosen land of his forefather-hunters and marry a sensible Eskimo girl; then even the countries with crops and flocks and money to be earned cannot hold him back. At the beginning of the present century when some Labrador Eskimoes were taken for exhibition to U. S. A. they were childlike, easily led astray (478). Their experiences did them no good: they were a little shaken in their sense of right and wrong and they had acquired ideas that unfitted them somewhat for the life of an Eskimo, but restored to their own village they confessed that Labrador was, after all, the best land in all the world.

To sum up, we shall here recall the opinions of two men who have known the Labrador Eskimo better and loved them more than most of those who have come into closer contact with them. The Rev. PERRETT, the experienced missionary, one of the grand old men of Labrador, once bore witness as follows: During my service in Labrador I have come into contact with many grand characters among the Eskimoes, upright, honest God-fearing, truth-loving, saintly souls (478). And HUTTON, their medical man, after having lived five years amongst them stated: As I look back over the years many who have now entered into their rest seem to rise up before my eyes: fine, true Eskimo, keen and clever at the hunt. They form a strangely attractive folk, with children's fears and childhood's quaint ideas, and childhood's whims and fancies and unreasoning demands, but with a manly bravery in the face of pain or danger and a manly mastery of the terrible rigours of their daily work that call for admiration. Auch in den Hütten giebt es denkende Wesen, ja Dichter und Philosophen, und ein munteres Menschengeschlecht mit Witz und Frohsinn tummelt sich auf den Schneefeldern; ausdauernde, kühne, energische, wohlwollende und mildherzige Leute gepanzert gegen äussere Anfechtungen, wrote KOHL.

The mind of the Eskimo must in many respects remain inscrutable. Their outlook on life differs strangely from our own; their thoughts and ways of reasoning also. Impatience and fatalism there go hand in hand. But when we see a little deeper into their mind, we shall find the characteristics of the fluctuating mind of a child. Just big children, say the missionaries. But good children they are, easy to teach and guide and keen to develop.

The future of the Eskimoes on the Atlantic Coast.

I have tried to describe these kind folk and their conditions of life as I found them, and the picture given shows clearly that the assertion so often met in travel descriptions that the Eskimoes on Labrador are still in thought and action very near the people of the Stone Age implies a cardinal exaggeration. In life and knowledge they are no whit behind the so-called settlers in the same coastal districts.

Now, many will certainly ask whether the new cultural stirring we have seen in the Eskimoes has really been for good for the people themselves? I believe we must look at the question from the following points of view.

When we divest the life of the Arctic primitive people of the charm which the unusual, the savage, and the picturesque can at the first moment have for an untrained observer, even the hardiest savage's life will seem a motley web of repeated episodes of which severe want and suffering, starvation, cold, and sickness form the warp. If we consider this I do not think that anyone will answer the above question regarding the Labrador Eskimo otherwise than with an unconditional: Yes. The Eskimo themselves are grateful to the missionaries for having helped them step out of the magic circle which ossified tradition generally lays around the inherited culture of primitive man. HAWKES also confirms that the general attitude taken up by the Moravians towards the Eskimo of a not-too-familiar kindness, and the founding of their authority on it instead of on force is particularly interesting because of the success with which it has been attended. It may be said that the missionaries have saved their bodies as well as their souls.

Regarding the material culture, the Eskimoes of our days still adapt their economic life to the surrounding natural conditions, though now with modern equipment. And the modernisation process continues. The present day Eskimo goes about in rubber boots and sells kamiks, says LINDOW. While their neighbours in the West, the Indians, still persist in extreme primitiveness, as will appear in the next chapter, the Eskimoes have entered upon a period of rapid development. Nor does anything indicate that this development will stop at its present stage. On the other hand there is no sign that the Eskimo will abandon their inherited way of living upon capture to go over to new kinds of occupation.

The northern half of Newfoundland-Labrador holds little promise for the white settler, and it is to the Eskimo that one is inclined to look for population. The way he lives makes an Eskimo self-supporting there. He can continue to be so as long as the game is sufficient. He can even extend his hunting area in the interior if the Naskaupee Indians disappear (cf. pp. 592 sq., 680, 695 sq.).

— However, as regards the supply of the game in the sea I think it is good to repeat here the urge to consideration which Sir WILFRED GRENFELL expressed some decades ago: See that the game is not exterminated! I myself put the question to the people, to what extent, for example, the reckless mass-murdering of the Greenland seal off Newfoundland can affect the seal-hunting of the Labrador Eskimoes. In Hebron they replied to my astonishment that there was no reason for fear. But this reply is of little significance in my opinion, for the people show by it that they have not at all noticed certain slow changes that are taking place in the animal life of their coast. For a great part of the year, of course, the game of the sea is effectively protected from industrialised capture by the impenetrable barrier which the ice-edge of the ocean lays round Labrador. It would, however, be a mistake to be too sanguine and trust that the game will not run short in the future if the hunting continues as now while the sea is open. The seasonal wanderings of the animal along the coast has caused the annually repeated industrialised killing of seals down to Newfoundland, and will certainly in the long run exercise a growing, harmful influence on the supply of seal also on the northern Labrador coast. When the hunters use modern fire-arms it has been shown that he can unwittingly do much evil. Already it is said that the walrus is very rare in Labrador and practically speaking exterminated in Newfoundland. The number of whales has strikingly decreased it is said; whale-factories have been closed. According to GRENFELL the supply of seals has decreased so much that he is inclined to see in this a cause of the depopulating of certain districts. The supply of salmon has declined very much, it is said. The capture of fur-bearing animals has, in spite of the improvement of equip-ment and the increased number of hunters, become relatively much less year by year, and in any case has not increased. The great auk and the Labrador duck have been completely exterminated, and other web-footed birds have declined considerably in number. The woodland caribou have become very rare, if they even exist any longer; I could discover only surprisingly few caribou tracks in my flights over the woods (cf. also MacGREGOR, p. 29). If people continue to hunt the great sea animals with steamer, motorboat and aeroplane, as they do now, they also must come to an end. — Cf. 952.

If we assume that a guardianship of hunting with an effective adaptation of the principles of protection developed in Canada will be set up in the future both in Newfoundland and Newfoundland-Labrador, it can also be said with full assurance that t h e m a t e r i a l c o n d i t i o n s f o r t h e c a r-r y i n g o n o f t h e E s k i m o e s' i n h e r i t e d e c o n o m i c l i f e w i l l b e s e c u r e d a n d a t t h e s a m e t i m e t h e i r s o c i a l g r o u p s w i l l c o n t i n u e c o m p l e t e l y s a f e g u a r d e d o n t h e A t l a n t i c c o a s t. The psychological conditions for nomadism also continue in the highest degree; the hunting instinct of the Eskimo can indeed be compared most nearly with that of the predaceans.

The church and the school have thoroughly opened up the Eskimo's world of ideas. When he has learnt to read and write and calculate his power of imitation can be developed without limit; it may be said that he has entered world economy as a commensurable factor. The competitive ability of this people in business and trade has thus been considerably developed. Some Eskimoes have even already understood how to break away from the control of the local trade, and proved themselves skilful in managing for themselves; but there are very few of them. To facilitate this for the others some regulations are needed.

The development cannot be stopped. The Eskimoes have entered upon a stage of fermentation and it is not least the trade which supports and furthers this continued progress. The burning point of the question regarding the continuance and welfare of the Eskimoes' social groups seems to me under these conditions to lie in the financial sphere. If the local trade is not regulated in such a way that it de facto safeguards the Eskimo's economic autonomy he will, like the Labrador Indian, gradually sink deeper and deeper into economic dependence and perpetual indebtedness with all their evil consequences! A reform is necessary and should start from payments *in cash* for the Eskimoes' products, instead of as now, *in trade*, say neutral observers; for in the latter case the Eskimo has quite another idea of the importance of economic values and competition. They will then enjoy all the fruits of their own toil, this will stimulate them and they will be able to build up their own welfare individually and also help their small societies to progress as in other parts of the world. Perhaps they will gradually be denationalized and assimilated into the future stock of Labrador. As yet it is impossible to say when that day will dawn.

This is not the right place to discuss such questions and make out at whose door the responsibility must be laid. Yet it must be stated that the Eskimo in general still lack the ability to demand the necessary changes for themselves, as will be clear from what has been said. If so, it is necessary that the authorities should intervene between the seller and the buyer and regulate matters

If this happens I see no reason why the Eskimo of Labrador should be depressed to a worse position in the world than their kinsmen in Greenland and Alaska. It would even seem easier for the Labrador Eskimo to become relatively wealthy and useful citizens for their ancestral country than it has been for the Danish Greenlanders, who live in a distant and poorer country than Labrador.

In spite of their limited possibilities the Innuit are an intelligent people. With a fine instinct they have remained as nearly as possible on the lines that nature intended for them. They possess, moreover, a physique that must rank high among the species of mankind, for they have fully adapted themselves and their mode of living to the exigencies of a cruel environment, and their physical robustness has enabled them to defy the onslaughts of the cold. They are also a peaceable, law-abiding folk, flagrant breach of order or discipline is very rare; there has been no prison and no police and practically no serious crime, HUTTON states. Is not this a race well worth preserving? PRICHARD exclaims! Cf. however p. 472 sq.

It has been considered (GRENFELL, PADDON) that the Eskimo half-breeds will form the future population of the coastal districts because the Eskimo themselves are dying out, they say: except for those fortunate tribes which have been under the fostering care of the Moravian missionaries, the Eskimo race has disappeared from Labrador, where at one time there were doubtless many thousands (325). Yet there is systematic medical experience (HUTTON, SUK) which does not confirm the lofty thoughts of the half-breeds' influence on the development on Labrador. Following this the half-breeds seem to live beyond the limits of their natural habitat, their physique degenerates and they fall victim to diseases unknown amongst the Eskimo. They can never compete successfully with the Eskimoes, who are so persevering in utilising the local sources of livelihood. Moreover, the Eskimoes at the Moravian settlements are fairly educated, very few being unable to read and write; in this they are far ahead of many of the white settlers, even PRICHARD found (cf. also 891).

At the Moravian settlements the population just about held its own; Fig. 215. Were it not for the diseases which have been criminally introduced there, they would have shown a substantial increase (see above and 17, 116). Let us therefore trust with GOSLING that the legislation needed to protect these real children of the coast will no longer be delayed, and that this deeply interesting race may again flourish on their native shores.

The Indians of Newfoundland-Labrador.

We shall now move away from the roaring surf of the ocean, from the archipelagoes and the fiords, to the soughing woodlands and the deserted rolling plateaus of the tundra in the interior of the peninsula: to the land of the Indians. It must not be supposed that all the statements made in this chapter with reference to the Indians are the result of my personal observation. Much of the information is gleaned from the officers of the trading posts, from trappers, from natives and from the literature (995).

Ethnological connections of the Labrador Indians.

The Indians of Newfoundland-Labrador speak a Cree language, an eastern branch of the widespread language stock to which the Algonkians belong. When Europeans began to take possession of North America it was the Algonkians who mostly occupied its eastern half.

This is why a Labrador Indian still understands the meaning of many of the Indian place-names from the Missisippi to Massachussetts and Manhattan away to the Rocky Mountains, with their expressive and finely drawn distinctions, forming a kind of geographical text-book. The language of the Labrador Indians is also very rich; for example in Father LEMOINE's French-Montagnais dictionary there are 12,000 words (564). It is a difficult language; LEMOINE gives a great number of inflexions for one single regular verb, and there are no less than fifteen conjugations.

From the original Algonkian language two dialects have developed and are to-day spoken in Labrador: M o n t a g n a i s and N a s k a u p e e. For a philologist their differences are not more important than their general similarities; the Indians themselves, however, told me that the differences can be so great that it is difficult sometimes for a Montagnais to understand a Naskaupee's speech.

Judging from their language the original home of these Indians lay in the west. This is also indicated by a story of wanderings preserved among the Labrador Indians (Naskaupee), which states that their forefathers came from the west, from a district to the North of a huge river [St. Lawrence River?], and that east of that home was an extensive area of salt water [Hudson Bay?]. When they arrived in the Labrador peninsula they found that only Eskimo had been there before them. Of course, experience has shown that stories of wanderings amongst primitive peoples must in general be treated with a certain amount of scepticism. In this case, however, there may be some truth in the story, for the language undoubtedly shows relationship between the Labrador Indians and the western Indian tribes. To some extent the story is also supported by historical information, as we shall see later; circumstances indicate that the former were separated out from the mass of Algonkian peoples at a comparatively late period, and then during the wanderings of the Indian peoples were pushed farther and farther to the east and north-east. Confirmation of the statement that the Eskimoes were on the peninsula before the immigration of the Indians is found in the fact that an older Eskimo tribe, the socalled *Tunnit*, were also living on Labrador before the immigration of the present Eskimoes (see p. 452 sq.); some place-names in Labrador are still reminders of the Tunnits.

What little is known of prehistoric conditions in Labrador does not allow of any far-reaching conclusions regarding the wanderings of the people in those districts. However it can be of interest to present here some information in this respect. But first we will review the

Distribution of the Indian population in Labrador.

In the north-eastern Algonkian culture there is a great cleavage line, the St. Lawrence River: to the north live the Montagnais-Naskaupee and to the south the Wabenaki (871).

The coast of Hudson Bay and its southern and eastern upland were already in prehistoric times occupied by tribes belonging to the eastern Cree Indians, or as they call themselves: M u s k e g o-w u g [= »the fen people»]. After their arrival in Labrador they divided into two large groups (855, 877). The N u t c i m i u i i u [= »the southerly inland people»] settled in the forests south and east of Hudson Bay and James Bay. North of the former, from Fort George to the north the K e e w a t i n-w u g [= »the north people»] took possession of the country. SKINNER has drawn the eastern boundary of the latter group over Lake Nichikun; see Fig. 222.

East and north-east of these are two groups of Indians: the N a s k a u p e e now wander in the north and M o n t a g n a i s in the south; the boundaries on Map 222 have been drawn in the main according to data given by SKINNER (855) and the information I myself obtained on Newfoundland-Labrador (see farther on); cf. Fig. 253.

When SKINNER compared his notes regarding the east Cree mentioned above with TURNER's information about the Ungava Cree, that is the Nas-

Fig. 252. Schematic presentation of the hunting grounds of the Labrador Indian bands according to FRANK G. SPECK (873).

kaupee Indians, he found a remarkable likeness in the material culture of these two groups. The east Cree themselves at Hudson Bay said that the Naskaupee are nearly related to them. SKINNER therefore concluded that the Naskaupee are a group of the east Cree who, because of their isolation in the wilderness from the influence of the whites were able to preserve their primitive traits fairly unchanged for a comparatively long time. Further SKINNER's southerly inland east Cree also maintained that a near relationship prevails between them and the Montagnais. Cf. Fig. 222.

From a cultural and dialectic point of view the peninsula according to SPECK falls roughly into three subdivisions:

37

1. East from a line from Ungava Bay over Lake Nichikun to the St. Lawrence Gulf at Seven Islands; the term Naskaupee (Naskwopi) has been applied
here quite generally in the past.

2. West of this line, south of the 'Height of Land' to a point above Lake
St. John and from here to a line a little above Quebec and north of St. Lawrence
is the traditional Montagnais area, divided by the Indians into the N o t-
c i m i' w i l n u t s [= people of the interior] and W i n i p e g' w i l n u t s
[= people of the Sea].

3. The north-western sector from below Rupert House as far as the uninhabited area north of Lake Minto and east to about longitude 72° W are the
M i s t a s s i n i   C r e e, at present a quite unknown group (871).

WISSLER in his map (1938) of the North American Indian groups, combines
the northerly muskego-wug with the Naskaupee and draws the limit between
them in the north and the Montagnais in the south from the Atlantic over Lake
Melville — Hamilton River — Lake Kaniapiskau and East Main River to
Hudson Bay. The boundary of the Montagnais in the south is formed by the
Gulf coast and the River St. Lawrence and in the west by the Ottawa River
and a line drawn thence to the north as far as James Bay.

According to WISSLER then only two large Indian groups need to be considered in Labrador. Apart from details as to the course of the boundary line
this last fact would agree with the idea which I found prevalent among the
white population of Newfoundland-Labrador.

However simple the conditions of the spread of these two tribes may
seem to those who only know WISSLER's map they are not by any means so
in reality; on the contrary they are quite complicated. To make this clear I
only need recall the fact e.g. that it has been possible to confirm from the
differences in the dialect that the Naskaupee were intermingled with the Montagnais until quite late times, for example at Musquarro on the Gulf Coast
(873); see Fig. 222. Descendants of the semi-settled Naskaupee are also to be
found at another place of the Gulf Coast, namely among the Montagnais on
Seven Islands.

These circumstances in themselves bear witness to the fact that in this
part of the country there must have been quite recently

Displacements of the Indian Tribes.

Not much is yet known of the oldest Indian population in Labrador.
JENNESS (490) on the basis of archeological facts suggested that eastern Labra-

dor was inhabited in prehistoric times by B e o t h u k s, an Indian tribe who
are mostly known to ethnology under the name of Newfoundland's Red
Indians (307 a, 455, 490, 583) (cf. p. 450 sq.).   It is said that they later withdrew from Labrador, presumably southwards to Newfoundland, where the
white population brutally hunted and exterminated them in the 1820's (596).
This idea is well supported by the results of excavations on the south-eastern
coast of Labrador, during which both LLOYD (583, 584) and WINTEMBERG
(1041, 1042, 1044) found ancient dwelling-places with remains from the
Beothuks.   Later a suggestion was made that the present day Indians of
St. Augustine River in Labrador (see p. 609) are related to the Beothuks
(873).

CARTIER, during his voyage in 1534 made some interesting observations.
He says about the land: it should not be named the Newland, but the
land of stones and rocks, frightful and ill-shaped. — — — Except at Blanc
Sablon there is nothing but moss and stunted wood; in short, I deam rather
than otherwise that it is the land God gave to Cain. There are people in
the said land who are well enough in body, but they are wild and savage
folks. They have their hair tied upon their heads in the fashion of a fistful
of hay trussed up and a nail or some other thing passed through it, and
therein they stick some feathers of birds. They clothe themselwes with
skins of beasts, both men and women, but the women are closer — and
tighter in the said skins and girded about the body. They paint themselves
with certain tawny colours. They have boats, in which they go by the sea,
which are made of the bark of the birch trees, wherewith they fish a good
many seals. Since having seen them I am sure this is not their abode, and
that they come from warmer lands in order to take the said seals and other
things for their living. (Cf. 819).

— The description is not applicable to the Eskimos; it seems most likely
that Cartier met the Montagnais Indians, who always came down to the
coast at the summer season.

Other reasons point to the probability that long before the arrival in
Labrador of the Europeans and the ancestors of the present Eskimo, hordes
of Indians were widely spread in the southern and eastern parts of the peninsula (cf. p. 450 sq.). However, there were scarcely any Indians far away in
the north.

Why did the Beothuks abandon Labrador, it may be asked. Many will
consider it had some connection with the immigration of the ancestors of the
present day Eskimo who, not content with hunting merely in the coastal
districts, forced their way far into the interior (cf. 873, 878, VII, XXXIX);
even to-day some signs of this can be found, e.g. in the Ungava peninsula north
of the Stillwater River (265, 588). It is easier however to seek a reason in

the movements traceable among the Indians themselves in that active new immigrants pushed out those tribes who had been there for a long time (74).

Yet any real information about the displacement of people in Labrador is only available from the last few centuries.

It is historically well known that the Iroquois began quite late to force their way from the south over the upper part of the St. Lawrence basin into Labrador. The Jesuit LE JEUNE saw in 1634 in Quebec the Montagnais torture some Iroquois prisoners to death in the cruellest way; he describes how an Iroquois chief defiantly went to death singing joyfully to the last; from this chief's body a Montagnais afterwards took the heart, cut it into pieces and distributed them to his children; now the enemy was finally conquered! The Iroquois now separated the Montagnais from the western Algonkians; the former were thus already at that time living north of the St. Lawrence River. Pressure from the Iroquois clearly caused the Montagnais to move slowly to the east; and these formed a link in the obstinate struggle with the Eskimo described above (cf. p. 480 sq.).

Both in the north and the south the memory of the Iroquois cruelties during this invasion is remembered by the Labrador Indians; not many decades ago Low heard that disobedient children were frightened by being told that the Iroquois would come and take them away if they did not behave. A mythical people, called the K ā t c i m ē d g ī z ū, are greatly feared by the Naskaupee, STRONG (909) also reports; they are said to come into the far interior in high-bowed canoes, and they steal Naskaupee children. While their whistling may be heard they are invincible to all but the Naskaupee conjurors, whose familiar spirits drive them away. — This would seem to be a northern version of the tales inspired by Iroquois raids in early times.

An Iroquois dwelling excavated by WINTEMBERG at Khegaska a little east of Natashkwan River shows that the fortune of war had driven them quite a long way eastward out on the coast. In 1660 there was a fight between the two Indian tribes at Lake St. John.

Another circumstance which helped to drive the Montagnais to the east was the invasion by the Micmac Indians in the 17th century to the Mingan district (cf. p. 479). In their birch-bark canoes they had come over the Gulf of St. Lawrence from the Gaspé Peninsula, and in the shelter of the great island Anticosti succeeded in forcing their way to the northern shore, where they defeated both the Montagnais and the Eskimoes.

It is therefore a historical fact that the Iroquois and the Micmacs again set the Labrador Indians in motion in the 17th century, and it seems that the slow displacement of the earlier immigrants, the Montagnais and the Naskaupee towards the east and north-east which we shall shortly study, was due to the invasion of just these Indian tribes. — Such slow movements of masses of people undoubtedly took place also in prehistoric times.

The struggle of the Labrador Indians with the Eskimo which stood in the way of their progress to the east has already been described; cf. p. 479 sqq. Here we shall only present some fragments to supplement the information regarding the Indians' advance towards the north-east.

N a s k a u p e e. The Jesuit LAURE who was the first to speak of the *Cuneskapi* [Naskaupee] met them in 1731 to the north of Lake Ashuanipi, that is on the Lake Plateau of Newfoundland-Labrador. The Indians' missionary, Father ARNAUD, visited the Naskaupee on Lake Manicuagan in 1853, and he says that their hunting-grounds lay to the north-west of that lake. According to HIND the Naskaupee's hunting-grounds extended in the 1860's from Lake Mistassini away to the Atlantic — a distance of 800 km (cf. 876). HIND also says that at the same date the Naskaupee from the interior used to come in the summer down the rivers Manicuagan, St Marguerite, Trinity and Moisie to the Gulf of St Lawrence to meet their '*robe noire*', i.e. the priest, at Seven Islands. All this indicates that the Naskaupee in the middle of the 19th century occupied at least the larger part of the Lake Plateau.

Further I think that the place-name Naskaupee River which falls into Grand Lake in the Lake Melville district can be considered as a proof that the Naskaupee also controlled not so long ago the area round the inner part of the Lake Melville drainage system. Some support for this idea can be found in TURNER's placing the southern boundary of the Naskaupee in the 1880's at Hamilton Inlet; at that time their trade was concentrated around the H.B.C. factories Rigolet and North West River. Low too says that in the 1880's the Naskaupee were hunting west and north-west of Hamilton Inlet; and there can be no question here of a temporary stay in the district, for about 200 Indians had dealings at that time with the factory at the North West River. BELL's map shows that in 1895 the Naskaupee Indians were in the area between the Naskaupee River and the Hopedale district. The Big Hill portage at Grand Falls was described to BRYANT in 1897 as »The Old Naskaupee Trail».

The Naskaupee had also come as far as the Atlantic coast. CARTWRIGHT relates that in 1774 he saw two Naskaupee canoes near a river mouth at Indian Tickle (350 miles south-east of the southern boundary for the Naskaupee in Fig. 253). Further he says that in 1771 he found traces of the Naskaupee near Denbigh Island and in other places on the Atlantic coast north of the Straits of Belle Isle.

Considering all this it can scarcely be doubted that the Naskaupee were formerly living a long way further south than their present limit which we find upon the map just referred to. Yet here a strict difference must be made between the information from the 19th century on the one hand and on the other hand that which we shall give later on from the latter half of the nineteenth century.

The setting up in 1824 of Hudson's Bay Company's factory Fort Chimo at Koksoak River near Ungava Bay was for the Naskaupee a kind of silent invitation to begin to do business at this place. Here it seems that an advantageous trade proceeded for a few decades and the Company had now also got the trapping of the Indians on their hand. But when 1843—1866 the factory was closed and another, Fort Naskaupee or Fort Petitsikapau, opened in 1840 on the north part of the Lake Plateau it seems that the Indians had already become so used to doing business with the whites that they were attracted to the new factory in its turn. Later on the trade became divided also among some other southerly H.B.C. stations, like Fort Mishikamau at the lake of the same name and Fort Winnikapau at Hamilton River, which were opened later. From the trapping lands round these factories the movements of the Naskaupee gradually extended still further south. HIND says that during his visit to the district he found that the Naskaupee had made their way for the first time to the Moisie Factory on the Gulf coast in 1849. The reason for this seems to have been that the caribou in their area had for some years moved along new paths so that a lack of game was felt and a general famine ravaged among the Indians. Crowds of them hurried by different ways down to the Gulf coast in the hope of finding there something to satisfy their hunger. Many of them remained behind after the famine period had passed and formed a special group, the so-called C o a s t I n d i a n s, who earned their living by seal-hunting and fishing, while others died of »the diseases of the whites».

Thus in the middle of the 19th century a tendency can be observed among the Naskaupee prompted by an accidental circumstance to move to the south in the interior of the country, both in families and as individuals.

M o n t a g n a i s. In this group on the other hand the displacement has, as far as is known, always gone towards the north-east and north. From the beginning of historical time down to our days Seven Islands has been an important centre of attraction for the Montagnais; there runs the winter road to Hudson Bay over Lake St. John with a canoe connection thence over the Moisie River (cf. p. 610). The Jesuit reports relate that in the middle of the 17th century there were Montagnais at Seven Islands, in 1660 even a large number of them. The missionary DE LA BROSSE opened a school for Indians here in 1669, he composed an alphabet and a catechism for them and also compiled a dictionary of their language. At the beginning of the 18th century the Montagnais had already wormed themselves at Blanc Sablon into the Eskimoes' old district (cf. p. 481 see above). In 1775 CARTWRIGHT related that two Montagnais families were living on the Atlantic coast near Sandwich Bay; two men with a wife and children came from the valley of the White Bear River to what is now Cartwright to visit him. He further stated that 160 Montagnais Indians were living on Nevil Island (?, Nevesek Island in Lake Melville?). Four Montagnais visited him in 1778 and some forty Indians spent the winter in the neighbourhood of Cartwright. The Montagnais were living at Hamilton Inlet until 1779. According to CARTWRIGHT there was then a population of about 500 people between Hamilton Inlet and the Straits of Belle Isle; as these people seemed to have earned their living mostly by seal-hunting, fishing, and catching wild fowl, it may be presumed that they were Eskimoes. During the 19th century the Montagnais gradually spread out over all parts of the extensive Lake Melville drainage system. SPECK expresses the opinion that the Moisie Montagnais moved to the north only after the time when HIND (1860) travelled up the river. When finally at the end of last century the mountainous district between the coast from Hopedale to Aillik and the George River was swept by great forest fires and deserted by the caribou and the Naskaupee this part remained empty for a long time. Later, when the reindeer lichen had grown up again and the caribou returned at the beginning of the 20th century, the Montagnais began to hunt this old Naskaupee land and gradually took possession of the district (120, 123). In this way the boundary between the Montagnais and the Naskaupee had been further shifted a good bit to the north and the boundary conditions had arisen which now prevail on Labrador, cf. Fig. 253. To this there may have been another contributory cause, namely the gradual invasion by the trappers of the best grounds for capture within the drainage area of Lake Melville; this we shall study later on.

Thus a tendency may be observed in the Montagnais to continually spread over the whole area of the taiga. Parallel with the displacement of the Naskaupee people in contrary directions individuals and families belonging to both tribes have ceaselessly been intermingled. This movement still continues (cf. the Davis Inlet Indians, pp. 608, 654)..

Reviewing this sporadic and meagre information from the literature in order of time, the conclusion will be as follows: The Indian tribes have contributed essentially to drive the Labrador Eskimoes from the southern and south-eastern parts of their former area in order to take over their hunting-grounds. The movement of the Montagnais to the north-east and north doubtless started from the pressure exercised by the Micmac Indians in the district of the Mingan Islands (see above), and further to the west by the Mohawk Indians (Iroquois); WISSLER's map should be consulted for the position of these last mentioned (1048). It is possible that the enormous forest fires on the Gulf coast have also furthered the displacement to the east of the Montagnais. As the displacement to the north occurred the Montagnais spread themselves over all those parts of the land which the Naskaupee had evacuated and which are mostly occupied by primeval forest. After the purely chance forward movement to the south in the middle of the 19th century the Naskaupee, who always seem to have kept to the north of the Montagnais, were pushed by the latter farther and farther to the north. Finally (after the H.B.C. factory Fort Chimo had been opened again) they found themselves out in the area round the George River and the Koksoak River on the north-eastern part of the peninsula so rich in tundra. The eternal struggle between the Indians and the Eskimoes was brought to an end at the beginning of 1800's by the united efforts of the Moravian missionaries and the English traders and a kind of peace settled down upon the peninsula. Now the Naskaupee reached even as far as Ungava Bay in the north; in this upland the place-names undoubtedly prove that the

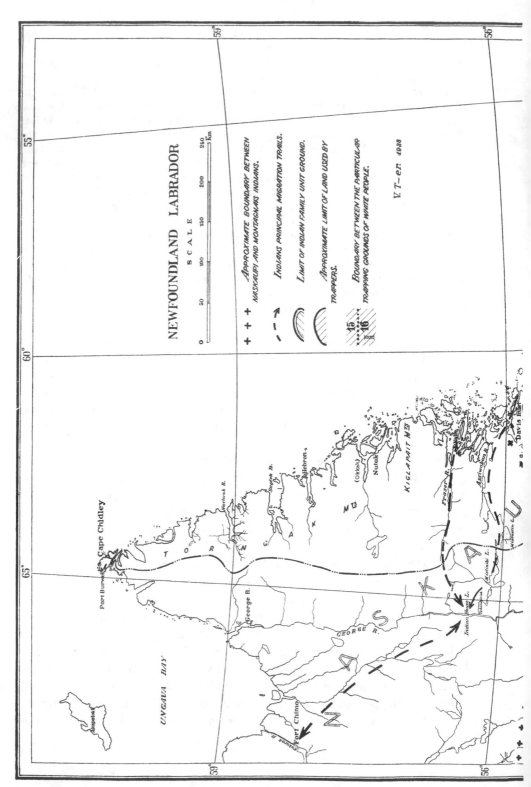

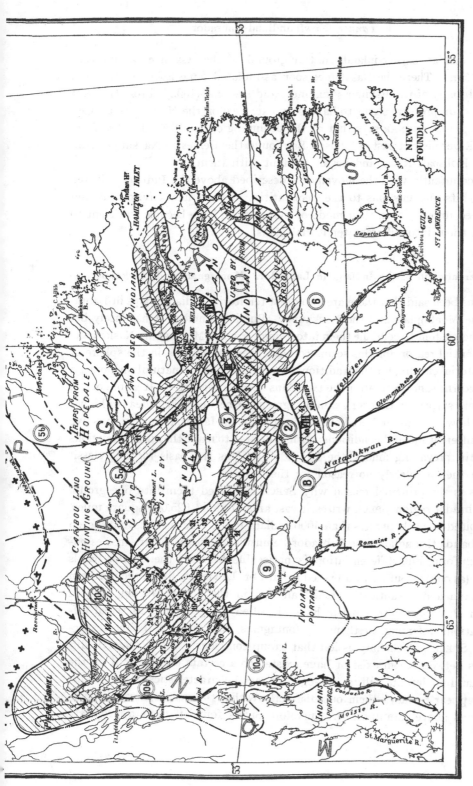

Fig. 253. The migration trails and the approximate location of the different Newfoundland-Labrador Indian bands (figures in double circle) as well as the trapping grounds of white people; cf. the legend on the map. Compiled on the basis of information gathered by the author from the trappers in Labrador 1937. Cf. p. 703 sq.

district was an earlier inland hunting-ground of the Eskimoes at no very distant time. There the Naskaupee soon assimilated some Eskimo blood and adopted many of the Eskimoes' manners and customs (1002). From the west too a certain pressure seems to have been exercised on the Naskaupee at least at times; thus TURNER says that at the beginning of the 1880's hunters from Whale River district east of Hudson Bay found themselves on Naskaupee land by the Koksoak River and began to hunt with them.

Concomitantly with the movements described above the Indians adjusted their habits and customs to the peculiarities of their new enviroment and developed differences in their mode of living which have become sufficient to classify them into different territorial clans.

Differentiation of the Indians in Labrador. Montagnais and Naskaupee.

It has been said that there are two great territorial Indian groups in Labrador: the Montagnais and the Naskaupee. But in ethnology it has lately become the custom to combine them into a single group: Montagnais-Naskaupee. The essential reason for this seems to have been phenomena in the sphere of non-material culture, for the traditional conception of supernatural phenomena remained the same even after the two clans became separated. It is, of course, extremely surprising that for example agreement between their ideas of higher things show such an unaltered outline in spite of the fact that both groups live under different natural conditions. If it is desired to make the magico-religious ideas still prevailing among the Labrador Indians the basis for this classification there is clearly no reason for the division.

This is the principal reason why SPECK combined them into one group. The Labrador region, SPECK writes, a vast area of about 625,000 square miles, is inhabited by probably less than 4000 Indians (871). The customs and speech of the bands throughout the Labrador Peninsula are in the wider sense essentially uniform, but differentiation is noticeable between the smaller bands located (a) on the southern watershed toward St. Lawrence, (b) on the northern or Ungava and Atlantic watershed and (c) in the portion facing towards Hudson Bay. Some ethnologists have classified the groupings as tribal differentiations, referring to them as Montagnais, Naskaupee and Cree respectively. SPECK got the impression that group-consciousness and dialectic differences between them exist or have existed to a certain degree. But on the other hand no definite culture analogies or racial characteristics (HALLOWELL, BOAS) appear to mark them out distinctly within the geographical districts mentioned above (MICHELSON). Therefore SPECK found it more appropriate to refer to these geographical groupings of Indians as Montagnais-Naskaupee (871, 877).

On the other hand SPECK (871) pointed out that in the Labrador area one may recognize a certain integration of subgroups referred to as 'b a n d s' (GOLDENWEISER). This means a group inhabiting a fairly definite territory with a moie or less stable number of families, mostly possessing paternally inherited privileges of hunting within the boundaries of the territory, often having an elected chief, speaking with idioms and phonetic forms by which they and outsiders distinguish themselves as composing a unit, often with a minor emphasis on this or that social or religious development, often with a somewhat distinctive styles of manufacture and art, and finally, travelling together as a horde and coming out to trade at a definite place on the coast. The bands are located in the drainage areas of the rivers and lakes whose names most of them bear. SPECK has moreovei determined the most of the Labrador bands, Fig. 252. But he points out that the remote north-western bands' areas on this map should be regarded as tentative only, because the information is derived from indirect native sources and not through contact.

To obtain the premises for an objective systematization the American ethnologists considered it desirable to direct their study of the Indians first to research work concerning these small sub-social units, 'bands', and then, having a thorough knowledge of the characteristics of each of these, it would be possible to re-group the bands into units of a higher order. Such an attempt by SPECK has been given in Fig. 252. There is no doubt that this scientist has entered upon an appropriate and practical path; I can say this with a certain knowledge of the subject, for I myself used just the same method successfully a decade and a half ago in working out the systematic divisions of the North-Fennoscandian Lapps (Fennia, Vol. 49). As regards the Lapps, it proved to be quite inacceptable to mix together habits, customs and ideas belonging to widely separated districts in the north and in the south, as the philologist-ethnologists of the north had usually done, for such a synthesis rested upon premises which from a geographical point of view are not commensurable. A consistent examination of the scattered Indian bands in the wilderness of Labrador will, of course, come up against many difficulties, and will demand a very long time for its completion. For, as SPECK himself has said, uncertainty will render vague the demarcation lines etc. of the different bands until each group has been examined without reliance upon an intermediary.

No real, thorough ethnological distinction can be clearly shown to exist between the two territorial clans, the Naskaupee and the Montagnais. The small differences which can be traced in their sociology and their material culture will probably be taken by many ethnologists to be only nuances of the basic characteristics. In spite of this I think that for most important practical reasons the two groups must be kept apart in the following presentation. A few fragments of information will explain more in detail the differences between the groups. If, for example, we take an anthropogeographical view of the questions which interest us here I think we can find an acceptable systematization.

The woods and the tundra offer quite different conditions for the hunter and make quite special demands upon his activity. Caribou are to be found in both regions, but their position as to game is different within them. In the tundra the importance of the caribou surpasses that of all other animals; it is the wild reindeer which almost alone supports the economic existence of the Indians there. On the other hand, in the forest areas the caribou has in these days considerable importance only in such districts where there are large connected treeless areas; in the forested parts the lack of caribou is compensated by different kinds of small game: the beaver, the porcupine, the hare, and birds. There are also the finest fur-bearing animals (see farther on), and the deeper and thicker the forests, the thicker, darker and more valuable are the skins; the skin of the marten, for example, comes to resemble sable. In Labrador the tundra is more connected towards the north, the forest region towards the south, and this circumstance naturally has an influence upon the kind of hunting; this in its turn is decisive for the economic and social configuration of the hunting societies.

As the poverty of nature increases the importance of caribou hunting rises (206 a). This is especially true of the tundra; the hunting must be extensive because of the biological habits of this chief game, and under these circumstances the hunting areas of the tribes form one single great whole. The economic system again develops on more or less distinctly communistic principles and yet the social structure remains loose and undeveloped.

In the primeval forests with mixed game and valuable fur-bearing animals the method of hunting becomes more intensive. For this purpose the hunting ground is generally divided into smaller areas within which hunting and trapping are carried on by a smaller group alone, a tent group or a great-family. The economic system in this way is stamped by the individual right of each family to own the booty. The social structure is already on a higher plane of development, and the society bears the stamp of the great-family. Within the c o r e s of these two areas which form Newfoundland-Labrador, the tundra in the north and the taiga in the south, nature is compelling the development of a different structure in the economic and social habits of the Indian tribes.

As the Indian peoples were slowly displaced from their good old hunting-grounds out towards the still more inhospitable extreme end of the continent in the north-east, they met ever worse natural conditions and existence possibilites. Cold and barrenness increased the farther they moved to the north-east, and they were compelled continually to adapt their material culture to this worsening; a rather difficult task at their primitive stage. Above it has been indicated that the Indian hunters are distinguished by their great mobility resulting in an ability to adapt themselves to any new natural conditions. From this again it follows that if there is peace and calm in an Indian society the local differences tend to develop permanent and more or less homogeneous groups of hunters, in this case the Naskaupee and the Montagnais, as well as local transition types between them. Trade with the whites (Hudson's Bay Company) showed itself during the 19th century apt to further a stabilization of the types. In nort-easternmost Labrador the change to new surroundings led to a degeneration of the Indians' material culture.

On the other hand, as regards the magico-religious ideas, the continuity remained fairly unchanged in the north and the south (cf. 878). The starting-point of my conception of the kind of differentiation lies just in this contrast.

The circumstance here quoted is confirmed to a certain extent by reality in Labrador (cf. 965). For example it appears that the hunter and the hunter's family who had grown up among the Naskaupee but moved over to the Montagnais or vice versa, in order to earn their living, had to give up the customs of their youth and adapt themselves to those which nature enforced upon the people in their new surroundings. In the boundary districts between the two great core areas different degrees of mixed types of course arise. Father COOPER told me about Indians west of James' Bay who, while hunting in the forest follow the principles of their family areas, on the tundra again the principles of hunting communism prevail. From a geographical point of view this is a boundary case due to the biogeographical conditions. Another such boundary case is to be found in the districts south of the east-west watershed in Labrador, where the hunting varies according to the season and the supply of game, these again varying with the geographical conditions. These exceptions, however, do not invalidate the main rule. This conception, which I found applicable to the Lappish half-nomadic hunters, seems of course not to have been unknown to SPECK (873) who writes: Two types should henceforth be recognized as prevailing in the Labradorian culture-area, the variation being traceable, I believe, to famine conditions and the natural history of the game animals.

It seems to me that the same more or less subconscious conception of, let us say, two different derivates of the hunting life is vaguely present in the minds of the white population of Labrador when they speak of the Naskaupee and the Montagnais. It should be specially mentioned that the managers of the H.B.C. stations, who may be reckoned as best acquainted with the Labrador Indians whose language they speak, continue to differentiate between the Montagnais and the Naskaupee. Also among them it seems to be just on the nature of the hunting-grounds and its consequences that the differentiation of the two great groups of to-day is based. The Indians themselves, of course, have no critical eye for ethnological distinctions, yet their intuitive apprehension seemed to me something to be reckoned with. They make for example a difference between the coastal people, i.e. those who in summer visit the coast of the Gulf of St. Lawrence, and the forest or inland people; both have some distinctive customs, but this is not all. Subconsciously the southern Indians make a certain ethnic distinction in for example the name N a s k a u p e e (cf. p. 657) by which the most northerly Indians do not like to be known; they call themselves N e n e n o t. During a conversation with the former and the present chiefs of the Montagnais' Mishikamau band, they told me that they call themselves Forest Indians, I n a′ p a′ o u′ w i l-n u t s, while they call the Naskaupee Tundra Indians, M o u s′ a′ u a u′-s h i′ b o w i l n u t s, and they seemd to consider the difference between the groups as categorical. As regards languages they said that certain difficulties arise in trying to understand the Naskaupee. The Naskaupee earn their living by hunting the caribou and some foxes, they said, on the other hand the Montagnais restrict themselves often to small game and valuable fur-bearing animals. They also said that the Montagnais were not so dependent on meat as the Naskaupee, but could manage quite well with »white man's food», while the Naskaupee consider that only flesh is nourishing food.

The differences in the northern and southern Indian groups' habits as described above, which are based upon the differing natural conditions in their respective areas, seem thus to me to justify the separation of the two terri-torial groups of Indians in Newfoundland-Labrador: the M o n t a g n a i s and the N a s k a u p e e.

Yet here it must not be forgotten that the two groups are not a permanent division. Because of the internal changes in both groups the differences are in our day in process of disappearing. The customs of the Naskaupee will clearly in the future agree more or less completely with those of the Montagnais (cf. p. 696).

Present distribution of the two Indian groups.

The approximate boundary between the two groups has been drawn in Fig. 253 from the information I received from the H.B.C. managers and the trappers. In our day this may be regarded as running along the Notaquannon River and from its sources westward to Resolution Lake, a river lake in the George River, and thence further to the west. The divide between the Naskaupee and the Montagnais is at the same time a divide between Lake Mishikamau and the George River, said WALLACE in 1905 (995). The drawing of the boundary was, however, quite a risky undertaking, for neither the information with regard to the extent of the Indians' hunting-ground in former or later times, nor the outlines of the map used as a basis proved to quite definite.

This boundary, or perhaps more correctly boundary zone, is, moreover, not an insurmountable wall; the Montagnais, for instance, are continually crossing it in our days, as will be shown later on.

From Fig. 210 will be seen how sparsely the Indian population is to-day distributed in these two districts.

Censuses.

According to KROEBER (539) the subarctic Algonkian Indians (Montagnais, Cree, and Naskaupi) number about 2,300. Yet those living within the frontiers of Newfoundland-Labrador may be reckoned as numbering only between 250 and 300; in 1935 they were 273 (222). Probably no reliable census exists for them; it would also be difficult to compile a register of a wandering people whose individuals are one summer on the Gulf of St. Lawrence, another on the Atlantic coast. The difficulties are also increased by the custom of the Indians to change names not only once but several times so that the malicious spirits shall not catch them when they fall ill; — HIND says: If a child is ill the parents will not infrequently change its name, under the impression that its life will thereby be saved. It is thus not easy, sometimes impossible, to identify persons who may make their appearance at different places bearing different names.

Now, it should not be forgotten that a considerable number of the so-called Canadian Indians (p. 609) pass nine months of the year in the southern part of the Montagnais area. The exact number of those is not known to me. However, to arrive at the number of Indians in Newfoundland-Labrador those persons must be added who spend the summer on the Gulf coast but in winter practise their trapping in our area; cf. p. 6C9 sq.

HIND (434) gave (following the Blue Book, 1857) the following statistics concerning the Indians visiting the Hudson's Bay Company posts in eastern Labrador:

on the north shore of the Gulf

Seven Islands	300	persons
Mingan	500	»
Musquarro	100	»
Natashkquan	100	»

in the interior of the peninsula and on the Atlantic coast

North West River	100	persons
Fort Naskaupi (Nascopie)	200	»
Rigolet	100	»
Kipekak (Kibokok)	100	»
	Total 1500	persons

BELL (56) gives the following numbers of Indians dealing at the different H.B.C. stations in 1895:

at Hamilton Inlet, Montagnais	125
» Davis Inlet, Naskaupee	230
» Fort Chimo »	90

According to TURNER (965) the number of Naskaupee Indians in the 1880's was about 350. Shortly after 1900 CABOT (123) reported that they consisted of some 40 to 50 families which, reckoning 4 persons to a family, means 200 individuals. In 1937 I was told that the Naskaupee in the so-called Davis Inlet group comprised some 20 families and those of Fort Chimo scarcely more. The REV. PAUL HETTASCH after his contact with them for several decades considered it certain in 1939 that they had decreased and at that date numbered only about 200. BUDGEL, JUN., the H. B. C. manager at Voicey's Bay, estimated the number who visited him in 1939 at about 100. It is clear that the Naskaupee are decreasing.

Information from the government given by TOWNSEND (950) shows that the Gulf Indians in 1908 numbered 694 and were divided as follows:

Natashquan	76
Seven Islands	377
Mingan	241

In 1912 some 25 Indian families came south to St. Augustine, according to BRYANT (108).

The census of 1935 (222) gives the number of Indians on the Atlantic coast as:

North West River 154 of whom 56 married, 8 widowers and 90 unmarried.
Davis Inlet   81 »   »   38   »   2   »   » 41   »
Voicey's Bay   38 »   »   16   »   2   »   » 20   »

Unfortunately, because of the present war, I have no possible access to records and other information concerning earlier population figures (cf. however the sources listed in the bibliography).

Judging from the scanty and heterogeneous statistics given above and the information I obtained in Labrador there is some degree of certainty in the statement that the Naskaupee have greatly decreased — comparatively speaking — in the last half century, while the Montagnais seem to have increased a little. They, too, were formerly at the mercy of every whim of Nature in the wilderness, but since they have become so to say the servants of the white traders their living conditions have become more even and starvation has ceased to plague them (of this more later) (cf. 539 a, 539 b). The figures in themselves give merely a superficial idea of the scanty population of the two territorial clans. The statement that to each individual Indian 400 km² of hunting ground or more can be reckoned is more revealing — the best parts of the area are of course exploited by the white trappers. — For the sake of comparison it may be stated that in the interior of Norwegian Finnmark in the 1550's about 115 km.² per inhabitant was the figure (according to SMITH quoted by SOLEM, 1933).

The population figures illustrate a general condition: the stage of cultural development usually bears some relation to the density of population. In Labrador the families are fewer in number the further we proceed toward the northern interior: the game is fewer and more uncertain in supply; there the social and hunting systems can never be organized in detail or the land better exploited. The Naskaupee can be considered pioneers on the extending frontier of migration.

The Indians of Labrador are certainly not so little mixed with foreign racial elements; HIND speaks of Montagnais half-breeds. The dilution of the Indian blood probably began already with the trade in furs when the so-called *coureurs-des-bois* roved about among the Indians, but I do not know of any detailed information in this connection (cf. p. 662). Most of the Indians who spend the summer on the St. Lawrence coast west of St. Augustine represent a type which must have more or less white blood in their veins through consorting with the traders and settlers. With but two or three exceptions the Indians of St. Augustine and the Atlantic coast are considered unmixed Indians (995). Yet paradoxically enough some members of the Davis Inlet band owe their inception to the mating four generations ago of a Scotsman (or Scots-Cree half-breed) with an Ungava Eskimo woman (897, 906). This family, called Rich, is completely Indianized; cf. Fig. 286. Some Eskimo blood has been mixed in the northernmost Labrador Indians.

Health.

I have not found any scientific information in the literature regarding the health and pathology of the Labrador Indians. But the late Dr HARRY PADDON told me of some results of his observations of the North West River group of the Montagnais. The Labrador Indians look healthy if mostly thin, powerfully built and sinewy. They are not protected by a armour of fat like the Eskimo, but they impress one as being a strong type; see Figs. 254—258, 214, 285—287, 299, even if they cannot be compared with the Eskimoes in this respect; the Indian can never carry such heavy burdens on his back, nor can he compete with the Eskimo in endurance as a boatman. Yet the Indian also is unbelievably hardened against wind and weather, and his imperviousness to cold is phenomenal. He lies down to sleep out-of-doors in winter in caribou-skin only, and examines his fishing-nets in a howling snowstorm, dipping his hands in the water now and then in order that they shall not freeze. His children run about in the snow barefoot. His whole body seems to be as hardened as his face. Part of his hardiness may be due to the practice of fasting which was formerly common among the Montagnais and sometimes lasted eight days; the main purpose was probably magical (435). The women, too, are strong; a few days after childbirth the mother is up and about, the baby is stuffed into a fur bag with moss as a napkin and on goes the caravan.

Many Indians live long, sometimes to the age of 70 and more if they do not die earlier from an accident, or of famine or disease (1050). Yet as a rule their lives are much shortened by lung diseases and stomach disorders; for these they have only the shaman's drum to help; seldom have they had any touch with a doctor. The Indian population in the north at least has been greatly decreased by the ravages of disease, for their resistance power to the white man's illnesses is also small (cf. Table on p. 592 sq.). It is said, however, that the Indians who live isolated in the interior suffer less from disease than the

Eskimo on the coast. One form of influenza appears fairly regularly in spring and autumn; spitting and uncleanliness foster the spread of germs. Venereal diseases are said to be absent from the Indians. On the other hand consumption ravages the youth and tubercular hip disease is common among the Naskaupee children. Carbuncles also occur. Infantile mortality is terribly high and diarrhœa further decreases the number of older children. It would be a profitable task to investigate systematically the health of the Indians in the same way as has been done among the Eskimo (475, 910).

Fig. 254. The former chief, Atvon Pest (to the right) and the present chief, Panace Pest (to the left) of the Montagnais' Mishikamau band. In the background is the village North West River; to the right Hudson's Bay Company's store and to the left trapper houses. Photo the author, July 21st 1937.

REMARKS.

The mode of life and the customs of the Indians undoubtedly present many interesting aspects for the ethnologist. Before describing life among the two territorial groups in Newfoundland-Labrador it may be well to state that while not a little is known of habits and customs among the Labrador Eskimo, only some scattered ethnological fragments are available regarding the Newfoundland-Labrador Indians, especially the life of the northern Montagnais in the taiga. They like to stay in their primeval forests and only make flying visits to the Hudson's Bay Company's factories, in winter to exchange furs for food and other necessities and in summer to trade, to meet their priest and receive

Holy Communion; cf. Fig. 259. If those who wish to meet these Indians do not take these opportunities, they can only hope for a lucky chance to catch a glimpse of them, for they soon retire to their taiga again. In Rigolet I was told that the sociable Eskimo say jokingly: The ape can talk just as well as man; but he is much more cunning than man and therefore he is silent; for if he began to talk it might easily happen that man would find a way of making him work. That is why the Indian is silent and hides away in the recesses of the woods!

The Indians show no inclination for contact with white strangers. Their shyness at a chance meeting is probably chiefly due to the fact that only a very few of those in Newfoundland-Labrador can make themselves understood in English — or perhaps they like to give the impression that they do not understand that language. Nor has it been necessary for them to learn it, for the intelligent local representatives of the Hudson's Bay Company — who are now really the Indians' 'big chiefs' — usually speak their language fluently. On the other hand, I met one man among the Montagnais who spoke French fairly fluently and with an excellent accent. It is clear, therefore, that it was in general difficult for me on my Labrador journeys to obtain direct information about these people, and I gratefully testify to the readiness of H.B.C.'s managers and some trappers to help me, partly by interpreting and partly by telling me what they themselves knew of their Indian neighbours. I also sincerely hope that what is here written will contribute to some extent to a spread of the knowledge of the Indians' life in Newfoundland-Labrador

The Montagnais Indians.

The name, of course, was given them by the French missionaries (943). The English-speaking population often use the name 'Mountaineers'. The Indians themselves use the name Tshe'tsi'uetin'euerno (59, 434).

SPECK has, in his comprehensive works (877, 871), given a fine account of the non-material culture of the Montagnais-Naskaupee. But no special detailed information of the habits and customs of the Montagnais in Newfoundland-Labrador exists (cf. 129, 168, 818, 819, 875, 916, 1067, 1068). Remarks and notes appear here and there in travel accounts and in some special ethnological works (see bibliography). I shall, with the help of my own observations and the information I obtained on the spot, describe them to the best of my ability.

Some general impressions.

The physical type and the dress at once indicate the difference between the Indians and the Eskimoes, who always (except those of Rigolet) go clad in their white sillapack.

My first meeting with Montagnais left no impression of having been face to face with the lordly sons of the wilderness; it was not »the last of the Mohicans»

Fig. 255. Atvon, the former chief of the North West River Montagnais (second from left in the middle row) and one of his sons (to the left of Atvon) who is considered amongst the Indians an expert in scapulamancy. Photo the author, July 7th 1939.

whom I met. Externally the 'Redskins' of Labrador are curiously prosaic. They are not red but fairly dark-skinned. They were called Redskins because once upon a time, like other Indians, they painted their faces in war time with a red colour obtained from roots, bark and ochre in order to terrify their enemies and also to please the spirits who control life and victory. LE JEUNE in 1632 mentioned that the Montagnais had painted their faces with red, black, and blue stripes when he first saw them. Their clothing was of skins, they wore no covering on the head and their long, greasy hair hung low over their shoulders. They were armed with bows and arrows, a shield, and a lance.

These same people also presumably wore in olden times beautiful white
chamois leather clothes richly ornamented with paint — cf. Figs 288, 289,
forehead bands and feather decorations; but they are now out of fashion

Fig. 256. Montagnais Indians arriving at North West River to see their priest. Photo
the author, July 20th 1937.

(cf. p. 640 sq.). Now they go about dressed in unimaginatively ugly, ready-
made clothes which almost give them a proletarized appearance, and the
stamp of poverty marks their destiny. The only thing reminiscent of a national
dress is the woman's pointed cap, made of wedge-shaped pieces of blue, black
and red, see Figs. 256, 257, which HIND as early as 1861 found to resemble

somewhat in shape the Frygian cap. The cap in conjunction with the cork-screw curls bound round a piece of wood on both temples suits the dark-skinned face. Some of the Seven Islands men also wear bead embroidered caps, Fig. 268. But this headwear and the moccasins were the only ethnographical distinctiveness in the appearance of the modern Montagnais. On the other hand it must be said that some of the girls had got hold of some second-hand headgear, and a few young wives made a display in dresses of thread-bare, dark-blue shantung.

Another difference is that the Indian is taller and slimmer than the Eskimo in general. The faces of many — both men and women — are attractive; the eyes are usually wide open and have a soft look. The skull is broader, the face not so long and the nose more prominent (see table, p. 445); Figs. 255, 256, 257. The hair is black, thick and straggling, and the men generally brush it back over one ear while the women roll it up into two big corkscrews, one on each side of the face; Fig. 257. The beard growth is very weak. HIND was much struck by the general aspect of the Montagnais women: their eyes are inclined, their noses aquiline and their jaws square, the last feature being also the distinguishing character of the men. Both men and women generally have thick lips. The wild expression one expects in an Indian: an eagle glance from under coal-black hair framing a broad face with tightly closed lips, is only exceptionally seen among the Montagnais. The figures of the younger women are nearly always slim and lithe and they are especially winning in looks, with fairly regular features, and hands so well formed that few of their sisters in society circles in large towns can compete with them in this respect.

The Montagnais impress the observer as being healthy, families are large, the children fit and jolly; see Fig. 258. Yet infant mortality is said to be high and tuberculosis frequent among the children and young people. One other thing I must add: never did I see the brand of 'fire water' on the face of a Montagnais.

They are a very peaceful people. As one learns to know them better one soon finds something attractive in their calmness. I myself can fully confirm the opinion of one of my white trapper friends: that they really radiate friendliness and goodwill when they see a person for the second time. The indi-vidual is very closely bound to the community. It is not any public authority which keeps the hunter band together, nor the chieftaincy, says LIPS (580), but tradition, custom, and public opinion in the band itself.

I much regretted not being able to talk to them in their own language, and most of my information was therefore obtained through an interpreter. Yet the local population also told me much of the mental characteristics distinguishing the Montagnais from the Eskimo and that merits repetition here because it throws some light upon the Indians' psychological state to-day. The mentality of the Indian, they said, is stamped with the nature of the wilderness which so rapidly changes from wild fury to dreamy peace (cf. 303). That is why, say the trappers, it is so difficult to understand the primitive Indian's

Fig. 257. Montagnais visiting North West River in July 1937. Photo the author.

train of thought. He is mostly a child of impulse, and that is why he generally lacks the orderliness, strength of will, and endurance of spirit which is found in the Eskimo. Already his ideas and way of expressing himself have a child-like touch; many a time it was a great trouble for my interpreter to paraphrase my question to fit it for the simple understanding of my Indian informer. Their world of ideas is clearly restricted, mostly to hunting and the life in the wilderness; they seem to be quite indifferent to the mentality and impulses of the white man, however loudly they laugh and nod at the speaker. They can scarcely apprehend an abstraction, and it must be very difficult for a priest to get them to understand the mysteries of the Christian doctrine; the only thing they seem to have really grasped is the notion of »the Great

Spirit» which still can scarcely fit into their original animism (cf. 177). The psyche of the Labrador Indian is clearly very differently attuned, it may be said, from that of the lively and speculative Eskimo.

The Labrador Indian impresses the 'pale-face' in that he is so care-free with regard to the morrow. People say he is like the wild animals of the woods; he is in good condition in the autumn but thins off towards the spring, and the reason is his improvidence in the matter of food. The Indians collect little or nothing, when hunting is at its height, for provision against the hard times

Fig. 258. Thoroughly healthy Montagnais' children in the camp of the Mishikamau band at Grand Lake. Photo the author, Aug. 14th 1937.

which mostly repeat themselves every spring. A speculative observer thought that the people were indolent when they noticed that the old basis of their economic life was gradually disappearing and did not make any effort to substitute a new one; but this scarcely touches the reality. The Indian's attitude to his needs was the same at the no very distant time when the forest still swarmed with caribou and perhaps moose so that the hunter took his food just when he needed it, so to say direct from Nature's own storehouse. In these days, when the number of hunters has probably increased while the supply of game has somewhat diminished or remained the same, the Indian persists in the same care-free attitude from custom it may be said, for if there is a main thread running through the life of the savage it is the power of custom. The trouble of collecting and carrying about food in the wilderness seems to loom large for the Indian and dull the urge of sensible prudence. Nothing worries him as he enjoys the charms of summer.

Thus no preparations help to make life easier and pleasanter during the hard winter and early spring. The Indian only laughs at the silly pale-face — a ridiculous figure who gives himself so much trouble, for example, to live in a large house which needs so much timber-felling to be kept warm. Why should one buy and carry about heavy stores? MERRICK (p. 95) on the 11. November met a great Indian family in the Hamilton River valley; their whole store of food was not so large as that for two white trappers for hunting till January, he says. In this way time and again the Indians have the bitter experience of finding their food come to an end half-way to their hunting-grounds after their stay on the coast. *'Sham sheevan'* [= very hungry] is there-fore their chronic state in winter (cf. p. 624). When the Indians are hungry they must take what they can; the whites call it: stealing; they examine the trapper's 'tilt' [blockhouse] and eat up the flour, while their dogs sniff up the larder under the snow, say the trappers. Need knows no law, say the Indians in response: we are starving. But the Indians are always hungry, say the trappers; year after year it is the same thing (104). It must, however, be observed that to the Indian this is no criminal action; if he takes without permission something from a white trapper's store he has not used violence against him; he has only taken something of what nature designed for the maintenance of all. He is in his attitude to life a naive communist. WALLACE in 1905 wrote: No Labrador Indian north of the Grand River will ever disturb a cache unless driven to it by direct necessity, and even then he will leave something in payment for what he takes. — Yet improvement may be noticed to-day in regard to provident action: some Montagnais have begun to realise the value of carrying with them on their tabansks [toboggans] small bags containing dried meat and pemmikan. —

Ease, not to say laziness, seems to dominate the greater part of his life. It is said that the smell of food can smother his objection to regular work. But if he gets his wage in advance he will slip away when the longing to hunt urges him. It is useless to try to force him in any way, for if he does not want to do a thing he will not, and if he is offended resentment can simmer in him for a long time. He ought, for example, to be most useful to explorers and cartographers with his skill as a guide. Yet WALLACE in 1905 stated: It is impossible to engage Groswater Bay Indians to guide you; their strength holds out well enough but regular work bores them and then they begin to do all

they can to annoy the pale-face, and are said even to mislead him intentionally. This is why many have found them more trouble than use on expeditions.
But of course there are exceptions. In the same district I heard much of a
legendary Montagnais called Loue who in his time was of great service to the
Hudson's Bay Company, as MERRICK (p. 60) relates. Yet, for all the disadvantages, it is good to have an Indian in one's company, for no one can 'read the
country' so well and if other Indians or Eskimoes are met on the journey he is
a good introducer. Another thing that is useful is his capacity for drawing
maps (872, 995). HIND tells of maps constructed by Montagnais at the request
of Father Arnaud in 1859 and showing the route from the Hamilton River via
Petitsikapau Lake and the Ashuanipi River to near the sources of the Moisie's
eastern branch and running down to the Gulf. This map well illustrates the
Indian's remarkable ability to delineate the general features of the country
through which he has passed; HIND says it was singularly exact and accurate.
He also tells of a »letter» he saw stuck in a cleft pole overhanging the river
bank; it was 'written' on birch bark and was a small map of the country with
arrows indicating the writer's direction, some crosses to show where he had
camped and a large cross marking his intended first winter headquarters (cf.
580). Near by was a small birch bark roll containing a little tobacco. The Montagnais is thus a master at drawing detailed maps on birch bark; Low's excellent
map of the enormous peninsula of Labrador is largely based on the Indians'
'pot hooks'. This skill is quite surprising and original, for when he draws a
caribou he likes to begin at the tail. The only other artistic expression I
came across was in their beaded embroidery of which the patterns are
traditional· and which reveals a great sense of colour harmony.

Most illustrative of his mentality is the Indian's love of sitting on a slope
and talking all day long (cf. 168). It is said of the people of Labrador: The
fishermen think in split cod, the Eskimo in seal and the Indian in fur-bearing
animals; start this last subject and the vocabulary of the Montagnais will
not be exhausted until hunger drives him to get up.

The Indians are all hunters and some of them fishers too. Their world is
thus different from our own, and one gets quite another impression of his
energy and endurance when one sees him hunting. When the Montagnais gets
'caribou fever' he becomes the true savage. His insatiable passion has found
an outlet, and he can never let anything in the way of game pass him without
qualms of regret. It is then hard for him to suppress his instinct to kill (995).
But beyond hunting he seems to have no ambition. It is the same with his
sense of order; in this as in much else he resembles the gipsy, not least in his

nonchalance and the begging of his old women. Again, it is safe to say that there is not a truthful Indian in Labrador; they will be most plausible when it serves their purpose better than telling the truth. But they will not steal, says WALLACE; when they are staying among the whites they are allowed to go in and out as they choose; and the H. B. C. manager at North West River told me there were never any thefts during their visits to the village. But the villagers shook their heads; some of the trappers know from experience that many things disappear from their tilts in the lonely forests once the Indians have discovered it.

Of course the white population differ in their opinions of the Indians, but the latter naturally differ in character and talent   Those who have only seen the Indians of the Gulf coast cannot easily appraise the northern Montagnais; and the Montagnais in the wilderness is quite another fellow from the one who visits the trading-posts. When the local inhabitants assert that the Indians are unreliable and dishonest, doomed to eternal pauperism because of their laziness, they may be right. On the other hand many of the North West River trappers are also right in their opinion of the Montagnais; they speak their language fluently and often have the most intimate association with them in the wilderness. They say that the Indian would never hesitate to give the white man food if he was in distress out there.   One of my puritan trapper friends who was evidently unusually captivated by them and who had, as I found, won their hearts, told me he considered they merited our deep sympathy and full admiration: 'the Indians are such good people that the Christian ideals of goodness cannot be imagined more fully realised anywhere on earth than in their communities' and he added: 'their language contains no swear words'. Two centuries ago (in 1731) the Jesuit LAURE, who was indeed the first educated white man to have contact with the Naskaupee, said that an outsider could not imagine the goodness of these Indians. These statements undeniably give a stranger much food for thought when he is unable to form an opinion directly because of his ignorance of the Montagnais language. It is also said that the Indian was honest until he met the whites, and is mostly so still in the wilderness; there they live according to their lights which are those of the untutored savage who knows nothing of the civilization in the world outside (995). I do not know if there is a single Montagnais in the North West River group (see farther on) who can read or write.

Let us now consider more closely the life of the Indians in the interior where they show another side of their nature. There they hunt and trap with the utmost keenness. There they show great reserve towards strangers and prefer to be alone. BRYANT (1913) believed they considered the interior as their particular hunting reserve and that their aversion from strangers was due to their desire to prevent them from spying out the economic secrets of their country and establishing industry where there should be hunting; they had little sympathy with the new form of civilization of which the main task was to search for ores, paper wood, and sources of hydraulic power. If I remember rightly it was CABOT who experienced in a very practical way the Indians' objection to his, to them, mysterious journeys; under cover of night one of them would creep down to his canoe and bore in it small holes through which the water flowed in when it was launched; this was done several times, and the owner of the canoe understood that this was the Indians' tactful way of informing the stranger that his room was more welcome than his company. It is, of course, the bitter experience of the cruel whites of former days, 'coureurs des bois' and traders, that urges the Indian to suspicion, caution, and even aversion when he meets the white stranger in the wilderness. Old injustices are not readily forgotten by a savage society.

Yet an unwritten law of sparsely inhabited parts all over the world is hospitality and the Montagnais tent, too, is always open for any hungry Redskin or trapper who wanders in over the threshold, and if there is anything to eat he is offered it. But one must know with whom one is faced. When in the wilds I met Indians I had met earlier I noticed no trace of reserve, unless among the children, who are not so accustomed to seeing strangers. When I visited an Indian camp unexpectedly the youngsters seemed to take me for the 'Man of the Woods' or some other horrible creature; they sprang up into the trees like cats and gazed at me from behind the branches, whistling and clacking their tongues to each other. But after I had talked with their elders a while they came down again and joined the ring around me, silent and abashed; sometimes one of them would even come and feel my coat.

Thus the Labrador Indian is something of a psychological paradox to the white stranger, who can never compete with him in endurance and agility in the wilds. But in mentality he seems to have stopped in the forecourts of civilization at the stage reached by a Northern boy of at most 14 (cf. p. 447). Certainly this must be the standard of comparison for many of the curious sides of his habits and customs. Cf. however 876.

Distribution of small Indian groups over the taiga.

The geographical character of the taiga (p. 339) is only very roughly known. Quite a lot of mystery always surrounds the existence of the Indians isolated on the plateaus. 'On their hunting-grounds in the interior the Indians wander hither and thither during the greater part of the year so that it is not easy to get to know their customs and abodes', I was told by the whites on the coast. The statement is very suggestive, though implying an exaggeration; it is certainly true that the organized hunting nomadism of olden times is on the wane but the basic principles of their occupation can still be traced fairly clearly. The hunters still spread out in small parties quite systematically so that every section of the wide forest is exploited. Here is an attempt to describe the system.

Fig. 259. Father O'Brien, the Indian missionary, performs the Latin Mass for the Montagnais Indians in the Roman Catholic chapel at North West River. Photo E. H. KRANCK, July 20th 1937.

In that part of Newfoundland-Labrador held by the Montagnais they are divided into two large groups called: the Gulf Montagnais and the North West River Montagnais. An off shoot of the latter are the so-called Davis Inlet Montagnais who now form a separate group. The names are taken from those parts which the Indians visit in the summer to meet their priest and do business.

THE NORTH WEST RIVER GROUPS.

These Indians normally visit the trading-station of this name once in winter to sell their skins and buy food, and once in the summer about the 20th July, when they gather to meet their spiritual advisor who arranges church services for them and administers Holy Communion to them, and also to supplement their food stocks. On this second visit the hunters are accompanied by their families; cf. Fig. 260. It was just during their summer visit to North West River that I met them for the first time.

This group comprises 35 to 40 Montagnais families, or rather perhaps tent parties, who visit the station in summer. In winter their numbers are somewhat larger, for then some Montagnais Indians also come who hunt far away in the interior and who therefore go in the summer to different places on the Gulf of St. Lawrence between Seven Islands and the St. Augustine river for their spiritual nourishment and their business.

When at the end of July these people move back to their hunting-grounds the families spread out in the following way; cf. Map, Fig. 253:

1. About 10 families go to the Mealy Mts — Kenamu River district; there they live on caribou and any other edible game. They wander as far as to the sources of the Eagle and Eskimo Rivers until they come upon the trapping grounds of the white trappers of Dove Brook, from which they wisely

Fig. 260. The season's first Indian tents pitched at North West River July 4th 1939; looking from north-east. In the background in the forest is the Indian Chapel. Aero-photo the author, 1939.

keep away. From the end of October they set their traps for the fur animals in the wooded valleys round the mountain group.

2. About 4 or 5 families go to the surroundings of the Hamilton River but at the farthest to the mouth of the Minipi River.

3. About 5 or 6 families move to the mountainous country round Goose [Bay] River.

4. About 7 to 10 families go to the district round the Beaver and Susan Rivers, wandering thence away in towards the district around Mishikamau Lake.

5. About 5 to 8 families move up the Naskaupee River valley. Here they separate into (a) those families who continue north-west towards Mishi-kamau Lake and the source of the George River, while the others (b) make their way during the early part of the year over Lake Snegamok to the mountains west and south from Hopedale.

These 5 b families have gradually been increased by those who formerly spent the winter to the west and south-west of this district so that sometimes they number 15 or more. Some of the added families have belonged to the Gulf Indians, others even to the Naskaupee from Ungava: some of them still spend their summer on the Gulf coast and their route now lies over the upper course of the Hamilton River to Ashuanipi Lake and thence, as before, along the Moisie River to Seven Islands. It is, however, now more usual for these families, too, to visit the Davis Inlet factory in the summer to trade. Tent parties of the early pioneers are said still to visit the North West River sometimes in the summer.

— As appears from the map, Groups 5a and 5b keep to what is well known to be old Naskaupee territory. Group 5b is specially interesting as showing how easily natural causes give rise to an internal displacement of people among a primitive hunting folk. Just before the end of last century a number of Montagnais tent parties used to move for their winter hunting up the North West and Naskaupee Rivers to their sources and those of other rivers rising in the neighbourhood but flowing out into the Atlantic, thus the area between Hopedale and the George River sources south of the Notaquanon River. This country had been to a great extent destroyed by enormous forest fires and was for a long time quite empty of game, while in those parts not destroyed the caribou were very uncertain and even the fish few. For this reason the Naskaupee had abandoned it although it was their ancient possession. When, however, the Montagnais discovered that the Naskaupee had definitely abandoned the just mentioned lake area, and that the caribou were returning with new plant growth, some families moved thither from North West River. Rumour soon spread the news and the number of these families increased from half-a-dozen to some fifteen or even more because some of those who had regularly moved from Seven Islands to the Lake Plateau (Mishikamau) now joined the pioneers from North West River. Thus arose a new group: the D a v i s I n l e t M o n t a g-n a i s also called the D a v i s I n l e t b a n d. Several times in the winter and late summer they would visit the trading posts on the Atlantic coast and then they beat out new tracks through the Notaquanon River valley. The tent party who used to visit Seven Islands and had first kept to Mishikamau Lake followed this new track over the Hamilton River to the Gulf coast; it was easy and protected from the wind, running wholly through the woods. For some time the pioneers' parties continued to move to North West River in summer.

Group 5a usually goes under the name of G r a n d L a k e or M i s h i-k a m a u I n d i a n s, or M i s h i k a m a u b a n d.

The Montagnais now never come to the areas east of the meridian running through the eastern end of Snegamok Lake; one family certainly hunted in the district between Double Mer and Makkovik in the winter of 1938—39, but that was an absolute exception. They have deserted this part of the country because the white trappers and the Eskimoes have begun to take possession of it for their fur paths.

Very seldom or never did any of the tent parties 1 to 4 and 5a move down in the summer to the Gulf of St. Lawrence, though one or other of them have visited the trading-post at Davis Inlet. It was also very rarely that any of them wandered out to the barren grounds, the tundra caribou there might be an attraction but then they would miss their valuable fur animals.

The new development of 5 b (cf. p. 654) appears to some extent as a transfer to

THE GULF [OR MUSQUARRO] GROUPS, ALSO CALLED CANADIAN INDIANS.

6. About 15 families now move up the St. Augustine (or Menshen) River to their hunting-grounds in the great forests in the tract between the source of this river and those of the Paradise and Eskimo Rivers. For many years the St. Augustine River has formed the main canoe route of the Montagnais to the Kenamou River falling into Hamilton Inlet; by it the Indians can journey from the Gulf of St. Lawrence to this inlet in seven days. The St. Augustine River Indians also moved to the Eskimo River's sources in 1938—39; on the other hand the land to the east is said to have been abandoned by the Indians long ago (cf. p. 611).

Some of these families would cross the watershed and move some way down the valley of the Traverspine River. WALLACE (995) stated at the beginning of our century that this group traded both on the Gulf coast and at North West River. SPECK (873) describes the hunting as usually carried on in companies — a few hunters in family parties. According to CABOT (123) the large group divides into smaller parties towards spring. In the 1860's PACKARD (716) affirmed that these Indians were wholly dependent on the caribou; and also LOW (588) says that in the 1880's the caribou was the chief object of this group's hunting, and fur-trapping of much less interest.

7. About 20 families follow the Olmanoshibo [= East Romaine] or Menshen River up to the hunting-grounds round their sources, whence they spread a little way over the watershed down into the Minipi River's valley track.

8. About 25 families hunt around the upper courses of the Natashkwan and Minipi Rivers but they may extend their movements even to the Hamilton River. In the winter the men visit North West River to trade, leaving their families in the interior. Natashkwan was formerly a great seal resort because of its gently sloping beach; for this reason some white families formed a settlement there and this attracted the Indians. It is said to be the mission which now draws these families to Musquarro. According to SPECK these Indians hunt in the upper course of the river in large family parties till late in the winter when they divide into separate families.

9. About 50 families keep near Mingan in summer: the islands of Mingan have long been a great haunt of sea mammals and attracted Indians. When moving inland they follow the Romaine River to the portage at the watershed towards the Lake Plateau and Attikonak Lake, thence to the hunting-grounds lying near Ossokmanuan Lake and the Hamilton River.

10. About 35 families are now to be found in the summer at the mission and trading centre at Seven Islands so often mentioned. When moving inland they follow the Moisie or Big River, the Indians' Miste'shibo [= Great River]. For centuries it was a veritable artery in their life and still forms the natives' main line of communication between the coast and the interior. On the other hand it has lost its earlier importance as an inter-Indian trading route, when the Indians moving up the Moisie River continued as far as Ungava Bay; cf. HIND's map and description; see also Low.

According to information given to SPECK some 10 families (10 a) live in winter round the upper course of the Moisie and hunt away towards Mingan and Attikonak Lake. Some of them even hunt as far as to the sources of the Hamilton River and round Ashuanipi Lake, while some hunt south of the water divide. Here they carefully observe family rights in the hunting areas.

The other hunter families move along the sources of the Koopasho and over a portage to Ashuanipi Lake and the Lake Plateau as well as further north.

(10b) About 6 families remain in the west and south-west parts of the Lake Plateau (SPECK's Petitsikapau Band).

(10c) About 13 families move on to the northern part of the Lake Plateau and wander between the Hamilton River and Mishikamau Lake (SPECK's Mishikamau Band).

(10d) Finally a number of families move on as far as the tract between the sources of the George River and Hopedale (cf. 5 b above).

Seven Islands Bay or Chi'sche'dec Bay has always been a great Montagnais rendezvous because of its strategical situation. It has been a nodal point between two great Indian lines of communication with the interior and even across to Hudson Bay. In 1671 there were routes from the Saguenay and Moisie Rivers to Hudson Bay. The bay is connected by a long, deep valley with Lake St. John 300 miles to the south-west, a great meeting-place of Indians in olden times during their winter moves. It is also not far from the Moisie River which once formed part of a canoe route to Hudson Bay. In the spring and at the approach of winter myriads of ducks, geese and swans used to gather here and even sea mammals. CARTIER, who visited the Bay of Seven Islands in 1535, found it one of the most sheltered anchorages on the north shore of the Gulf and tells a marvellous tale of fish shaped like horses [walrus] spending the night on land and the day in the sea; LESCARBOT in 1609 called them 'hippopotami'. At Mingan there is a Walrus Island. The walrus, now never seen in the Gulf, was once common up to the mouth of the Saguenay (435). The Indians when visiting the Gulf coast during the summer months lived on the harbour seal; their flesh is not unpalatable, HIND noted. The Montagnais hunted in the Bay for centuries because of its swarms of wild fowl.

HIND reports that in the summer of 1859 500 Montagnais had pitched their tents at Mingan. They came there from all parts of their wintering grounds between the St. John's River and the Straits of Belle Isle, some in canoes, others in boats purchased from the American fishermen, others again on foot. He says that the rioting and debauchery resembled the war dances which centuries before were performed by victorious warriors around their tortured prisoners.

Only a brief acquaintance with these groups is needed to convince one that their subdivisions are not distinguished by stability; the numbers of the hunter families vary from year to year with the movements to and fro. But their territorial occupation can also vary somewhat by reason of the varying supply of game.

The Gulf Indians move in their canoes up to the sources of the rivers lingering much on the way. In the lower parts of the valleys running into the Gulf they can no longer earn their livelihood, for the white trappers have now taken possession of those areas. In this way a kind of conventional boundary line has arisen there and in those valleys running into the south-eastern part of the Atlantic coast. North of Hamilton Inlet the line is vague and irregular but the tendency is for the white trappers and the Eskimoes to extend their grounds further and further up the valleys as will be described later (p. 649 sq.).

In Fig. 253 I have tried to indicate the movements of these different groups, but it was naturally impossible to avoid their being very schematic.

It must be considered surprising at first to find that on the map (cf. also p. 650) the Montagnais no longer spread over the south-eastern part of New-foundland-Labrador south-east of the Mealy Mts district. From what I learned in Henley's Harbour it is by pure chance that the Indians come down even to the mouth of a river some fifty miles west of Forteau Bay (Napetipi River?) whereas the St. Augustine River is visited regularly. Trappers who carry on 'furing' in the wooded valleys running into the Straits of Belle Isle's north-west shore, e.g. Chateau Bay, have told me that they know the Indians do sometimes come down to these tracts; they have there visited the trappers tilts and the latter have lost their winter stores. In the summer of 1937 also, there was an Indian at Hope Simpson on the Atlantic coast trying to find work, which showed that he was not quite unacquainted with the neighbourhood. Yet there is no regular Indian trapping here. The reason is said to be partly that the caribou are very sporadic, and partly that people from Newfoundland

come here again and again with their traps and snares and greatly decrease
the supply of game.

The Gulf Indians too sometimes, pay short visits to North West River
to buy food if they need it. But there is nothing regular about this; the
H. B. C. like them to sell their skins at the stations on the Canadian side
and the buyers in North West River (Thévain, Faquit, Revillon Frères, and
H. B. C.) have agreed to pay similar prices, so the Indians have no advantage
from selling their products at this station.

Annual cycle of economic life.

The country occupied by the Montagnais is poor and anxiety for earning
possibilities drives them ever restlessly forward; it may also be that desire for
change or even old habits are contributory causes; it has been said 'the wander-
lust was in their blood; their birch-bark canoes were their second home, they
read the secrets of the woods like an open book' (514). Their stoppages were
mostly short; if they stay in one place more than a week it becomes so dirty
with refuse and excreta that discomfort warns them to move on. If someone
has fallen ill at a place they are only too anxious to move away, to escape
the evil spirits, they say; so they roam again, month after month, year after
year, even if the game does not compel them to do so. Yet for most of the
camp families the main tracks of their movements remain much the same.

Let us now try to imagine the annual cycle of the Montagnais Indians'
life.

The Gulf Indians.

I shall consider Pierre Gabriel's movements as a typical migration; he belongs to Group 10; cf. Fig. 253. I met him at Kalle'koane'kau on the Lake
Plateau and he told me of his movements etc. in fairly fluent French.

Move to the hunting-grounds in the autumn.

They move to the hunting-grounds in late summer. When the care-free days
at the coast with their feasting and loafing near their end the Indians begin
to make their purchases. After many discussions they get ready for the
journey. Pierre's group generally breaks up at Seven Islands about August
10th. They move up the Moisie River and usually two families go together;
thus they are sure of mutual help in case of sudden illness or accident, when
a small family would be unable to procure food or even to watch their nets

and traps so long as »the evil spirit is raging» and the men of the camp shake with fever. It is also well to have the help of others in carrying the baggage at difficult places. Though once it happened that Pierre and his family travelled quite alone. The number of individuals in a camp is carefully thought out and must not exceed what the supply of game can feed. That is why Indians are only found in large numbers during their summer visits to the coast or when gathering in parties to hunt the caribou.

The Indian is a poor man, yet the outfit he takes with him on his journeys (tent, bed and other clothes, food, tools and other gear, new traps etc.) amounts to quite a fair amount at the start. Though his main food still consists of nature's gifts — fish and game — yet he must now also have flour, lard, sugar, etc. with him, and they are quite heavy. To carry all this equipment up to the lake plateau a family with, for example, three children will need 2 to 4 canoes according to circumstances. The loads must be reduced to the minimum possible; the portages will be arduous, and the larger baggage must be moved by stages; the canoes bear a load to the end of one stage of a few miles, leave it there and drop back for another. If anything is abandoned it is part of the food store. It is no picnic to ascend the Labrador rivers.

In the open season travelling in canoes is more or less comfortable; the Indian does not walk much in the summer, the forests are too thick and full of rubbish, he thinks: he just steps into his canoe which in the wilds takes the place of both cycle and motorcar. Since childhood he has known the rapids, the dangerous rock snags and the foaming eddies; in time he becomes daring in shooting even difficult rapids. It is fine to see him sitting on his heels in his canoe, paddle in hand, diving into the whirling spray. He himself delights in this and his experience from boyhood up has taught him how the power of the water increases rapidly with its speed. A slight change in an eddy may swamp the fragile craft, bring it under the fall or break it on a rock; indecision may prove fatal. Yet I never heard of a serious accident. It is not the easiest thing to balance a canoe, but after an Indian boy has hung on to the upturned keel a few times, with the food and other baggage floating around him he soon becomes a real boatman. Again, to lose the canoe would be almost equivalent to losing the lives of the whole party, for it would be almost impossible to reach the coast dwellings on foot in summertime. In winter most of the difficulties of such a journey disappear for then the route lies over frozen lakes and rapid progress can be made.

All this shows that Gabriel and his company have no easy journey. The canoe must be paddled, poled, or tracked. Tracking lines are attached to the bow of the canoe and are drawn by a man ahead, floundering forward, but the heavy growth of willow lining the banks soon forces him into the cold water, where the swift current makes it very difficult to keep a footing upon the slippery boulders of the river bed. If the canoe hits a submerged rock the travellers may easily find themselves in the rushing, seething flood rolling down through the rocks. It is a cold business in late autumn to push the canoe, inch by inch, up the opposing, rock-filled waters; it demands a hardy fellow. The rapids seeth hither and thither between and over stones and heaps of rubbish in which the canoe rope can easily be caught. Inch by inch he who draws the canoe struggles along against the current, bent almost double, fumbling for foothold. Sometimes he falls into a deep hole and the canoe towing-rope is locked firmly. But the Indian boatman well understands how to avoid the many sand bars and the partly submerged stumps and logs here and there; boldly he twists his canoe between them with his paddles. When the stream becomes too full of obstacles he picks up his canoe and carries it. Again, it is no light task to bear the canoe over the steep portages where the path is only roughly hacked out in the thick brushwood of the forest, in many places over high mountains. However, along the Moisie River there are many well-worn portage paths round the falls and rapids proving the antiquity of this highway route from the Gulf to the table-lands of the interior (map in 435). But the Indian is strong; it is said that he finds it no trouble to carry a 198 pound barrel of flour over a 3 to 4 mile long portage, and even heavier weights over shorter stretches: 'just for badness'. He also has learnt to solve the carrying problem; the centre of gravity of the burden must follow the middle line of the body and this is well achieved by placing a carrying strap round the forehead. Soon habit becomes second nature.

Often the equipment of a company amounts to half a ton or more and must then be transported in lots from one stretch of slack water to another. If a storm arises in a river lake there is nothing for it but to land and wait till it is over, for with only a few inches free-board from the surface of the water it will soon ship so much that it capsizes or sinks, and certainly no Indian can swim. Yet on the whole the canoe part of the journey goes quickly; the daily rate including stoppages is estimated at up to 20 miles per day.

To economise the stock of food taken Pierre's family during the journey live mostly on fish; they use ordinary gill nets some sixty yards in length.

Only after a month and a half of paddling and poling does Pierre arrive at his first big stopping-place: Ashuanipi Lake; there the family takes a rest. They now feel at home and there is no reason to hurry.

Thence they move comfortably further north along the watercourse from lake to lake to Petitsikapau Lake, though sometimes they are delayed against their will by the crossing of the lakes, which can be troublesome: Lake Mishi-kamau especially has a bad name for heavy seas and high waves which may keep them on its shores for long periods because no canoe could weather them (995).

Now the wilds lie open to them and they can move about it at will only avoiding collision with the white trappers. In talking with different Montagnais I learned that nowadays the different families do not, as formerly, feel them-selves confined to certain hunting tracts which were theirs of old-established

Fig. 261. Camp of the Indian Mishikamau band about three km south from the outlet of the Naskaupee River in Grand Lake. Photo the author, Aug. 14th 1937.

Fig. 262.   Tents in the Indian camp of the Mishikamau band.   Photo the author, Aug.
14th 1937.

or inherited rights and respected by the others. For the greater part of the
time the Indians now live as separate families or groups of some few families;
on the 10th January 1931 a group as large as 5 families came to Mud Lake
(644).

There is no semblance of a regular camping-ground. The shore of some large lake with islands is preferred, and what they like best of all is an island with an extensive view so that the game can be seen far away from the tent; perhaps also this choice is a tradition from when they needed to be on their guard against surprise by an enemy. A similar reason may govern the selection of the edge of a great bend in a river. The traces of an Indian encampment are not easily distinguished in the landscape; the tent poles rot away after a few years and even the tufts of grass which soon grow up look just like an otter-meadow. Sometimes a *lobstick* (a spruce tree with the lower branches lopped off) is seen, but I did not discover that this had any special connexion with a camp. The chief thing in choosing a camping-place nowadays seems to be the possibility for someone always to be able to watch for booty, when the hunter lies on his back on a caribou-skin, sleeping or smoking, and his squaw works on the ground surrounded by the dirty children.

The Indians sometimes stay in the t i l t s or blockhouses erected by the trappers on their hunting-grounds, but it would never occur to them to put up a fixed dwelling for themselves. The only protection against the elements which the Indians have is the linen tent — Figs 261, 262 — in summer a cool and airy dwelling but in winter as cold as a cellar even after it has been covered with deerskins and then with thick layers of snow. Sometimes they content themselves with an improvised *meetchwop* (p. 639).

Nor do the Montagnais systematically lay in a store of food; they build no *caches* like other hunters which would remove the need for carrying large amounts of food on their long journeys; it just never seems to occur to them to do so.

Life in autumn.

From Petitsikapau Lake Pierre's company bends south-east to Sandgirt Lake, and on this stage, too, they lay out their nets in the fish-filled waters. At this period trout is plentiful but later the fish retire to deep water beyond reach of the nets. Hunting in the woods also decreases in results as the leaves begin to fall, though fortunately there is the ptarmigan to relieve the necessities of the band.

From Sandgirt Lake the Indians using Pierre's route spread out in different directions: away to Mishikamau Lake and even to the Notaquanon River or the Hamilton River. While waiting for winter to set in Pierre himself generally paddles away to the mysterious hunting-grounds around the sources of the George River which he still considers as his family's hereditary possession; but sometimes he goes trapping to Mishikamau Lake or down the Hamilton

River. A round like this usually takes some months, but he tries to get back to Sandgirt Lake before it freezes.

When the cold starts Pierre sets out his traps and snares for birds and other small game — hare and porcupine. Hares — in Labrador called rabbits — and porcupines formerly existed in large numbers all over the country and the Indians told HIND that the former used to be one of their most reliable sources of food and were very easily caught both in winter and summer. Their disappearance has been largely instrumental in causing the Indians to leave certain districts. Yet the hare is poor as food and, compelled to live on it alone for months, they soon become weak, emaciated and prone to disease.

In early October the caribou — *athen* — collect for the rutting-season at places lying beyond the water-divides and round the sources of the George and Kipekak Rivers; the hunters therefore go there and kill many animals; (from the airplane I could see that the country in the former area was more closely crisscrossed with deer paths than I have ever seen anywhere else in the world; clearly there are tremendous migrations of the animal). Especially Mishikamau Lake has been a regular autumnal rendezvous of Indians since very old times. Tent poles of all ages, log caches, 'sweat holes', cairns, signal fire places, etc. are still there in great numbers (995). The country is low and swampy and surrounded on all sides by bare mountain summits. In open country rows of branches or brushwood, in forests strong fences forming the sides of an acute angle lead the caribou to an enclosure. In this at certain intervals are openings where snares of babiche are set up, in which the deer, trying to escape through the openings, is captured. The other animals run blind with fear into the enclosure, where the hunters watch and kill then with guns or lances. — It is just the same converging rows of brushwood, sticks with a head of peat, piles of stone, etc., which formed the *hangas*, that were in use for the same purpose still a century ago in north-eastern Lappland. In the rutting time the bulls are decoyed towards the hunters.

After returning from hunting the caribou camp is pitched for a short period while waiting for the lakes to be definitely frozen, which generally happens in early November. Now there is much to be done in a short time; the winter gear must be examined and mended, the traps set, the canoes laid up and covered with brushwood. The snowshoes always need some repairs and the sledges and *tabanasks* [= toboggans] generally require patching. All these are the men's jobs. The squaws dress the caribou skins with the animal's brains which are kneaded into them; then they are dried, washed in warm water and stretched and afterwards worked with a bone chisel to scrape away the hair; if the skin must be made into moccasins, she must also smoke it.

Tanned, unsmoked skins are cut into thin strips — *babiche* — the Indians' universal material used instead of belts and string; it forms the plaited work of the snowshoes, it binds together the sledge bottoms, it forms the nets and fishing-lines. The sinews of the caribou's back are dried, split and wound round the thigh-bone to be used as sewing-thread, *sinew*. It is difficult to imagine how the squaw manages to do all this in the bustling, hurrying time.

Life in Winter.

When winter has really started the long journeys on foot begin. In the endless forests where no surveyor has ever tramped the Indians' work in transporting their baggage is worthy of esteem. The continual moves from one place to another can, in the course of one year, amount to 1500 or 2000 miles, and all the means of transport at their disposal are the same as in ancient times — canoe, tabanask and snowshoes or his own back.

The winter life is varied by constant change of camping-ground; a few weeks in one and then a move to another perhaps two or more days' journey distant.

When snow covers the trails travelling is not so easy; the family must take to snowshoes and drag all their baggage on the tabanasks. This little sledge is made of boards from birch or larch bent up in the front and bound closely together with strips of babiche. A tabanask to run on a thin covering of snow is 16 inches (40 cm) broad, a quarter of an inch (0.6 cm) thick and 8 feet (2.5 m) long. When the ground is frozen and hoar frost or a little snow is lying the tabanask can be used just like the Lapps' *akja* and the *veturi* of the ancient Finnish hunters; it glides like a ski in deep snow; the baggage is collected in a skin or sailcloth sack and bound fast to it. Dragging a loaded toboggan over snow on cumbersome, circular snowshoes is not easy. A white hunter would never try to drag such a heavy load as the tabanask carries — a trapper told me, sometimes it amounts to 100 kg. It is considered amongst the whites a hard day's work for one man to pull so much over dry, floury snow for 10 miles; yet a rather weak Indian can be expected to carry about 70 kg. When travelling in mid-winter nearly every member of the tent party has a separate sledge. At the head of the caravan tramp the strong men on their creaking snowshoes, then come the women, the old ones and the girls of 12 to 14 years all have their own tabanasks and all dragging with all their might. The old men come next, some of them dragging sledges; last of all come the youngsters. But ahead of them all, half a kilometre in front of the coil of sledges, tramps for all he is worth a 14—15 year old boy, proud to be 'the leading dog in a span of dogs', it is said. The mist of his sweat often hides him somewhat but the track shows a certainty as if he could see through woods and mountains, and he does this all day and in all weathers. In this way the Indians travel from morning till night without stopping, day after day, winter after winter. No meals are provided: Eat light to travel far, say they. Often the family has to go to sleep on empty stomachs when darkness falls, and often they are compelled to leave their hard night camps without having broken their fast. Generally a hunter starts off half-an-hour before the caravan in the hope of procuring some game, but if he fails there is nothing to be done but suffer the hunger. But it is said that this soon affects the muscles and gives rise to a drowsy slackness when trudging hour after hour in snow half a meter deep with the freezing land monsoon piercing one's clothes. Yet give the Indian a drop of tea and a lump of dried meat or deer-fat and he is all right again and quite pleased with things. INGSTAD said that the Indian of Canada is fortunate enough to be able to stop thinking, and this statement may also be applied here.

Of animals it is said that when it is really cold they creep down and lie in the snow, and fresh tracks are rare. The Indian on the other hand does not bother about the temperature or wind, his insensitiveness to them is extraordinary. The times of the moves are decided by the leader and he is obeyed. When a future hunter makes known his desire to join the party, they stop for a day or two, but shortly after the new baby has seen the light they go on again. When the great cold makes the snow rough a sealskin is slipped under the tabanask and it runs as if on oil.

The Montagnais' little, long-nosed 'cracky dog' is invaluable to the hunter; cf. Fig. 262 (870). Yet many people may be surprised that the Indians have no draught animals. Large dogs are not kept by them as a rule, partly because of the food they require and partly because the cunning huskies soon learn to rob the traps. Only a few Montagnais are said to keep a span (1—2) of dogs, but the Naskaupee are said to be better-off: they move about in districts where caribou meat is plentiful; many of their families have a komatik with a team of huskydogs. The use of the dog as a beast of burden is unknown in Labrador. A tamed reindeer — Fennoscandia's 'ship of the winter wilderness' — would be most useful to the Indian.

Life in winter is governed by hunting, which is considered the only worthy occupation for an Indian. No law prevents him hunting and fishing where he chooses, though in these days he must observe the close seasons.

A new great tour is made when the snow cover has formed and the fur-animal traps are set around the lakes in a line which may be thirty miles long and take a week to visit. The family first lives chiefly on the flesh of these animals, and on hare, porcupine and birds caught in the snares and on ice-fish. At the height of winter — the best hunting time — they also get caribou meat, and food is in better supply. Sometimes there are feasts when caribou are killed, and the men eat so much that they are half-stupid the next day. If again the weather is bad the men remain at home starving and devote themselves to making and repairing tools and utensils.

The whites say that the hunting of caribou is still the mainstay of the Montagnais' existence, as HIND also once wrote; he included the Naskaupee in his note and compared the importance of the caribou here to that of the buffalo to the Prairie Indians. I am not convinced this is right nowadays (cf. above and p. 422 sqq.).

The American caribou is considered the fiercest, fleetest, wildest and shyest of all reindeers, so much so that they are rarely pursued by white hunters or shot by them except by good fortune. Indians on the other hand have the patience and instinctive craft to enable them to crawl up to them unseen and unsmelt, for the nose of the caribou can detect the slightest taint upon the air of anything human at least two miles up wind, says HIND (435).

Nowadays the hunt mostly takes the form of a drive which requires a special strategy. When the snow lies thick and soft in the forests the Indian on snowshoes hunts the caribou with good results; the animal sinks deep into the snow, the hunter trails the animal and runs it to exhaustion. Different rendezvous are chosen from which a good lookout is possible so as to see the signal fires by night or smoke pillars by day which announce the presence of a hunt or a herd. When the caribous are discovered some young men are stationed at certain points to drive them into a pass where the hunters form a shooting-line, thus making certain to kill some of them. — All the boys above the age of 14 to 15 take part in these hunts.

I do not know whether pitfalls are used when hunting caribou.

February is said to be the best hunting month. When the snow lies meters deep and is covered with a slight, icy crust the caribou fall through it and have difficulty in escaping from the hunter on his snowshoes. The Jesuit chronicles tell of several cases when famine has ravaged large areas because there has been little snow.

The result of the hunt may be hundreds of caribou. It is the property of the whole band and is divided between the hunters' families according to traditional rules; in this way communism prevails in regard to collective caribou-hunting. In respect of other game the rule seems to be that those present at the death share in the booty, while all that is taken in snares or traps belongs entirely to their owner.

Sometimes one hunter alone, or with a single comrade, starts off to hunt caribou. He picks up and follows its tracks—most easily in February or early March when a few inches of fresh snow has fallen, which also is a good support for the hunter who can from the leeward side creep up to the pasturing, wantoning deer. When the hunters have stolen upon a herd they either finish an animal by a sure shot, or the game takes alarm and starts off on the jump, when the hunt must be given up in despair.

Other methods of the Labrador Indians in hunting the caribou can be found in CARTWRIGHT's writings (134). It is most interesting to note that a similar strategy, similar methods and the same periods in hunting caribou were to be found among the Skolt Lapps in North Finland a century ago.

Co-operation in hunting the bear has also been usual among the Indians. But this animal is not so often pursued even at the present time, perhaps for fear of upsetting his powerful spirit.

Small game, such as hare, porcupine and birds are not considered worth hunting by the Indians unless there is a lack of snow and the caribou and fur animals are few in number.

The fur-bearing game on the other hand the Montagnais hunts with the passion of a wild animal. He has greatly developed his methods and equipment, for which I refer the reader to LIPS (579), CARTWRIGHT (134), and LOW (588). Fur-trapping is carried on between the end of October, when the skins are thickest and darkest, and the end of March. The Indians have taken more interest in it since trade in its products has been arranged by certain firms and the natural economy of the natives has been supplemented by the white man's goods. It has finally become at least as important as caribou-hunting; the skins are in demand at the trading-store, and in exchange for them the Indian can get ammunition, clothes and food.

The trapping-grounds of the Montagnais lie wholly within the forest area where the most valuable fur-animals are found. The beaver, for example, is never found outside the connected woods: in Davis Inlet the Indians never have beaver-skins to sell; on the other hand in some winters they can bring in 40—50 skins of marten and mink, something which is very rare in Nain. For comparison it may be stated that in North West River, which is the staple-place for skins from the forests of the central districts, not very long ago a thousand beaver-skins and several thousand skins of marten and mink used to be traded every winter.

Among the Naskaupee things are different; skins of the caribou, the fox, and the seal are the most important articles of sale to the northern trading-posts,

Nain and Hebron. The white fox is especially valued and is trapped by the Nas-kaupee in the barren grounds and the Eskimoes along the seashore and on the ice edge. Sometimes in the barren grounds there is a shortage of hare and small rodents and then the foxes abandon those parts and make their way to others where they can find their customary food. At such times the Naskaupee would be badly off if they had not the caribou which not only gives them food but whose tanned, shamoyed skin is a good object of barter.

The Montagnais are not so hardly hit if the fox deserts their hunting-grounds, for there are always other fur-bearing animals and game, and plenty of fish in the lakes so they need not starve if only they plan out their life sensibly. Lately, however, there has been a turn for the worse, for it seems clear that the valuable marten is definitely on the decline in North America. This animal lives in colonies and is easily exterminated when its tracks are discovered. As regards Labrador, in former days 1500 to 2000 marten skins would be bartered in North West River during the winter, but 1938—39 only 70 such skins were sold. The supply of beaver is constantly on the increase now after a close period.

I spoke with some presumable experts who considered that the hunting-grounds can still support the present day Montagnais quite well. The Indians themselves say they have not noticed any decrease in the supply of fur-bearing animals except the marten.

When game is scanty and the stomach empty the Indian must fall back upon fish. Lakes and rivers contain plenty of them especially when the frost begins to cover the surface of the former. But the fishing really does not interest the hunter very much. It is therefore the women's task to set out a net near the tent or bait some primitive hooks. Fishing nets are used all the year round; in the winter the nets are put under the ice, but then it is very difficult to dry the nets, and therefore hooks are mostly used. The catch is eaten then and there, the Indian does not think of preserving any of it for the future; he counts on getting caribou meat when the ice is firm and the ground covered with snow.

Yet one kind of fishing does interest the men. To catch larger fish — especially salmon — in summer, in daylight the fish are caught from the canoe with a thin wooden spear with a long metal point and sometimes with two barbs of wood on each side of it, at night he catches these large fish with a torch and a spear with two barbs; this method he calls w a s w a n o. His favourite resort for salmon-spearing by torchlight is the pool at the bottom of a fall.

In early winter fish are caught from the ice by 'jigging'. Here again it is the squaw who works with a primitive, wooden hook fastened to a variegated line on which a whole lot of fish may bob up; a cold method of fishing demand-ing the patience of an angel! Sometimes the hook is of metal but those of wood resemble the ones we know from the stone age in Fennoscandia; sometimes even a simple bit of dry juniper is used on the line when fishing for the greedy pike. I have heard nothing from the Montagnais concerning the use of hooks made from antlers. Imported hooks are mostly used nowadays. But ice-fishing is dependent on geographical conditions. Newfoundland-Labrador is very cold and the snow-covering of the ice generally thin so that the ice itself soon becomes six and seven feet thick and it is only with great difficulty that a fishing-hole in it can be kept open. For this reason ice-fishing is not so important for the Montagnais even in famine times, whereas the quadrupeds and the birds are their main support, for the stocks of flour, pork and molasses they take with them are but small and they never think of taking an extra sledge load of them. They just do not worry; if they don't find game to-day they will to-morrow, or some day in the future, is their happy-go-lucky way of thinking.

Someone has said that even when the Indian's stomach is empty, his temper is good. Indeed, if everything said of him is literally true there is in the world no finer artist in starving than the Indian of Labrador. Famine and privation are the main theme and reappear constantly in published descriptions of the Montagnais. In 1670 PATER ALBANEL related that hunger was their great enemy, and FATHER ARNAUD repeated this in 1859: In the spring when the snow had begun to melt the caribou and ptarmigan passed away. The interval between the disappearance of these animals and the arrival of the geese was yearly one of suffering to the improvident Indians of Labrador. If the fish also fail they have nothing to eat but *'tripe de roche'* (p. 343) and broth made from birch buds. When the goose is heard the melancholy faces brighten; the goose dance was therefore a time-honoured custom (943). According to LE JEUNE they had recourse to the inner bark of the birch and caribou moss in times of famine, which seem not to have been infrequent (497). Nowadays famine is never so disastrous as in bygone days; now they hurry to the trading-posts by forced marches and get the white man's food.

The winter life of these savages is thus precarious, and truly the aspect of the country which they delight to call their home is sufficient to cool the ardour of the warmest admirer of a life in the Labrador wilds, HIND concludes.

Life in Spring.

When the trapping finishes towards the end of March, when the sun begins to bleach the fur, the Indian goes on to hunt other game, chiefly the caribou and the porcupine, always supplementing them until June with fish when necessary; the great lake trout, kokemesh, which can weigh 20 kg., is caught by 'jigging', others are caught by spearing in the rapids. If there is a scarcity of food at the end of April and in May even the bear is hunted.

Generally in May a move is started towards the places where the canoes are laid up. The Indian likes to begin to use his summer method of travelling as soon as the ice has gone; and he must get to the storing place before spring really comes and the freshet sets in when it is impossible to ford the rivers on foot.

Arrived at the canoes the families prepare for the spring move. Some of them go to Seven Islands, others to North West River, to meet their priests in the early summer. The way to the Gulf coast is long and if spring is late these Indians make small sledges with runners — *catamarans* — on which the canoes and the baggage are laid and dragged along on the snow-track. Normally Pierre Gabriel can begin his canoe journey to the coast immediately the ice breaks up in the beginning of June. He then goes via Astray and Ashuanipi Lakes. On the Lake Plateau the portages are short and low, the lakes are many and the river stretches between are shallow and short; it is easy to travel. Canoes and equipment are carried over the watershed to Koopasho Lake [= Oath Lake]. Southwards from the watershed the portages are steep and take up a large percentage of the route; this is the 'bad country' of the Indians. The journey down the River Moisie, so full of seething rapids, is made quickly. Generally the Indians must hurry if they are to reach Seven Islands by about 20th June when the priest usually comes to serve them spiritually: read a Latin mass for them, celebrate Holy Communion, baptize, marry, or bury them.

On their arrival at the coast the annual cycle of Pierre Gabriel is at an end.

Indians moving to the hunting-grounds from the other trading-posts on the Gulf coast follow the same principles in the main.

From Mingan they ascend the St. John River, over a high portage to the Romaine River they proceed to Attikonak Lake and thence to the neighbourhood of the Grand Falls of the Hamilton River.

As long ago as the 1850's the Indians, according to FERLAND (252), used to begin their journey up the Natashkwan about 2nd September. Along this route the Minipi with its difficult rapids formed the next stage. According to TOWNSEND (950) the Indians used to come to the coast at the Natashkwan in May and move away in August.

PACKARD (716) relates how, in August 1860, two Indians were catching seal and ducks in the Eskimo River; they must have moved in to the hunting-grounds only in September.

Those families who in summer visit the villages on the Gulf coast east of Seven Islands move thither in much the same way, but as the rivers are shorter the journey is quicker.

At Mingan and Seven Islands there are churches for the Indians. There is also a church at Natashkwan, but the Indians there usually have their services 15 miles farther down the coast at Musquarro. Since 1903 the Roman Catholic Eudist Fathers, who were then expelled from France, have had spiritual charge of the Indians. In the village of Olomonoshibu, one of Hudson's Bay Company's most important trading-posts, there was, according to RICHET (799), a little church; he says that this place used to be visited by 300 Indians who came there to sell their fur stock.

It seems thus that the migrations described above hold good for all the Gulf Montagnais except Group 10 d (p. 610). It is as a rule only in the height of summer that the Gulf Indians are to be met with at the coast.

Variations.

Sometimes the Indians appear on the coast in winter from force of circumstances. WALLACE met some at St. Augustine in the first days of April. They had failed in the caribou hunt and had come out from the interior half-starved a week or so before (995). TOWNSEND tells of a characteristic event from Natashkwan in the winter of 1911—12; some Indians turned up suddenly in February with the corpse of an old woman who had long been ill and died in the interior. Instead of waiting, as they usually did, for the spring journey to the coast, they felt an urge to come down at once, for they feared that the disturbed spirit of the dead would run riot in their district, and the best thing to do was to get her buried in consecrated ground as soon as possible. Even after the burial it was more than a month before they summoned up courage to return to their hunting-grounds (cf. 109).

But other causes may also give rise to variations. The seasonal rhythm is anything but regular; winter may surprise the travellers before they reach Sandgirt or Mishikamau Lake; then they are forced to lay up their canoes on the spot while some of them go back for the sledges on which they can move further. This delay disturbs the hunting of fur animals and their earnings suffer. On the whole it may be said that the forest Indian's life is subject to the same whims of nature as we know so well in that of Lapland's people.

We now leave the Gulf Indians to consider

THE NORTH WEST RIVER INDIANS.

In doing this we can confine ourselves to the Mishikamau Group (5 a).

Shortly after the departure of the priest in July these Indians prepare to start from North West River. The lodges are taken down, canoes launched and their stock of worldly goods, including food from the store, placed on them for the journey to the wilderness. North West River is soon without Indians, who turn to their rivers and forests with instinctive love.

They move first up to Grand Lake where they camp somewhere at its north-west end. At the beginning of August, 1937, I visited them at the idyllic spot they had chosen — Fig. 261 — in the beautiful poplar forest on a headland on the northern shore, about 3 miles south of the mouth of the Naskaupee River. They generally stay some months in this neighbourhood to build canoes. There the families separate though usually not before the men have together been on an early caribou hunt in the district round Snegamok Lake.

The direction to be followed by the parties after they have separated from the camp is determined by the chief — *wuitsta'mau;* Fig. 254. The principle of circulation, already described (cf p. 632), seems to be followed in the main; each stopping-place on their move is occupied for a few weeks.

The Indians go up the small rivers with their many rapids in short stages. When after a very strenuous journey they reach Mishikamau Lake the watercourses are beginning to freeze.

Here they lay up and cover their canoes, take to their snowshoes and tabanasks and lay out traps for the fur-bearing animals. These Indians, too, get their food chiefly by hunting and ice-fishing. Then, already after a month or two, they go down to North West River to exchange their furs for food and other necessities, and then, dragging their heavily loaded tabanasks, they return to Mishikamau where the younger ones have been watching the traps. These Montagnais seem to be prudent and realize the value of having a good stock of food if the game fails; as they say, if this happens in March when the ice crust on the snow surface can bear, things go badly for people in this poor area. It is a tradition among the Indians themselves that the worst time for food and the greatest need for exertion is just before the close of winter.

When spring comes with a burst, or at the latest when the wild grey geese begin to make for the north, the canoes are taken out again and the band begins to move down in short stages to their summer camp at North West River. In 1939, at the beginning of July, I found there 18 tents with about 150 Indians, but one family party from Kenamish had not yet arrived; cf. Fig. 260.

This is how life passes among the Montagnais in Newfoundland-Labrador to-day.

Life in olden times.

PATER LE JEUNE (497), in Relations des Jesuits, 1634, gives the most interesting description of the life and habits of the 'wild' Montagnais of that time. From this and other Jesuit narratives one gets the impression that the Gulf Indian groups were living under conditions remarkably similar to those of the Labrador peoples of the present day — the same winter wanderings of small family groups living on the edge of subsistence and avoiding other groups during their migrations (879).

In 1786 CARTWRIGHT (134, vol. III) described the Montagnais with whom he seems to have had not so little intercourse. By birch bark canoe in summer and rackets in the winter they traversed the inland. They used guns and bows, but their chief dependence was on the gun and they were excellent marksmen. They were wonderfully clever at killing deer; in winter time if deer were scarce they followed the herd by the slot day and night until they tired them quite out, when they were sure to kill them all. They killed beaver by watching for and shooting them or by staking their houses and killing them with spears made for the purpose, or they took them in nets placed across a contraction in ponds. Hermit beavers were therefore always observed to be most numerous in those parts of the country which are frequented by Indians.

The Montagnais were also very dexterous in imitating the call of every bird and beast, by which they decoy them close to their lurking-places. And, as the destruction of animals is their whole study, there is not one whose nature and haunts they are not perfectly well acquainted with, in so much that one man will maintain himself, a wife, and five or six children in greater plenty and with a more regular supply, where an European could support himself singly, although he were a better shot. As the people never stay long in a place they live the year round in miserable wigwams covered by deer-skins and birch-rinds having no built houses.

Some eighty years ago CHISHOLM (435), who for forty years had close connection with the Gulf Indians, described their migration as follows: When leaving the coast for the interior, many families have particular rivers to go up by, and often in a large body. But once a certain distance inland, the whole party breaks up and disperses into bands of two and three families each to pass the winter, and seldom see each other any more until spring. But before taking their final leave of each other a place is appointed to meet [in the spring], and he or they who first arrive at the prescribed rendezvous, if having sufficient food to wait, keep

about the vicinity until the whole party collects. They then go to fetch their
canoes, wherever left when the cold sets in, and employ themselves, some in
making new canoes, others in repairing the old ones, until such time as the ice
breaks up in the large lakes and the waters subside in the rivers. They then
move off in a fleet of canoes towards the sea, and generally make their appearance
at the coast about the latter end of June.

It seems also to have been usual only some decades ago for the Indians to
collect in large groups at some of the great lakes on the south of the Lake
Plateau — generally Sandgirt Lake — before beginning their journeys in
smaller groups to the Gulf coast or North West River. It was a fine time,
seeing each other and exchanging news, or arranging their differences; probably
some came from far distant parts, perhaps even some Naskaupee, and there
was much to hear and to relate. It seems strange that they should settle their
business affairs at such a meeting-place, e.g. at Kanispiskapau L., and not
on the coast. But it was probably because not all the Montagnais went so far
at that time, and their economic life more closely depended on natural supplies
and conditions. Another reason for meeting thus may have been the opportunity for barter between north and south. Some kind of inland trade survived for many years; e.g. trade in canvas for canoes between the Gulf
Indians and the Naskaupee, whose country did not supply canoe bark (cf. 123).

To-day it is at the summer meeting on the coast that experiences are discussed and differences settled — all those quarrels which may have arisen
during the winter. Public opinion is a powerful force in the Indian community and on its attitude towards him the very life of the hunter depends.
He can go nowhere else, for to leave the community means to be outlawed
and doomed to starvation (580). The associated heads of families thus form
a kind of arbitration court.

Montagnais' summer visit to the coast is presumably of fairly recent origin.

It has been argued that the seasonal migration from the interior forests,
where their winter hunting-grounds are located, to the coast or to the great
bodies of fresh water to spend the summer there and return to their winter
grounds when the season of insect pests, forest fires and drought have ended,
is fundamental to northern culture (871). Thereby it is moreover thought that
conservation of the game is possible, it allows a quiet period for the purposes of recuperation. During the summer season at their camps beside water
these nomads subsist upon fish, water fowl, eggs and, where they come to the

sea, sealing. This undoubtedly holds good for the western Indians, yet I have good reason to think that our Indians in olden times never moved to the coast regularly. It is known that formerly the buyers of furs used to seek out the Indians in their hunting-grounds. Then the H. B. C. set up stations inside the country in order to control the fur trade more effectively. Within the present Montagnais area were established Fort Mishikamau, Fort Naskaupee or Petitsi-kapau at the northern end of the Lake Plateau and Fort Winnikapau in the Hamilton River valley. Under these conditions the Montagnais of Newfoundland-Labrador had no cause to make their way to the coast. It is said that in those days the life of the Indians in the interior all the year round was healthy and happy, further that the summer was a festive time. They lived by fishing with net or spear, by trapping birds with all kinds of snares, and by collecting eggs and picking berries. New canoes were made. The family vegetated without care for their maintenance or being cold. Rheumatism was less troublesome than in the cold season, and tuberculosis was unknown.

The missionaries, too, at least as late as the middle of last century, sought out the Indians where they collected in the interior (cf. p. 581). There both trade and church ceremonies took place; so why should they go to the coast? If the Jesuit chronicles can be believed the Missions in those days were fairly successful.

Then later when the Hudson's Bay Company closed the stations in the interior and opened new ones at the coasts the chiefs or some of the reliable Indians used to be entrusted with the conveyance of all the furs obtained to one of the Company's stations and the exchange of them there. If I was correctly informed the other Indians even at the beginning of this period lived permanently in the interior; only those whose hunting district was near the Gulf coast used to visit it, to hunt seal among other things.

What brought this Golden Age to an end I have been unable wholly to discover. Some information points to the fact that life in the interior became harder as the use of firearms and forest fires caused the game to decrease. After the middle of last century the habits seemed to change with some confusion as can be traced in HIND's descriptions; the caribou began to migrate differently and hunger drove the Indians down to the trading-posts on the coast, especially the Gulf coast. Here they found the *'oblate father'* or *'robe noire'*, the priest, who gave them Holy Communion and looked after them in many ways, probably giving alms to the poorest. What was the actual cause and what the effect I have been unable to judge; it may well be that the missionaries — as I have heard — found it much more convenient to let large numbers of Indians collect at certain times at certain places on the coast, as they now do, than to undertake the troublesome journey to their collecting places in the interior and just at the worst mosquito period. WALLACE (995) says that when the Jesuits in 1897 gave up the chapel in North West River this caused the Indians to go down south to the Gulf coast. I find it difficult to believe that the periodical migrations arose from such an event. CABOT (123) alleges that at the beginning of the present century few Indians regularly visited the east coast of the peninsula: For convenience to themselves the 'oblate Fathers' had prevailed upon the hunters who formerly traded at Hamilton Inlet to make the longer journey to Seven Islands. The fact probably is that for both these reasons the annual move to the coast soon became a fixed custom, and now nearly all the families do it unless perhaps some who have been unfortunate and have no fur to sell. Yet it must be remembered that as early as the 17th century many places on the Gulf of St. Lawrence were visited in summer — apparently spontaneously — by Montagnais Indians. But the question is still not clear.

One thing is certain that this custom is most unfortunate for the hunter people. The journeys — often forced — to and fro mean lost time and unnecessary trouble. Thirty to forty per cent of the year is wasted on them; how much work could they not do with their hands in that time? The hardships tell greatly on the well-being of the whole family, especially the children and the old people; for many of them the journeys to the coast are a veritable punishment; one or more goes under with the illnesses of the whites and others carry back to the wilds seeds of diseases which were formerly unknown among them and from which they will never be free — especially consumption.

HIND says he found that the Indians who stay behind on the coast rapidly lost the energy and bodily strength which characterized them when living in the interior, and which were absolutely necessary for the hunters to maintain themselves in a mountainous country thinly stocked with game. Once on the coast their habits soon changed; they learned to live on seal and fish, became very susceptible to changes in the weather and, during the spring, were liable to prolonged attacks of influenza; the young people became consumptive, the middle-aged rheumatic, and death rapidly thinned the ranks of this once numerous and singularly hardened race.

When in September the caribou's horns begin to harden the hunter must again be on the spot, it is said. The Indian is not one to do things in good time, so that when he travels he is always in a hurry and must hasten onwards whether the members of his family are old or young, well or ill, blind or seeing. The danger of accident is then greater than ever and often the lives of all for good or ill depend on the trapping skill and wide knowledge of locality of a boy or an old woman, though such knowledge does border on the phenomenal. The sick must accompany the rest at the same pace as the healthy. Only after a month's persistent boat-dragging, bearing, and canoe-poling up the narrow valleys do the latter open out into the district of the rivers' sources. Then the extensive, sun-bathed plateau country finally spreads out before the caravan and many of the members breathe more easily again.

The Indians themselves now know nothing of the time when their fore-fathers began to move down to the coast in summer; and little do they know of the meetings on the Lake Plateau in the olden days or the barter there, though some of the older people say they have heard of it.

The alteration in their migration habits is only one detail in a whole series of changes that have taken place among the Montagnais recently.

Some fragmentary notes on cultural changes.

Looking through the older literature on the Montagnais it is soon clear that their material culture has gradually undergone important changes though its main characteristics have remained the same.

FAMILY HUNTING AREAS HAVE DISINTEGRATED.

Among the nomadic hunters of forested regions it is usual to divide the country into 'spheres of interest' of which the exploitation is recognised as the right of a certain band of hunters, generally a 'great-family'. In such a division it is taken for granted that these hunters have the sole right of hunting and trapping there as well as of ownership of all booty. This custom, of course, implies a higher form of hunting culture than the primitive communism of the tundra people (cf. p. 674). There is in general a relation too between the faunal composition and the human institutions in the vast expanses of the north (878 a). Just because the natural conditions vary from the north towards the south in Newfoundland-Labrador, two different patterns of hunting are to be found: the communal hunting of the caribou in the tundra regions, and the segregated family activity in the forested area. (Cf. 177 a, 206, 207, 866, 868, 872, 878 a, 879, 895 b).

I was therefore extremely surprised to be told, by people who ought to know, that the Indian families of the present day ramble about the wilds according to their own impulses, without any old-established plan for the division of the hunting-grounds among them. In proof of this I was informed that a Montagnais family who one summer visits North West River or Davis Inlet will quite likely turn up at the Gulf coast the following summer, and in spite of the doubts I expressed, my informers maintained that there was not and never had been any family control of land among the Montagnais Indians.

Their statements, however, quite probably hold good of the Naskaupee who, it may be said, live as a kind of parasite on the caribou. As this animal always moves against the wind, and as the prevailing wind changes considerably with the climatic periods (p. 310 sqqq.), it may well be that the Naskaupee in following their game come to quite other portages, swimming-fords, and hunting passes in one year than they do in another. CABOT (119) also stated regarding the Naskaupee: the number of lodges on the eastern side of the country depends on the movements of the caribou.

Caribou migrations can last several years; the herd will suddenly move to far distant parts for no known reason and this means difficult times for the Indians; hunger is always at the door, except in summer when fish is obtainable. Then the hunters try to track single caribou — a very uncertain prey; they must leave the smaller game they still have to the family in the camp and spread out farther inland with their cracky dogs. With tough endurance they struggle through the forest day after day, but hunger and hardships tell upon their strength, they gradually become exhausted and finally realise that they will never be able to return to the camp; nor does anyone ask why they do not return; their fate is understood by all without words (119).

It would thus be difficult to maintain any division of the land among the tundra Indians; it would simply collapse with the absence of the caribou. But among the forest Indians, too, calamity may break down existing customs. For instance, extensive forest fires may, in parts of Labrador as in the Siberian taiga, completely exterminate vegetable growth and then fur-bearing animals and all other game desert them (101); with them, of course, the Indians who lived there must also move away if they are not to starve. We have already had an example of this (p. 608). The great Saguenay forest fire in the early part of last century (434) must have been a calamity which forced the Indians to seek out quite new hunting-grounds further inland, or to try to win a livelihood on the coast. Forest fires may have been a contributory cause of the slow displacement of the Naskaupee and Montagnais in the wilderness (p. 401 sqqq.); they have certainly broken up the divisions of the hunting-grounds, and brought other great misfortunes.

Migrations of fur-bearing animals and other game in the woods can also bring the Indians face to face with starvation, e.g. the fox's period of circulation in Labrador is 6—7 years (243). The migrations of the caribou can hit the Naskaupee quite hard, those of the fur-bearing animals the Montagnais. HIND reports such at St. Augustine. Then the families must move to quite new areas and perhaps never return to their own hunting-grounds (cf. 421).

The woods are generally the scene of distinctive individualism in the people's lives and, although circumstances can be quoted to prove the opposite, I am unwilling to renounce the idea that also in the Labrador taiga segregated family hunting existed, and that once the Indian had adjusted his equipment to the environment in those extensive family hunting-grounds he did not give them up so quickly.

Finally I received confirmation of my idea in the very exactly delimited great-family areas drawn for me by the Indians Pierre Gabriel and Mathieu André on my map. These are shown in Fig. 253. These two Indians said that there was a tacit agreement that in those areas trapping was forbidden to the hunters of any other family; the monopoly rights of hunting and trapping on such an area could be broken only if a stranger passed through it and was in need of food, he might then satisfy his immediate needs. Further they told me that the custom among the Montagnais was that on the death of the head of the great-family the area passed to the new head or the inheritor of his rights. — Later I found in CABOT's writings a corroborating statement [apparently it applies to Group 6]: most of the Indians are actually born upon hunting-lands handed down from their ancestors. At an early age each knows his own ground as the farmer's boy knows his father's farm (123). In another place CABOT stated even more definitely: the hunting-lands are held by individual (?) hunters, and are passed down from one generation to another by customs of inheritance similar to our own (cf. 879). Finally, to quote CABOT again: the hunting naturally descends upon some man of active age; if a daughter is married, the young husband may succeed to the lands. Parents, or even more distant relatives have, by common right, their place in the lodge. In fact, all must be taken care of in some way, in one lodge or another; about the hunters group the dependent ones, widows and orphans and incapacitated; none is denied his rights. This concerns particularly the Montagnais. Infringements upon each other's hunting-grounds are probably no more frequent, CABOT supposed, than the cutting of timber on another's land in civilization. Another remark of his is that the land is usually divided into three parts which are hunted in rotation year after year. — It is astonishing that I discovered exactly the same family tenure (now abandoned) of hunting-grounds among the Skolts-Lapps in Fennoscandia.

The above facts show that there was once a somewhat higher kind of social organisation among the Montagnais of the Lake Plateau even as now among those of the south-west (cf. 868, 872, 873, 876, 877, 878 a, 879, 206 a etc.). It is only a pity that CABOT did not define more precisely the districts to which he refers.

It has earlier been explained that in the area between the Height of Land and the Gulf of St. Lawrence, which SPECK (879) labels the L a u r e n t i a n r e g i o n, the segregated family hunting type of the Montagnais prevails most characteristically. In it the authority of the chief is at its minimum and family self-management at its maximum. The family terri-tory hunting system shades off into the band ownership of the caribou hunters, as may be expected ecologically, some groups even utilizing both methods under different cir-cumstances (878 a). On the other hand, north of the Height of Land only commu-nal hunting has been known. Now, from what has been stated above when considering the annual cycle we know there are, north of the east-westerly watershed, not only family hunting-grounds but also a gradual change from family to communal hunting (p. 588 sq.), and the same fact was reported long ago by SPECK (879) south of it. The varying seasonal ecological conditions compelled the hunter to adapt his activity to the supply of game, and that is why both methods have sometimes been in use by the same band.

Fig. 263. Indian youngsters of the Davis Inlet band. Photo the author, July 20th 1939.

In summer 1939 I met some older Seven Islands Montagnais and asked them about this matter. They said they knew the family hunting-ground principle but did not recognise the validity of any boundaries such as those of the areas claimed by the two Indians as described above. They said they could hunt within them whenever and to any extent they wished without any risk of punishment. The conclusion must therefore be drawn that the system had lost its authority on the Lake Plateau.

Thus the question of individual hunting-grounds in Newfoundland-Labrador to-day is vague. Though there are indications of the two examples of great-family hunting-grounds on the Lake Plateau, we have seen that this area was still in the hands of the Naskaupee in the middle of the 19th century, and they scarcely had separate hunting reserves in post-Columbian times. The conception of the segregated family grounds is consequently not very old and probably introduced when the Montagnais invaded the Lake Plateau. The Indian families south of Lake Melville still seem to have some such idea, but farther in the interior it is said that the majority of hunters do not respect any kind of family ground boundaries. It is of course true that the same hunters return year after year to the same area but this does not prevent other Indians even from great distances coming there to hunt.

Another argument in this connection is that the family hunting-ground question bears an intimate functional relation to the highly specialised economy arising from the fur trade which became important late in the 17th century (895 a) and is thus a post-European development (492 a). Many people may therefore think that the separate hunting-grounds arose in connection with the establishment of H. B. C. trading-posts on the Lake Plateau. SPECK (868) however has expressed the opinion that in other parts of America the family hunting territory system of the northern Algonkian was an aspect of aboriginal culture and also in Labrador (e.g. Lake Mistassini) he saw in it the marks of a deep stratum of circumpolar culture in one of its local phases. It is an undeniable fact that there the idea of family control cannot be traced to the sponsorship of the H. B. C. (878 a).

A further possibility is that in olden days the same arrangement was in force in Labrador (cf. 498, vol. 32) as among the hunters of boreo-arctic Fennoscandia: that the ownership of land was vested in the hunters collectively and not in individual families. However the right to use a certain area of forested land long ago became the monopoly of a great-family or a party of small families (cf. Fennia Vol. 49, no 4). Like other Jesuits LE JEUNE may have had the impression that the winter hunting of the Montagnais was more chaotic and unplanned than was actually the case (879); however, he noted (498, vol. 8) in the beginning of the 17th century that if these Indians became located they took their own hunting territory. It seems quite natural that a certain degree of permanency is necessary for a hunter to discover the character and resources of any given hunting district. I therefore fully agree with SPECK (879) that prior to white contact the dependence upon a migratory animal may have encouraged separate family use of territory. This would be precisely what we know from northern Fennoscandia. In this connection T. W. SCHMIDT's opinion (829 b, vol. I; 879) may also be quoted, that the family operated hunting territories are associated with an old phase of culture of the northern Algonkian hunters, which has been subjected to historical influences that have modified it, socially within and ecologically without, according to locality and ethnic background. The north eastern Algonkians tend to make the family, rather than the great-family the carrier of land tenure principles, and where this has been lost or weakened SCHMIDT thinks it is due to non-Algonkian influence. Certain environmental and cultural conditions among the lowest hunters are conducive to such a system (879). Moreover, SPECK and EISELEY concur in the opinion of HALLOWELL that a number of the characteristics of the hunting territory system are extremely variable. A more intensive study of local faunal variations seems warranted, along with a survey of their effects upon the property arrangements of the band (829 b).

All these facts and opinions lead irresistibly to the conclusion that far-reaching changes have taken place during later times in the hunting life of the Montagnais of the Lake Plateau; the old social system has broken down. What is the cause? Has the yield of the area become too scanty for them, in spite of assurances to the contrary? This thought cannot be dismissed without further study if one looks at Fig. 253, which shows clearly how the trappers have slowly encroached upon the family district of Pierre Gabriel and Mathieu André. For every new party of young white trappers who march out into the wilderness, the limits of the supposed family areas must give way to supply land for new fur paths.

In general in North America, as DAVIDSON (206 a) pointed out, it is to be expected that the family ownership of districts will assume less and less importance in the economic system of the hunting nomads as a result of a progressiveness in natural poverty the more one approaches the barren grounds. In east Labrador the reason seems to be excessive exploitation of the hunting-grounds.

It has already been stated that the trappers are hampering the Indians in nearly all the large valleys and I have certainly received the impression that the former's areas are being constantly extended. They spread out like tentacles and drain out all the resources of what was once the Indians' land, for when the fur-bearing animals escape into the forests in unfavourable weather many of them fall into the white men's traps. A glance at the map, Fig. 253, will make this clear as regards the Lake Melville watercourses (of this more later). Corresponding conditions are seen in the configuration of the area covered by the Gulf coast trappers. The Indians of course regard this intrusion as an injustice, but they resign themselves to it as their fate, afraid of 'the law', and the law, as elsewhere in primitive conditions is dictated by the majority with the 'right of the strong'. In this way I can imagine that the old social structure of the Montagnais has been loosened and changed.

The breaking-up and disappearance of the social organisation mentioned above in conjunction with other circumstances seems to me to have made

Newfoundland-Labrador a region of deculturisation, degeneration in social respects for the Indians. This idea seems to have been vaguely present already to HIND: It is natural that a semi-civilised people when unable to live in security and driven into a less favourable milieu betake themselves to a ruder and more migratory life, and they are gradually reduced to a lower state of barbarism, he wrote.

CHARACTER OF THE CHANGES IN MATERIAL CULTURE.

Different kinds of stone objects, scrapers and such like, are still in use among the Indians, but the stone heads on their arrows and spears have been abandoned. Splinters still lie on the surface where tools have been made, the moss and grass have not yet grown over them. It can be said that the Montagnais emerged from the Stone Age only a few centuries ago.

Many generations of these primitive people have adapted their tools and hunting gear to the needs of their habitat, and in many cases not even the new inventions of the present day can replace them. Examples of such are the canoe, the tabanask, and snowshoes (204 c), all purely Indian inventions and in use among them from time immemorial. The Indian is very clever with his hands; with only his axe and his peculiar 'crooked knife', *mok'a'tau'kan*, he constructs his boats and sledges as well as his tools. In olden days the forests along the river banks yielded all the material necessary for the canoe: poplar for its ribs, birch bark for its outer covering, the roots of the juniper to sew the parts together and black spruce whence came the resin for the seams (181, 317). Nails were never used, the materials were fastened together, both in canoes and sledges, either with threadlike roots or caribou sinews (as earlier among the Skolts in north Finland); in that way they became elastic. Great pains are needed for this work; both canoe and sledge are subjected to great strain; the former must be able to take a load of 450 to 500 lbs, not including the two paddlers. The Montagnais' birch bark canoe was exceedingly light and graceful; a man carried it easily on his shoulders overland, and it sank only five or six inches in the water. Their length must not exceed three fathoms owing to the very curving brooks and the mountainous character of the country. Lately the birch-bark has been replaced by painted sail-cloth (317), and this is a practical change as the latter is more easily repaired. In our district the Montagnais gave up the birch-bark canoe, *wueskivi-ush*, nearly a century ago. Old white trappers at Lake Melville however remember the time when the Indians of that neighbourhood used birch-bark canoes lined with caribou skin.

The *tabanask* (cf. p. 619) seems to have undergone little change. The snow-shoes (204 c) are still of the type of olden times; they are very broad and much shorter than those of the western Indians (e.g. 19 × 33 in).

The *tepee* (1008), the sharply conical tent covered with caribou skin over which was laid a birch-bark cover, was still in use among the Montagnais only about forty years ago. The *meetchwop* [= shelter], the dome-shaped lodge, is to be seen to-day; in winter it is covered with caribou-skin on which snow is piled. A half-tent or wind-screen of birch- or spruce-bark has also been seen. On the other hand the two-fire wigwam is unknown among them (cf. however p. 655).

There was formerly no stove in the tent, only a stone hearth was built in the middle and on it a fire was made; even after the embers had lost their glow the stones retained their heat and kept the air warm until far into the night. Since the coming of the white trappers the Indians have learned for more than half a century to use calico instead of skin and birch-bark — Fig. 262 —. When the cloth-tent is well stretched it is as watertight as a skin one, and much handier to transport; as a wigwam, too, it is just as comfortable. The sheet-iron stoves are also copied from the whites; they are about a foot square at the end and 2 feet long; Fig. 262. This was a most important acquisition as it enables the Indian to camp almost anywhere, even outside the forest if necessary, because twigs and pine cones are good enough as fuel for it. A quite warm and cosy *meetchwop* can rapidly be set up. One makes a half-oval framework — a cage one might say — of bent willows (reminding one of the kuatt-framework of the nomadic Lapps) and on this is laid the sledge covers, the tent-cloth, and caribou skins. Two logs form the threshold to prevent the snow driving in. The floor is covered with branches of balsam spruce and in the middle are placed two iron stoves which are soon glowing hot and not only spread warmth but give light. The baggage, sleeping things etc. are set up against the sides to keep out the draught. These new articles have clearly been a great advantage in camping.

Certain other Indian inventions are still retained unchanged. The s u d a-t o r y   b a t h (539) is still in use to-day; a favourite pastime or 'medicine' frequently resorted to. It does not take long to build a bath-house. At almost every portage and old camping-place HIND saw, when ascending the Moisie River, the stones which had been used by the Indians in preparing a vapour bath; and BRYANT found all along the Hamilton River a number of curious wicker frames, on the same principle as that used for a meetchwop, which the natives had used in bathing. When in use the framework is covered

with caribouskins and placed over a hole in the ground, into which a number
of large heated stones are rolled; the bather then goes inside and water is
thrown on the stones, creating intensely hot vapour; (I found the same kind
of bathing-lodge in use in Russian Lapland). When the vapour-bath was taken
for curative purposes the Indian 'patient' might remain in it for hours and
then plunge direct into a near-by lake or river.

It is the favourite remedy for sickness cherished by all the ramifications
of the great Cree nation, and has played an important part in their ceremonies, too. The conjuror, sitting within the tent, frequently pretended to see
the feeding haunts of the caribou or the winter lair of the bear, or the coming
of the geese in the spring. According to LE JEUNE the Montagnais of his
time had the utmost respect for
their conjurors and were greatly
afraid of them. The picture, Fig.
264 drawn by Mr WILLIAM HIND
during his brother's ascent of the
Moisie River (434) shows the Indian's vapour tabernacle and the
credulous spectators squatting
outside waiting for the message
and accounts of the prophetic
visions of the impostor.

Fig. 264. Montagnais Indians squatting outside
the conjuror's vapour tabernacle. Drawing by
WILLIAM HIND in 1860 (Cf. 434).

As far as my information goes
no Montagnais in Newfoundland-
Labrador has erected a fixed dwelling-place, of wood or anything else; and
this is indeed surprising in view of the presence in the country of the white
trappers' numerous blockhouses, the so-called 'tilts'.

In 1676, PATER ALBANEL relates, the Montagnais were entirely clothed in
caribou skins. They did not understand the use of firearms but were very
skilful with the bow and arrow. From CARTWRIGHT's writings one understands
skin was the only clothing of the Indians from head to foot. On the other
hand HIND noted in the 1860's that when they came to the coast the thin wild-
looking Indians threw off their caribou-skin coats and leggings and were
dressed in second-hand suits of European make. They then had a dirty half-
civilized look, the agreeable wild air of forest children which they once possessed had disappeared, they became a half-civilized dandy shorn of his natural
grace and dignity. Nowadays, as has been said before, their dress is usually
composed of the ready-made clothes sold at the trading-store. Old Montagnais

at North West River said they certainly had seen Naskaupee in national costume but never anything of the kind in their own circles, not even the long chief's coat shown in Fig. 288.

Moccasins never seem to go out of use.

In hunting the gun superseded the bow and arrow as early as the 17th century. But still to-day, when the Montagnais runs short of ammunition, he uses the long-bow, a four-sided bow as tall as a man and arrows thick and blunted at the end so as not to injure the fur of the hunted animal, e.g. the musk-rat. Even in hunting the caribou the bow is sometimes used and a metal point of 15 cm is then fixed on to the arrow; this point was formerly of stone, sometimes rock crystal, and was also fixed on to a spear. The unbarbed stone arrow head broke off in the wound and worked its way on into the deer. The bone point on the other hand was securely attached. The three-feet-long arrow, spoken of by CARTWRIGHT, I did not manage to trace.

Pipes of stone, probably soapstone, *calumet*, are said to be still in use by the Davis Inlet band.

My observations regarding these matters were by no means systematic. Yet the impression has remained with me that in spite of great changes in the equipment of the Montagnais the main characteristics of their ancient culture as a whole are still there; the changes are mostly in details and have certainly been of practical use in their hunting life.

MUCH OF THE ORIGINAL NON-MATERIAL CULTURE STILL REMAINS.

It is of course impossible for a stranger to penetrate the recesses of the thought life of the Indian in a few improvised conversations, or to approach the secret chambers of the soul of a child of nature. Yet I have obtained some information from persons who have known the Montagnais since childhood which supplement and illuminate the account given here.

As has already been said the Montagnais are still pure hunters and fishers; agriculture is unknown and the amount of plants (berries) gathered is insignificant. However, many changes have undoubtedly taken place during the two centuries of merely peaceful contact with European civilization; they have lost a lot of their old customs and beliefs (514); yet one must be careful not to over-estimate the depth of the metamorphosis of their psyche due to Christian culture.

In 1636 GABRIEL SAGARD represented the Montagnais as the lowest order of the Indian races then known (435). In 1660 the Jesuit missionary, MENARD, spoke of the Indian neophytes of Seven Islands. In the 1780's we read in CART-

WRIGHT's work: the Montagnais profess the Roman religion but know no more of it than merely to repeat a prayer or two. At the beginning of the 19th century all the Montagnais are said to have been baptized, whereas many Naskaupi were still pagans (421). However, in 1860 HIND spoke of »the half-savage Montagnais and wholly savage Naskaupees» with whom he came in contact, »and a man does not get rid of his heathen notions by being touched with a drop of Manitou [holy] water». »Whether these Christian children of the forest», he continues, »have a real knowledge of sin and redemption, and of the world to come is doubtful. Have they indeed overcome the dread of the malice of the evil spirits, or do they still endeavour to appease them with offerings?» Many people still ask themselves this question when they meet the Indians, and generally the answer is uncertain (177, 869) — Regarding the Roman Catholic mission work see (130, 191, 198, 226, 250, 251, 454, 496, 498, 569, 793, 809, 904, 944, 1080, 1127).

All the Montagnais are nominally Roman Catholics, stated BRYANT, but, as the ministrations of the priest extend over a period of only one or a few days every year, there is time in the long interval without his care for the precepts of the church to be forgotten. WALLACE (995) observed in the beginning of our century how his Indian servant Pete offered sacrifice to the Manitou; when caribou was killed some bone, usually a shoulder-blade, was hung in a tree as an offering to the caribou spirit that he might not interfere with future hunts and drive the animals away. Certainly the Indian's keenest response is still to the inherited traditions and conceptions which land and water and the shadows of animals gave their ancestors thousands of years ago. In the forests he is still driven forward in haste by quite other impulses and another notion world than that of the Christian Church, it is said. In the interior the conjurors are reported still to have much influence over their followers. The authority of this savage hierarchy is said to be rather strong, and the old incantations and prophecies are still practised in the forest wilds as on the barren grounds. The Indians of the far north of Labrador especially showed but little evidence of contact with white men (995). Another thing is that a careful Indian reinsures himself in the Christian religion. The Catholic religious ceremonies must appeal forcibly to his eye and his imagination, and tend to awaken sympathies in the hearts of uncultivated natures naturally prone to the grossest superstition and ever willing to believe in superior influences visibly represented — wrote HIND.

The Roman Catholic Indian Mission in Newfoundland-Labrador during a period which will soon be a century (435) seems to be strictly confined to the two Hudson's Bay Company's trading posts, North West River and Davis Inlet, where there are small chapels. The missionary remains with his wandering flock a day or two. The religious ceremonies are strictly performed during that time: mass is celebrated in Latin in the chapel, confessions are said to be made, marriages celebrated, burial services read and baptism administered, etc., until the priest starts for the other trading-post. After that the Indians disperse and do not see the missionary again until the following year (cf. PATER BABEL's report reprinted in 435). The devotion of the Indian congregation in the little wooden chapel in North West River gave me the impression of deep sincerity; they seemed to me to exhibit the whole zealous faith of the Catholics. The priest there was also to pray for the sick, bless their guns, and collect money for this magic.

When later I met the Montagnais several times in the forests, far from the priest's watchful eye, they quite openly revealed many purely pagan instincts. It is quite certain that the drum is not only a musical instrument for them, but is also used in magic; e.g. to find out in which direction the game should be sought: some object is placed upon the drum and they watch to see in which direction it moves when the instrument is beaten, said the old chief Atvon; Fig. 255. Scapulimancy is used with the same aim; Sine Pest was pointed out as an expert among the Mishikamau band when they revolved a shoulder-blade to find out where the game was at the moment. The Montagnais, like the Davis Inlet Indians, were blindly convinced of the infallibility of scapulimancy.

The former also make no secret of their conviction that each animal is composed of a body and a spirit. Man must honour the spirit to have success in hunting. For example, to show his reverence for the spirit of the mink its skull is still always put up on a pole; if this is not done the spirit may be angered and the hunter fail in the chase. With the same object the caribou horns are placed on a scaffolding; if they were left on the ground and the dogs got a chance to gnaw them it might call forth the greatest misfortune for the band as the animal's spirit would drive the herd away from that part. Gnawed bones of the caribou, the beaver, porcupine, and bear are always thrown into the fire so that the dogs shall not get them and anger their spirits. There was a singular observance whenever a bear was killed and brought into camp. No young children or girls or young married women who had not yet become mothers were permitted to remain in the lodge, either during the cooking of the bear or during the subsequent feast. Bear flesh was only allowed to be eaten by adult males and mothers (435). — This is an interesting custom amongst the aborigines all along the circumpolar outskirts of the forest belt (393, 420) up to Lapland where the ceremonies are said to have been performed as late as a century ago among the Skolts.

The Montagnais also believe that there is a living spirit in plants and

other objects and the hunter must avoid anything that might provoke
them.

Illness is brought about by malicious spirits who have taken up their
abode within the sick person and must be driven out in secret by incantations.

Numerous terrifying spirits still range the forests, *'Man of the Woods'*,
'Winnebago', or *'Windigo'*, the giant cannibal, twenty or thirty feet tall, who
lives on human flesh they think, and there are others. In winter such beings
are less dangerous, for their tracks in the snow reveal their whereabouts so
that people can be careful and keep the fire in the tent burning; then they
do not dare to approach the camp.

When hunting has been good the Montagnais arrange a *magushan:* apparently a ritual feast in honour of the beasts' spirits. Then the drums are beaten,
a ring is trampled out in the snow, the people take hold of each other from
behind either by the waist or the shoulders and then the tribe begin to tramp
hard in the ring, both men and women losing themselves body and soul in
primitive joy. The drums continue without a break and the dancers may go
on until dawn. I have been unable to discover what is really intended by this
dancing, and my informers either could not or would not tell me. As a matter
of fact all Labrador Indians seem to use the drums a great deal; sometimes the
drums are brought out in North West River immediately the priest has departed, though more seldom there, said the Indians, for they do not like the
white people streaming out to their camp during this performance.

Left for well or ill to the caprices of life in the wilds it is no wonder that the
Indians fall victim to superstition. Dominating mountains they endow with
supernatural attributes; when passing near one on no account can they be
persuaded to point or refer to it, believing that such ill conduct will be the
inevitable forerunner of stormy weather and misfortune. Such a taboo mountain is, according to BRYANT, the Pootakabooshau, ten miles north from Lake
Melville, which dominates the entire surrounding country at Mulligan. Grand
Falls they refuse to approach; there is a superstitious dread of it among them,
they believe the place to be the haunt of evil spirits, and assert that death will
soon overtake the venturesome mortal who dares to look upon the mysterious
cataract. BRYANT stated that it would be out of the question to look for aid
in this quarter amongst the roaming flock of the Indians. There runs a tale
among the Indians that many years ago two of their maidens, gathering firewood near the Falls were enticed to the brink and drawn over by the evil
spirit of the place. During the long years since then, these unfortunates have
been condemned to dwell beneath the fall and forced to toil daily dressing
deer-skins; until now, no longer young and beautiful, they can be seen betimes

through the mist, thrusting their white hair behind them and stretching out shrivelled arms towards any mortal who ventures to visit the confines of their mystic dwelling-place (cf. 105).

And to this very day the spirits seem to have unlimited control over the Montagnais' life in the forests, surmise those who know in North West River.

The reader is further referred to the descriptions by LE JEUNE (496), CARTWRIGHT (134), HIND (435), SPECK (877) and others of both early and late times concerning superstition among the Montagnais. In these a striking resemblance is to be found not only to the superstitious beliefs of the Labrador Eskimo but also to the ideas of the nomad hunters of north-west Eurasia a century ago (burial methods, significance of the back door of the tent, bear cult etc.). The similarities have not extended to the material culture.

This complex of superstition is however the basis of many of the most beautiful sides of the Indians' ethical culture. For instance the ancestral rules of mutual assistance are still fully valid and are one of the most interesting traits of this primitive society. When the conventional signs calling for help are seen near the trail by any passer-by, he rushes to the tent of the sufferer and gives emergency help even if it is for his most hated enemy. If he did not do this — i.e. intentionally disregarded the sign — he would be unsuccessful in his hunting and the following morning the 'Man of the North' might order his wind to cover the game tracks with snow (cf. LIPS' elucidating study in 580).

It is interesting to find that chieftainship — *vuitstamau* — still survives among the Montagnais; but the power of the chief does not extend beyond the local band within which it seems to be inherited. On Fig. 254 we have the Mishikamau chief in 1937, Atvon (Pest), farthest to the right, and his son, the present chief, Panace (Pest), farthest to the left; cf. Fig. 267. The chief exercises his right *de facto* but has no special nomination by the governmental authorities. Among the Montagnais his great task is in representing the regulating will of the band. One of his prerogatives where the office is invested with authority in social and economic concerns is that decisions in respect to hunting locations of families and individuals are referred to him (879). Much discussion precedes the great hunts; no one may remain with the women in the tent, it might mean failure in hunting; the discussions are unending. The drums beat in the monotonous rhythm which rouses their primitive instincts, and the wise men try to see how to arrange things by studying the blackened shoulder-blade. Then it is the chief who puts an end to it all by deciding and giving orders. Now many formalities must be observed; hunting is not merely a way of obtaining food, it is part of a cult. To bungle at it is regarded by the Indians in the same way as creating a scandal in our social gatherings, whereas a skilful hunter is honoured among the Montagnais as a great warrior used to be among the southern Indians. There are still many rites and directions to be observed before the hunting can start and these are the duty of the chief. Ravens, wolves and dogs may not be killed, nor may a bear be shot unless in self-defence; clearly th'ey fear that the spirit of this king of the forests will take revenge.

Conclusions.

In speaking of the progress or decline of a people it is always in relation to something else, and that something should, if possible, be taken from the milieu, from the natural surroundings and from the neighbouring people who are leading about the same kind of life; never should such a question be judged *in abstracto,* or else it is one's own taste which becomes the standard.

If we now remember the metamorphosis of the Labrador Eskimoes described above (p. 528) and compare with it the changes in the Montagnais society, the progress of the former and the backwardness of the latter is clearly apparent.

The Montagnais have been unable to develop at the same pace as their nearest neighbours, the Eskimo. The whole life of the former is dictated by the hunt, he dreams of nothing but the game and their spirits; here are his greatest achievements. Hunting elicits a special talent and personal skill; the sole ambition of the Indian is to be as perfect a hunter as possible, and this becomes routine in his life. Because of his geographical position and his mental stage he seldom comes into contact with the rest of the world, and no representative of modern civilization has helped him to break out of the charmed circle of prejudice and tradition in which he lives. It has already been said that I failed to find out that any Indian in Newfoundland-Labrador could read or write or count, whereas the Dominion Governor, MacGregor, writing nearly half a century ago stated that every Eskimo child above the age of seven could do all three. This explains why there has been no development of the Indian national consciousness.

If we think of the enthusiasm of the Eskimoes in learning the Christian morals and the vital interest they take in the advantages in daily life obtainable through the Europeanized culture we must ask ourselves: what have the representatives of Christian culture given to the Eskimo and what to the Indian? To an outsider it must appear that while the Moravian Missionaries have *de facto* excellently solved the delicate problem of the spiritual peace, the civilizing and the development of the competitive capabilities of the

Eskimo, the results of the Catholic Missionaries' activities in these matters among the Indians in Newfoundland-Labrador have been poor; they have never been able to create the atmosphere necessary for development.

LIPS (580) in his study of public opinion and mutual assistance among the Montagnais-Naskaupee came to the conclusion that the unwritten constitution of the Indians is by no means rigid, but regulated for instance by changing economic conditions. As a result of the dealings with the Hudson's Bay Company entirely new and definite rules have developed, have been sanctioned by Indian public opinion and embodied in a new law, 'merchant law, and law of contracts'. Under devoted conscientious leadership perhaps the Montagnais could also be raised to a higher plane of spiritual and material culture.

Eighty years ago CHISHOLM, after living for forty years in contact with the Montagnais (434) described them as honest, hospitable and benevolent. But they lived with much superstition, he stated, which can never be erased from their minds for want of education. These words are fully applicable at the present day. There is no doubt also much truth in HIND's statement: Indian customs never change; they are like the Indians themselves; and he adds: they will all go to the grave unchanged as long as they remain heathens.

The only people who seem to have any deeper interest in these children of nature appear to be the managers at the H. B. C. stations. And I believe the more progressive Indians would themselves recognize their great indebtedness to them. Perhaps this should be explained in a few words.

European civilisation began to penetrate to the Montagnais as early as the 17th century, at first like small streams (coureurs des bois), later as a mighty torrent (Hudson's Bay Company's fur trade and other fur firms). It was the fur trade that opened up the peninsula to the trader and the missionary, and with them a new historical phase for the free sons of the wilderness: their economic freedom gradually disappeared.

Several stages can be noted in this trade; all are dominated by the same principle: the exchange of the Indians' furs for the goods of the whites, chiefly weapons, ammunition, tobacco, molasses and all kinds of fancy articles and small trash. These last especially attracted the Indian; if in the humour to barter he might give his last skin in exchange for mere rubbishy things or tobacco and molasses. The consequence was that when it came to the obtaining of his winter stocks he was quite without anything to exchange and the station had to supply him on credit. The habits of the natives have altered little with time; they are still like children in yet another matter; they treat their debts with great nonchalance. According to old custom the debts of a hunter were cancelled by his death. It was therefore necessary for the trader to have a credit contract with him if business was to be stabilized; this needed guarantors and they were usually the managers of the Hudson's Bay Company's posts; in this way the Indian was prevented from making use of credit in one place and exchanging his furs in another. This custom developed and resulted in the clever managers gradually having all the Indians in his hand. SKINNER alleged that the Hudson's Bay Company had aimed at persuading, for example, the East Crees to spread farther and farther to the north and west to exploit the fur resources there: it seemed to him fairly impossible that the game resources were sufficient for the Cree and they were thus obliged to obtain their food from the Hudson's Bay Company and give them furs in exchange; it was apparently therefore only recently that they had migrated to that part of the peninsula. Af first it seems perhaps improbable that a trading company should be able to control and direct the lives of the Indians in the limitless wilds; yet we have already seen the same phenomena in the east of the peninsula in more recent times (p. 581 sq.); the Hudson's Bay Company directing the Naskaupee now to the north, now to the south as they considered best.

This policy has come to mean that the exploitation of the hunting-grounds is in many respects regulated by the fur trader; in the district under discussion the Hudson's Bay Company. With some exaggeration it may be said that in this way the Indian hunter is becoming only a kind of intelligent hunting-dog for the trader, kept alive by him to bring him skins. But I wish specially to state that as far as I could discover the company has on the whole acted in a manner absolutely humane. It must be remenbered that they take great risks in giving the Indians credit, not least the managers personally (811).

For the Indians' civilisation, the development of their mentality and talents, the Hudson's Bay Company has naturally not had any interest. On the contrary it was a necessity of the fur trade to disperse the Indian families among the creeks and rivers (492 a), and since this began everything possible had to be done to keep the Indian a nomad. As a good trader the Company therefore also opposed the desire of the Jesuits to concentrate the people in permanent settlements and introduce agriculture, etc. (868).

There is something to be deeply deplored in the fact that these children of the wilderness are compelled by the white man to extract a scanty living from the woods in the same way as their ancestors, for the Indians are undoubtedly in danger of being ruined in the trapping country nowadays occupied by the pale-faces.

The silent struggle for the hunting grounds.

Life in the wilds trains the judgment and produces generally a good and noble human type. Nearly every day the trapper is reminded of his own littleness, conditional existence, and subordinate position in creation, so he becomes humble and helpful towards his equals, and in return seeks protection from his neighbour when matters become precarious. He learns to love his neighbour almost as himself, and therefore the white trapper has sympathy for the Indians. Most of the trappers speak the Montagnais' language fluently from childhood. In the woods they are in intimate contact with the Indians. They thus know each other. Meanwhile people accustomed to European thought processes cannot help being surprised when they hear the trappers express the most vivid sympathy for their competitors, the Indians. Although the latter have given them unpleasant experiences in the disappearance of clothes or food from their tilts, the trappers seem to resign themselves easily to such happenings and enmity rarely arises, at least not a permanent one. This seems to be chiefly due to the fact that the trappers who are mostly Anglo-Saxon and very religious have realised that the Indians' ethics and conception of life and its demands are quite different from their own, but at any rate have their good sides. They generally accept the Indians in a brotherly fashion, gladly share with them their small surplus, often to the despair of their wives when they lightheartedly give away household utensils and other home necessaries which cannot be fetched from the shop just because the Indians need them. Many of the trappers even admire the natives (cf. p. 604). The Indian on his side looks upon the trapper as his big brother; disputes occur but again due chiefly to the difference of ideas as to what is right and fair. On the whole the intercourse of the two races is marked by sincere and conscious friendship. This is the background which makes so deeply tragic the silent drama now being played out in Newfoundland-Labrador.

Now as to the conclusions to be drawn from the map, Fig. 253. It is clear from this that the Montagnais are to-day being driven by the white trappers from those forest valleys which are most rich in fur-bearing animals, and just those hunting-grounds where their forefathers roamed. Moreover the white man's area is rapidly and *progressively* increasing, the map shows how his tentacles spread deeper and wider over the country. Of course the Indians may still hunt there but it is gradually becoming harder to make a living at it and the unhindered development of this state of things is clear.

The Mealy Mts district is now closed in by the Rigolet trappers from the north, the Dove Brook trappers from the south-east and the North West River trappers from the north-west and west; the last two groups will soon completely surround it in the south-west and then tighten the ring round the bare spots on the Mealy Mts. Here the game will dwindle and the Indians be compelled to abandon the area.

Looking at the Indians' country round the sources of the Rivers Susan and Beaver and further away to Mishikamau Lake we see how the Hamilton River trappers have already closed it in from the east, south and west, and the Naskaupee River trappers from the north-east. As the trappers have also taken possession of land east of Mishikamau Lake the future development is apparent: in a few years the ring will be closed in the north and then the contraction begins; the Indians will be forced gradually to abandon this area too.

A similar fate seems to be in store for the Indian district lying between those of the trappers from Rigolet, the Naskaupee River and Hopedale; there the white man's trapping-grounds are pushing in wedges from other parts of the coast although in the absence of sufficient information they could not be shown on the map.

Here we have had some instructive up to date examples.

The map also shows how the trappers who have fur paths farthest to the west, the so-called *'Height-of-Landers'*, have begun to penetrate the Indians' last great-family areas in the Hamilton River basin, and they have to some extent broken them up. In this process I should like to see strong support for my idea that the Indians' wanderings during the hunting season round the basin of the Lake Melville water system (p. 637) — mostly a few families in company — is due to the fact that an earlier, higher social order has been disintegrated in consequence of the white trappers' invasion of the Indians' great-family areas in the forest valleys; the people in this way lost their roots and it was easy to expel them from their inherited trapping grounds.

This development is slow, but the final result on the Lake Plateau can be imagined as resembling that now to be seen in the south-east of the peninsula, south-east of the Paradise River and east of the Eskimo River. The map shows no Indian migration traces in that sector, and it is not surprising. The Indians come regularly to the mouth of the Augustine River but never farther than a river mouth about fifty miles west of Forteau Bay. The trappers of that part certainly say that sometimes the Indians come to the forest valleys there; in Chateau Bay one trapper told me it happened sometimes that up in those valleys the white men's winter stores vanished from the caches (p. 611). But in any case the Indians no longer regularly visit this end of the country.

It was stated above that the occupation of the trapping-grounds by the whites is progressively developing to the disadvantage of the Indians. Perhaps this demands some more explanation.

The fact must not be disregarded that the Montagnais' material culture, like their social structure, is undergoing transformation through their contact with the whites. They do not lack adaptation capacity (cf. the Davis Inlet Indians) even if they have not gone the length of building fixed dwellings at their chief stopping-places. Yet it has just been stated (p. 646 sq.) that the Montagnais have at bottom lagged far behind in the general development of those parts (cf. 916). So long as there was sufficient game and fur-bearing animals there was of course no need for the Indians to alter their habits and their relations to their surroundings; their daily needs demanded no further development of their material culture; the Indian hunter had already reached the climax of that stage. But the forests' resources have changed for the worse and the Indians' position with them. When the fur trade was organized and the trapping equipment — traps and guns — improved, the stock of fur animals and other game became more exploited than the barrenness of the country could bear. The Montagnais were pressed back to the thinner forests.

The white man's providence and better organisation have given them further advantages over the Indians.

We shall see later (p. 704 sqq.) how systematically the white trappers prepare for and carry out their work. The whole matter is for many reasons different for the natives.

The Indian is always accompanied on his migrations by his family. He does not like to leave them alone in the wilds, though this happens only during collective caribou-hunting. As the Montagnais does not lay up stores of food he must supply his family by more or less continual hunting. During the trapping season he may suddenly be obliged to leave his traps and go after game. When that is done and he can again go to his traps he may find that beasts of prey have been at them and spoiled the valuable coats of the booty. Of course they know the habits of the animals and the strategy of the hunting as well as the whites (579), but the latter are so much better equipped in the matter of food, and can therefore regularly visit their traps. It may be added that year by year the percentage of fur animals caught by the white trappers increases while that of the Indians correspondingly diminishes. It has happened that the Indian, when this was brought to his notice, has in childish desperation laid out his traps along the trappers' *fur paths* (644); but there is always trouble afterwards, at least temporarily, and it always ends with the Indian withdrawing to desolate parts off the beaten track which are poor in game There he can wander about with his tabanask undisturbed by the whites.

The difference between the ways of livelihood of the white trappers and the Indians is thus about the same as that between hunting and a hunter's life, CABOT has so aptly remarked. The white man's family stays in the warm, comfortable house in the village and suffers no lack of food in the absence of fathers and their sons; the Indian must drag his whole family and his wretched possessions with him on his trapping journeys which are often disappointing and unfruitful; most of them lead a dog's life most of the time and every spring starve to the edge of the impossible.

All this shows how deeply the trappers' activities have interfered with the life of the Indian hunter on Newfoundland-Labrador. Perhaps the trapper realizes it; perhaps that is what rouses his sympathy, and his helpfulness to the aborigine hunter may be unconsciously based upon it. The Indian, on the other hand, does not fully understand his situation yet; he continues his life as his ancestors did centuries ago. While the Moravian Missionaries have interested themselves in solving the problem of civilizing the Eskimoes and caring for the future of his possibilities of a livelihood in Labrador, the Catholic missionaries seem to have been quite indifferent to the Indian's subsistence and future life in this world; they seem to think that they have carried out their responsibilities for his weal or woe by dealing out some meagre alms. — (Cf. 315).

The future of the Montagnais Indians.

The question now arises, how is all this to end? The crisis has already continued several decades. CABOT noticed that the Montagnais had passed the Rubicon of children of nature: at the beginning of this century they were already in permanent dependence on their white neighbours; if they had suddenly to stop buying from the stores their lives in the spring would be very precarious. It is the same here as everywhere else on the earth: when the aborigines cannot or will not manage in the old ways with the natural resources at their disposal and have not yet risen to a higher stage of culture there is nothing but to continue their declining mode of living. In 1636 GABRIEL SAGARD stated that the Montagnais were the lowest of the Indians in the St. Lawrence area (434; cf p. 641). During the subsequent three hundred years they have not succeeded in raising themselves above the early hunting plane.

One thing seems quite clear: they are now, as they were formerly, hardly capable of adapting themselves to the white man's practice of living in permanent habitations.

The Indians are timid and they think the wilderness is limitless. The peaceful Indian hunters constantly give way to the whites until one fine day they will find themselves in the most distant and most inhospitable outskirts of the country, poor alike in fur animals and game. They can there live undisturbed as their forefathers did. Yet every new generation of trappers encroaches more upon them; it cannot be long before the white man has reached the Indian's last refuge. When this happens the latter's hunting life will be extinguished for ever in these parts (cf. 315).

Most observers consider this to be the course of development in Newfoundland-Labrador. For my part I do not think the matter is quite so simple. We have found (p. 608) that the Indians possess a not inconsiderable ability to adapt themselves to new natural conditions; we really know nothing of their ability to adapt themselves to new social conditions.

It is therefore too early to paint too darkly the future of these Indians; it may happen that great changes occur in the trappers' ways of living. A large sawmill is being planned in North West River. It may be that in a few years one of the greatest power stations in the world will start working at Grand Falls in the Hamilton River; then already a railway will connect the iron ore mines on the Lake Plateau with the North American network. Such events cannot but react upon the lives of the trapping communities. It is difficult to say how great will be the attraction for the younger generation of fur hunters of this great industrial development with its certain preference for white employees; and it is therefore impossible to make any responsible prophecy as to the future fate of the Montagnais.

— In August 1939 I had a conversation with one of North America's foremost experts in the sphere of fur animals' trapping. It then appeared unnecessary to be too pessimistic about its future in Newfoundland-Labrador. It is probable that the trappers in the east of Labrador will be compelled to give up the country to the Indians and take up others methods of earning their living as has already happened in many parts of Canada, by which means the natives have had every possibility of continuing their capture of the fur animals; they are more accustomed to the country than the whites. Then too experience in Canada has shown that Indians conform better than the whites to the regulations issued by the authorities for the purpose of

protecting certain fur animals and are there better able to hunt them systematically, said the expert (1082 a). He continued: it is probable that it
will still be the Indians after five generations who are following a profitable
trapping occupation in Labrador's endless wilds. And farther into the future
than that we need not look.

Davis Inlet Indians.

This mixed group of Indians is named after the Hudson's Bay Company's
station where they barter their furs. The nucleus of this band — cf. Fig. 252 —
came from North West River originally as we know (p. 6C8), and it preserves
in parts its linguistic and cultural particularities. Intermarriages and shifts
of residence have tended to keep the band well mixed. The Davis Inlet people
have thus later been much influenced by the Barren Ground band (cf. p. 672)
with whom they have intermarried and associated for several generations,
according to STRONG (909); formerly, when the caribou were abundant there,
they often lived together on Indian House Lake. The description on page
667 sqq. of Barren Ground-Voicey's Bay band among the Naskaupee should
also be compared here. Also between the Davis Inlet band and the Naskaupee of Chimo band there was much intercourse.

In this band the constancy of number of individuals is lacking which is
distinctive of the others. In 1926 STRONG found it to consist of 36 persons.
They mostly stayed in the interior and only now and then were seen at the
trading station; Fig. 265. The census of 1935 shews 81 individuals of which
38 were married, 2 widows and 41 unmarried. In the middle of August 1939,
when I met the band at Davis Inlet — Fig. 266 — and had a talk with their
chief, Susapish or Joe Rich, — Fig. 267 — it comprised 107 persons. This
rapid increase in population clearly shows that the part of the country where
they move about has a special attraction for the Indians. Its composition
was peculiar, for besides the predominating descendante of the Montagnais
from the North West River group there were three Seven Islands men with
their families (from the Mishikamau band) — Fig. 268 — and, surprising to
relate, two Naskaupi from the Ungava or Chimo band — Fig. 269 — (one by
marriage with a Naskaupee girl). Chief Susapish said he was sole ruler over
them all.

WALLACE met with some of the Davis Inlet Indians in the middle of September 1905 at George River, at about 55° 20′ (995). The encampment consisted

of two deerskin wigwams. One was a large one and oblong in shape, the other of good size but round. The smaller tent was heated by a single fire in the centre, the larger one by three fires distributed at intervals down its length. Chief Torna occupied, with his family, the smallest lodge, while the others made their home in the larger one. WALLACE's party was discovered long before arriving and was met by the whole population, men, women, children, dogs and all. The reception of the picturesque group was tumultuous and cordial. The swarthy-faced (at least eleven) men, lean, sinewy and well-built,

Fig. 265. Hudson's Bay Company's trading post at Davis Inlet; looking north. Photo the author, Aug. 1939.

with their long, straight, black hair reaching to their shoulders, most of them hatless and wearing a red handkerchief banded cross the forehead, moccasined feet and vari-colored leggings, the women quaint and odd, squatty and fat, the eager-faced children, little hunting dogs and big wolf-like huskies.

These Indians still lived in the crude primordial fashion of their fore-fathers. They had not accustomed themselves to the use of flour, sugar etc., their food was almost wholly flesh, fish and berries. They had adopted shot-guns, but the bow and arrow had still its place with them; the boys were very expert in its use. Some of these people had never seen the priest, WALLACE reports (995). Already then the Eskimo sledge had obtained an entrance among these Indians. From Atuknipi on George River there seemed to be a trail to Davis Inlet (995).

Nowadays, when the hunting period begins in August the Davis Inlet band divides into smaller groups of about twelve persons who move into the interior. Some go via Sango Bay, some via Adlatok Bay, and it is round the sources of the Adlatok River that the latter prefer to remain in the winter, though they sometimes go as far as Snegamok Lake. It also happens that they look in at Makkovik and Kipekak. Often the hunting goes even nowadays as far as the George River.

In the valleys running down to the Atlantic the hunters of this band also

Fig. 266. Part of the Davis Inlet band Indians' camp at Davis Inlet; looking south.
Photo the author, Aug. 1939.

come upon areas where the Eskimoes have laid out traps, and this is a great trouble to them as it compels them to make their way farther into the country to find unoccupied land and thus avoid conflicts. Circumstances (place names in the islands; cf also p. 582) indicate that the Montagnais in Labrador formerly extended their trapping-grounds to the ocean shore, just like the Beothuks in Newfoundland in olden days. I was unable to find out how the Indians lived at that time; CARTWRIGHT alone gives some information on the subject; they seemed to have been in their canoes out to the most distant islands to collect eggs.

From what was to be heard at the trading-station of Davis Inlet it may be concluded that the hunters of this band are in general not very enterprising;

when they were starving they joined up with the Barren Ground or Voicey's Bay band (Naskaupee).

Their habits seem in no vital respects to differ from those of the other Montagnais; yet the caribou is their chief spoil and they prefer caribou meat to all the food of the whites.

The Naskaupee Indians.

While the Montagnais can be most expressively termed the taiga people, the wild men of the primeval woods, the woodsmen, the Naskaupi are the children of the sub-arctic tundra with its wooded valleys and the river landscape, and are therefore sometimes called 'the river-men'. Their life is made up of wandering in the tracks of the caribou. With an allusion to the Eskimo, the raw meat eaters (p. 454), the Naskaupee have also been called: 'the caribou eaters'.

The name Naskaupee is differently spelt by different writers: E s c o p i c s (CURTIS), C u n e s k a p i (PATER LAURE), N a s q u a p e e (HIND), N a s c o-p i e s (McLEAN); here the spelling N a s k a u p e e (SPECK) is used.

The origin and meaning of the denomination is not quite clear. FERLAND (252) says that the heaps of stones 9—10 ft high set up along the Gulf coast to serve as navigation marks or to indicate the position of dwellings used to be called naskaupees. Another explanation, however, is more probable. It is well known that many natives, especially of Indian stock, go under the name given them by their neighbours or their enemies and this has often an opprobrious meaning of which the bearers have no idea. The name Naskaupee, now generally accepted, seems to have begun as a term of abuse; an expression of the Montagnais' contempt for their wild northern brethren; some say it is equivalent to 'pagan', others that it indicates a 'false, unreliable creature', 'helplessly lost in poverty', 'whose mental gifts are so little developed that he even lacks the conception of individual property', or on the whole an ignoramus with a tinge of falseness. The Montagnais, alleges TURNER, are supposed to have given their former allies this name after they had acted traitorously towards them in a common feud against the Eskimoes of the east coast. HIND translates it as 'a people standing upright'. Their neighbours in the west call them simply *Msuhau-au-eo*, which means 'tundra people'.

The Naskaupee themselves will not pay any attention to this name; they call themselves N e n e n o t which has been translated as 'the real' or 'the ideal red man', or 'our own people'. They use the denomination naskaupee for an evil spirit which leads people astray in a snowstorm.

42

Some outstanding features.

I was prepared to find these Indians very primitive and I was not disappointed.

Already their outward appearance indicates poverty and primitiveness. Never before have I had so overpowering an impression that I stood face to face with people of the stone age, or perhaps even cave people; it was at the end of July 1937 at a camp of mixed Montagnais and Naskaupee beside Davis

Fig. 267. Susapish, the chief of the Davis Inlet band, with his family at Davis Inlet. Photo the author, July 1939.

Inlet. It was an attractive, selected party who had come to meet the Catholic priest: thin, supple men, Figs 270 and 271, with unbelievably long legs, slim women with flashing glances — Figs 272, 273, and 274 — and shy children — Fig. 275 — all appeared in their savage, primitive dirtiness; they are not nearly so clean as the Montagnais. Although it was the height of summer many of them were clothed in caribou-skin parkas, a bloused gown of caribou-skin with a hood, Fig. 275.

Their features were mostly hard, the glance searching and almost scornful when looking at a stranger; Fig. 276. Yet many of them had rather delicately formed features, small hands and feet, a large and rather soft eye inclined towards the nose; the hair was intensely black, coarse and thick, the teeth regular and beautifully white (435); Figs 272, 273, 277, 278, 283 and 279.

At first they were rather shy and silent but that soon wore off and they laughed and talked gaily; the natural joyfulness of primitive people broke through; they shook hands in a most friendly way, laughed and joked and all talked at the same time. When they raised their voices the men spoke in a curious, flat-toned falsetto. The impression was one of surprising attractiveness in contrast to my first meeting with the mild, melancholy, but repulsive Montagnais. From the first moment it was clear that I was meeting a fragment of mankind who had to a great extent been preserved from the influence of Europeanizing culture.

The primitiveness of the Naskaupee is a natural consequence of their isolation from the influences of the civilized world.

I have found nothing in the literature regarding the work of missionaries among the Barren Ground Lake Naskaupee. During the time when the Montagnais were being already strongly influenced by the Jesuits the Naskaupee Indians were not at all known to the white. The Lake Plateau Indians certainly had no contact with these Roman Catholics before the beginning of the 18th century (LAURE, 1731). On the other hand, only lately have the northern

Fig. 268. Montagnais of the Seven Islands band (note the embroidered cap) who has joined the Davis Inlet band at Davis Inlet. Photo the author, Aug. 1939.

Naskaupee been visited by a missionary. BABEL (27) was with them in 1866 and he is said to have been the first priest to travel from the St. Lawrence River to Ungava Bay. In 1877 the Indian missionary PÈRE LACASSE went to Ungava, but he had no success; he arrived just in a year when the hunting was very bad and the Indians starving, and the latter believed that the presence of this priest was the cause of the disasters. No one was converted. In 1905 the Rev. Samuel M. Stewart of the Church Missionary Society of England was at work at Fort Chimo (995). Today the southern Naskaupee at least are christianized.

Hudson's Bay Company's people in Fort Chimo and other trading-posts have probably had more opportunities of influencing the Indians than either missionaries or travelling investigators. But it has scarcely been in the interest of the fur trade to raise their level of existence; cf. p. 812. Mc LEAN (610) after his long experience as manager at Fort Chimo said in 1849 that the Naskaupee were selfish and inhospitable in the extreme, they demanded payment for everything, but he adds — and that is interesting for us —: that this they must have learned from the white man as they were very generous to each other. FATHER ARNAUD, who in 1854 passed a winter among the southern Naskaupee, reports that they were very good to each other; they liked to meet and help each other.

Fig. 269. Naskaupee of the Ungava band who has joined the Davis Inlet band at Davis Inlet. Photo the author, Aug. 1939.

There has been little intercourse with their easterly neighbours the Eskimo until quite lately; this may seem surprising when their districts are only separated by a comparatively low range of hills, but is probably due to century-old feuds, (p. 441 sqq.); much is related of them in their folk-lore. At the beginning of the 19th century TURNER said the Eskimoes rarely ventured into the interior, unless along some of the great rivers falling into Ungava Bay. There, however, it sometimes happened that some Eskimoes camped beside some Indians. The old Indian men and women left behind with the children while the others were out on the winter-hunting were more than glad to have jolly Eskimoes so near from whose stores they could profit until the hosts themselves were quite destitute of food and moved away to a quieter place.

In any case not so few similarities can be observed to-day between the northerly Naskaupee and the Eskimo, both in the sphere of material and non-material culture (1002), and this must be traced to the cultural borrowings. The *ulo*, the woman's knife (*cagwi'igon*), is also used by the Naskaupee in tanning, this is no doubt a loan of late years (p. 529).

The old life customs, clothing, tools and dwellings then remained until recently among these Indians.

When CABOT visited these people at the beginning of the current century their clothing in winter — and apparently all the year round for some of them — was of caribou-skin, the hair inside, and the outside of the finely-dressed chamois-like skin decorated with various designs in colour: blue, red and yellow. Following tradition they wore an under-shirt of the silky skin of unborn caribou, and over this was a parka. Moccasins and leggings of the same caribou-skin were worn on their bare flesh. The beautiful chief's coats of white shamoyed caribou-skin, with quasi-magical (?) ornamentation in red and blue painting were masterpieces of primitive art; Fig. 288. When the garment had a hood it was sometimes the skin of a wolf's head, with the ears standing up and the hair outside, giving the wearer a startling and ferocious appearance. Tight-fitting deerskin or red cloth leggings decorated with heads, and deerskin moccasins completed the costume (995). Also the woman's dressing of the same materials was most skilful; cf. Fig. 289.

CABOT says that among the south-eastern Naskaupee the wearing of trousers only began about the beginning of this century, when many of them wore them under their leggings. But he alleges that the people of the George River still preferred the old way. Sleeping-bags were generally made of caribou-skin and their covers of hare skins. But even then the George River people were beginning to abandon their national dress; their taste of olden times had deteriorated.

Their artistic sense of bygone days has now disappeared or at least is not to be compared with that of the Eskimoes. One can see moccassins neatly ornamented with coloured beads or porcupine quills worked with silk into flower designs, and caps embroidered with glass beads, but that is all. Nowadays it seems that ready-made clothes have got the upper hand, at least in summer-time. Many of the men I saw had a kind of sports cap on their heads, whereas formerly they wore no hat, only pulled the hood over the head. This hood is similar in shape to the Eskimo's kulitak.

Measured by economic and social standards the Naskaupi must be considered among the most primitive peoples of the earth. But this opinion must not be regarded as a disparaging one in their own environment, for in the godforsaken regions where they live an unusually high degree of buoyancy and inventiveness is demanded even to keep alive, and these qualities they certainly possess. The isolation from contact with the white people and the conditions of nature under which they live, have evidently been contributory causes of the retention of their primitiveness which, as we shall see (p. 697), once resulted from a process of deculturisation.

It might be thought that the race could be kept pure in the wilds where the Naskaupee live. Yet they are not so unmixed as their appearance may indicate at first glance. According to WAUGH's notes the original Naskaupi element should be a mere remnant now. His list of names showed twelve men

who could be considered Naskaupee, the remainder, four, being Montagnais.
A number of the women also were Montagnais. The language spoken showed,
I believe, WAUGH continues, some confusion with Montagnais. Two at least
of the Ungava additions to the band showed unmistakable signs of Eskimo
admixture; one of these men spoke Eskimo quite well. Some of those who
visit Davis Inlet are known with certainty to be mixed with the Montagnais.
Three or four families are descended from half-castes (RICH, cf. Fig. 267), but

Fig. 270. Naskaupee and Montagnais of the Davis Inlet band waiting for their priest at
Davis Inlet. The hood is drawn because of the plague of mosquitoes. Photo the author,
July 1937.

in the third generation they are completely Indian both in speech and habit;
some of them can speak a little English but this does not conceal their Indian
type.

As far as my information goes these Indians and Eskimoes intermarry, but
this may be rare in spite of the tolerance prevailing to-day (cf. p. 443 sq.). Only
in a few cases have I found that an Indian has taken an Innuk to wife, but
never an Eskimo a squaw. Yet among the Voicey's Bay band there is at
present an Indian with Eskimo blood, and many of the others can speak the
Eskimo language. It has happened that Naskaupee, especially children, have
stayed for a time in an Eskimo camp, but they have never been assimilated
and have returned to their people.

The Naskaupee's health conditions seemed to leave nothing to be desired, Fig. 284. They keep to their traditional domain and thereby avoid contagion by the epidemics which now and again circulate along the coast. The Naskaupee are still said to be hardier than the Montagnais. They bear cold as well as the Eskimo. The little children show astonishing indifference to temperature; they sometimes run barefoot in the worst cold. The birth rate is said to be good and the natural increase of numbers is satisfactory; the children are brought to birth as easily as the young of wild animals. However, they could not avoid the diseases completely. For instance in 1919 they suffered much from smallpox and even scurvy due to want of fresh meat. It is these epidemics and the awful starvation which has reduced the population (cf. p. 592 sq.).

Fig. 271. Naskaupee Indians at Davis Inlet. Photo E. H. KRANCK, July 1937.

These Indians, of course, know nothing of what we call hygiene; the wilderness requires nothing for the health but food and clothing. They scarcely ever wash themselves unless sometimes the face; the mosquitoes bite and the cold stings worse when the face has been freed from its natural fatty secretion. Instead they use the vapour bath frequently, they really love it. It seems to be not unusual that when all arrangements for a move have been made after long discussions, the head of the family suddenly feels he must have a bath and then everything else must stop until he has finished.

When painful warnings are felt of rheumatism or other illness they like to have a bath. For pains in the limbs, etc., often a needle or similar sharp instrument is stuck into the veins, to bleed the person. The Naskaupi have also many medicines of their own composed of different herbs, among them an effective means of artificial abortion. However on the whole the effect of such remedies is probably more magical than physiological (see CABOT on this point). — Most of these customs are forcibly reminiscent of old popular customs in the subartic regions and are also known in northern Fennoscandia.

It has already been said that the distinction between the Montagnais and the Naskaupee is rather superficial. Fifty years ago TURNER stated that the language of the Ungava Naskaupee was a little different from that of the Montagnais. According to what some Montagnais told me, the language of the Barren Ground Naskaupee differs so much from their southern kinsmen that they had sometimes difficulty in understanding each other, a fact showing that they represent different clans which have lived apart for a rather long time. HIND, on the other hand, stated that the southern Naskaupee, who then lived at the Lake Plateau, could communicate with the Montagnais without any difficulty.

Fig. 272. Naskaupee squaw with children. Photo the author, Davis Inlet, July 1939.

The Naskaupee speak very quickly and with a curious modulation of the voice reminding one of a fretful child, and they often shriek out what they have on their minds; an outsider might think they were quarrelling and highly excited when they are discussing ordinary subjects (995). Their voices are penetrating and carry far; TURNER says that on calm days they will stand and talk to each other, one on each side of a broad river. They can also imitate the cries of animals so well that the latter do not discover their source until it is too late; they can even modulate their voice to lure the shy geese from their hiding-places.

It was only possible for me to collect on the spot some comparatively scanty information regarding the Naskaupee and their way of living, but that was from some of themselves and some Hudson's Bay Company functionaries. But if this is supplemented by the observations of McLEAN (610), TURNER (963, 965) and CABOT (119, 123), WALLACE (995), WAUGH (1002) and STRONG (909) during their stays among these people it appears that a good deal is known of their condition and customs, and from that it is possible to form a conception of them generally both in former times and at present. The above reports I have freely used in the following account.

The Naskaupee have divided themselves into three bands.

The once vast extent of the country hunted by the wandering Naskaupee may be conceived when, 180 years ago, one finds this people side by side with their allies the Montagnais on the Saguenay, and 100 miles west of the Straits of Belle Isle, places from 800 to 900 miles apart. Since then their country has been gradually reduced and they have finally found themselves in the most inhospitable tracts of the Labrador peninsula.

The Naskaupee, Algonkin Indians' outposts in the north-east, are divided into three territorial clans, or perhaps rather population groups, in agreement with the geographical character of the Labrador Peninsula; cf. Figs 222 and 252:

in the west: White Whale River or Mackenzie River group, whose district borders on Hudson Bay,

in the north: Ungava or Chimo group in the upland of Ungava Bay,

in the east: Barren Ground River [= George River or Mushauan Shibo] or Voicey's Bay group in the Atlantic upland.

It is wise to remember that many of the accounts which have been written of these people refer to bands living many hundreds of miles apart, and it is therefore no wonder that differences occur in the descriptions of their habits and customs.

Fig. 273. Naskaupee squaw with children. Photo the author, at Davis Inlet, July 1939.

McLEAN and TURNER and WALLACE tell us about the Naskaupee who hunted in that portion of the peninsula which is styled Ungava; CABOT, STRONG, and WAUGH met the Barren Ground Lake people, who traded in Davis Inlet and Nain; PATER LAURE, PATER ARNAUD and HIND had intercourse with the southernmost Naskaupee who once lived at the Lake Plateau; CARTWRIGHT gave some scattered notes about those who more or less occasionally seem to have appeared on the Atlantic coast between Hamilton Inlet and the south east corner of the peninsula.

Of the above bands the first and the second live outside Newfoundland-Labrador; but as the second band in many cases resembled the third in olden times we shall also consider it here. Naturally the main interest is centred in this last group.

An interesting characteristic which more or less vaguely appears among the Naskaupee is that the bands are still composed largely of kinsfolk, most often a man's sons with their families; a postmarital residence may be either patrilocal or matrilocal. For this reason the principle of the related clans can be thought to be the basis of the division of the bands, as well as the geographical conditions. In recent times however there has been a shifting of families from group to group within the band, or from bard to band. PRICHARD reports that the Naskaupee of the Barren Ground band were for many years in the habit of coming out to the Atlantic coast at Davis Inlet to barter. But in 1910 for some reason they made a change, a trading party of fifteen, of whom some four or five were Montagnais, came out to Voicey's Bay, and, abandoning their canoes there, they hired the resident settlers to take them by trap-boat to Nain to transact their business at the Moravian Mission's store. They had never before brought their women to Davis Inlet but three accompanied them to Nain. WAUGH noted that among the Voicey's Bay band (see later on) in 1921 were also some Naskaupee from the Mackenzie River. In the Voicey's Bay tribe there are actually some Ungava Bay people and at least one Montagnais. This circumstance makes any generalisation impossible.

Fig. 274. Naskaupee widow visiting Davis Inlet. Photo the author, Aug. 1939.

It seems to me that here is not only a progressive change in the composition of the population of the small groups, but also a symptom which betokens the breaking-up of the old social order.

Changes in the material culture confirm this. As a whole the old hunting culture as it was at the end of last century and even a few decades ago takes us back to a considerably older stage of development than the one we know among the Montagnais. Since then, however, habits and customs among the Naskaupee have also been subjected to thorough changes at a forced pace and now they are beginning to resemble those of the Montagnais. This transition would be well worth a special sociological study.

Barren Ground Band, later called Voicey's Bay Band.

Few of the Indian bands show so clearly as these the spontaneous changes in customs which come about in the lives of aborigines and revolutionise their way of living when they come in contact with white people. The line between an older phase, B a r r e n G r o u n d p e r i o d, and a younger phase, V o i c e y's B a y p e r i o d, seems to be about the year 1912. (Cf. 1002).

THE BARREN GROUND PERIOD.

Fig. 275. Naskaupee boy in kulitak of reindeer-skin with the hood drawn against mosquitoes. Photo the author, Aug. 1939.

In the Barren Ground Naskaupee at the beginning of our century we meet a true hunter people dependent upon game only for their livelihood. In those days they were still managing to live upon the country's natural products. Among the objects of their hunt the caribou was incomparably the most important. Its flesh was their food, their clothes were fashioned from the pelts, the hair worn inside, they sewed with the sinews, the lodges were covered with hides of stags. Shamoyed, slightly smoked caribou skin was their most important object of barter. Yet the herds of caribou were usually accompanied by white foxes and these, with the red foxes, were hunted in the winter and so gave the Naskaupee some extra products to exchange. The prevailing opinion however seems to be that the Barren Ground Naskaupee were never very interested in trapping. Nor did they like the long journeys to the trading posts. It was the caribou that absorbed their thoughts, and they never seemd to care about any great effort to procure furs to exchange for the necessaries which it was possible to obtain from the Hudson's Bay Company stations.

As stated above (p. 582 sq.) the Naskaupee were pushed north by the Montagnais as the latter retreated before the onslaughts of the Iroquois about a century later than the occupation of Canada by the French. They retreated through the wooded south, and their remorseless pursuers were not shaken off until the tribe had been pushed up on to the rolling, to a great extent timberless plateau of the interior of Labrador with its innumerable lakes and mountains. After many wanderings the Naskaupee must have found a veritable Land of Promise in the valley where the Barren Ground Lake is. The country was virgin and caribou plentiful. Spruce, alder, tamarack and birch still grow here in the sheltered hollows, woods and marshes are about evenly distributed. The tangled bush housed coveys of willow grouse, while the very rocks yielded

Fig. 276. Naskaupee mother and child. Photo the author, July 1939.

the crouching ptarmigan. There were also hares, Canadian geese, ducks, foxes and wolves. Game on the hills, fish in the rivers and lakes, wood for their fires and tepee poles — what more could they desire?

They learned to live here by the caribou, and while the herds kept to the old migratory routes all went well for the tribe, living secure behind their impenetrable rampart of wilderness, gorging and starving alternately. Thus they existed until the dreadful crises came when the deer and the game failed, as for instance in 1893 (761).

As late as the eighteen-ninety's it seems that this group alone had appropriated the hunting within the wide area north of parallel 54° 30" and south of parallel 57° 30″, reckoned from the ocean away to a western boundary which seems mainly to have coincided with the watershed west of the George River. I have however found no proof that they also extended their hunting to any notable extent west of Indian House Lake. As already stated (p. 608) the band gave up the southern part of this area after it had been destroyed by great forest fires, when the caribou herds abandoned it, and now the boundary can be regarded as stretching approximately from Resolution Lake along the Notakwanon River away to the Atlantic. But as we shall see the boundary is fictitious.

The centre of gravitation for these Indians at the turn of the century was the so-called Barren Ground Lake or Indian House Lake [= in older descriptions Lake Erlandson?], an expansion of the George River; cf. Figs 182 and 280. It is somewhat more than fifty miles in length and only one to two miles wide and is shut in on either side by high hills which on the west usually lie a little back from the lake. Long wedge-shaped points — eskers — reach here and there out from the west shore and in some places slope back to the hills in high terraces — fluvioglacial deltas —, the highest more than 150 feet above the lake. On either side of the George River towards the south rise high, rolling ridges dotted with spruce, willow and tamarack. The whole land surface there is seamed with the old trails of caribou that had

Fig. 277. A happy Naskaupee squaw with her baby at Davis Inlet. Photo the author, July 1939.

pursued their age-long wanderings in this desolate region, PRICHARD observed, and each promontory showed remains of deserted Indian encampments.

One of the chief places on the route of the immense caribou herds had been, from time immemorial, at the narrows of the lake, and therefore the Indians always collected here in late summer; it seems to have been such an important centre and for so long that Hudson's Bay Company set up one of their trading stations on the northern shore of the lake: Erlandson Lake Post [?]. As late as 1905 WALLACE wrote that Indian House Lake was a general rendezvous for the Indians during the summer months when they congregated there to fish and to hunt reindeer. In the autumn they scattered to the trapping-grounds, where the fur-bearing animals were found in greater abundance.

PRICHARD noticed that about August [1910] the Naskaupee used to pitch their *tepees* deep in the interior. The chief event of the year, the autumn killing of the migrating caribou, should now take place. From ancient times they had gathered to slay the deer near Barren Ground Lake while the animals crossed the lake on their mysterious journeyings. The hunt took place around the lake and south of the lake along the George River valley as well as east of the river and lake. The hills near the lake were especially cut up by caribou paths and the banks covered with the shed hair of their spring coats. The country round Indian House Lake is reported to have been a veritable place of slaughter (758); all the travellers in this part of the country have noted great piles of caribou bones and antlers.

Fig. 278. Naskaupee beauty of the Davis Inlet band. Photo the author, Aug. 1939.

Let us now look at the great annual climax in the hunting life of these people, the great caribou hunt which took place every year during the latter part of the summer or at the beginning of the autumn when the great caribou migrations of the type still known in north-west Canada began. Both the time and the place for the migrations varied somewhat from year to year and a certain strategy had been developed, forming one of the most systematic hunts I had ever heard described.

About the time the migration might be expected all was activity in the Indian camps. Scouts were sent up and down the George River and when the leading caribou made their appearance signals were given by fire or smoke to the other watchers, for the first deer must not be touched, otherwise the herd would turn away. All must keep silent and quiet until a fair number of the advance guard had crossed the river. As mentioned above the principal hunting-passes were near the narrows of Barren Ground Lake, opposite the large esker landscape with much moss on the hills and grass in the swampy hollows. On all strategic places on the western shore of the lake large wooden fences had been built, thus forcing the herds to reach the shore in definite places, from where they began to swim across (cf. also CARTWRIGHT, 134). On the eastern shore opposite these places were the old men and women, who all day long sat watching like falcons, with pipes hanging from their dribbling mouths. When the caribou finally came in sight at the opposite shore they gave the hunters a sign. There was a whisper: *athen* (= caribou), and the primitive hunting passion was released. Everyone became wild with excitement. All control was lost and the result can be imagined.

The leading caribou approaches the lake and plunges in, the herd follows. When the deer are swimming in the lake, the Indians in canoes divide into two parties; the one behind it cuts the flock off from the land, the other spreads round the sides of flock, encircling it. The herd caught between the two is confused and mills round and round; the flock is driven backwards and forwards. Right into them, even up on the backs of the terrified caribous the hunters drive their canoes. They spear right and left, directly into the kidney of the animal and thus stabbed it gives a few kicks and dies at once. The Indian in his lust to murder is said to surpass the hereditary enemy of the caribou, the wolf: It is a case of killing in passing. It is said that they slay as many as a thousand, a terrible massacre of the helpless swimming creatures. The slaughter is also idiotic because not all the flesh is saved; from some animals only the tongue is taken (cf. 1009).

The chief means of support for these hunters was thus that of the prehistoric period, although they were well off for guns. In favourable years the spear alone furnished anough caribou for their main support.

At one time hundreds of caribou carcases lay piled up and decaying at the traditional camping-place of the Barren Ground Naskaupee at the mouth of the Tshinutivish Brook, Fig. 281, which runs from the east into the great lake, writes CABOT. PRICHARD wrote that the whole country near the George River was covered with caribou bones. Another noticeable fact was that rarely could one look round the sky without seeing flocks of ravens, he added. WALLACE, too, (1905) wrote that the George River all the way down to the Barren Ground Lake had been in past years a veritable slaughter house.

Fig. 270. A fine Naskaupee rascal at Davis Inlet. Photo the author, Aug. 1939.

There were great piles of caribou antlers, sometimes two or three hundred pairs in a single pile, where the Indians had speared the animals in the river. Abandoned camps were distributed at frequent intervals. Countless multitudes of caribou have no doubt been slain in this manner in Indian House Lake. The hunters might kill a whole herd, leaving nothing for breeding purposes. No judgement was shown, and yet they must know better than anyone else that to clear the country of these animals is to face starvation. The band has also learned that Labrador is a bitter mother; there were few years when one month or another did not see the Naskaupees face to face with famine, this is more especially the case just with those Indians who pass their lives in the most remote and poorest part of the peninsula.

The caribou are erratic in their movements, as has been stated several times, and can never be depended upon with any degree of certainity. And should the Indians fail in their hunt they are face to face with starvation, as was the case in the winter of 1893—94, when fully half of the people perished from lack of food (955). Nor is it easy to judge how the caribou will move during the year; many causes may alter their direction. According to information obtained by CABOT when visiting the Barren Ground Indians at the beginning of the present century, their northerly migrations have for a long time extended to near the coast. After the great forest fires mentioned above the migrations shifted towards the interior and the Indians had to keep watch on several other hunting-passes. One such was in the great Lake Mistinipi (cf. CABOT's

Fig. 280. Looking upward over Indian House Lake from about the Tshinutivish Brook (761).

map in 119) where the sound over which the caribou often swam was a third of a mile broad, and six to eight miles from the east end of this lake.

But the Indians also collected to hunt caribou at other times of the year. For instance the country round about the head-waters of the larger riverways were common meeting-grounds in the old deer-hunting days at the fording season of the migratory caribou.

Even then, it was said, the game was unlimited and therefore Indians from Ungava Bay also joined them sometimes though they had fine hunting-passes on the Koksoak River, and it is also not surprising that such attraction drew thither some of the Davis Inlet Group, especially the so-called Shango Indians, or the Rich family; they lived almost as parasites on the Naskaupi (p. 657). There was plenty of room for all; about 300 square miles per person. Furthermore this area could easily be doubled if the extremities of their recorded wanderings be taken as indicating their territorial limits (909).

Small parties of caribou were also hunted in summer. The horse flies and mosquitoes which are such a plague to the white man are friends to the Indian hunter, for they cause the caribou to gather in small parties.

Deer decoys were often used, i.e. groups of large erratic boulders, each boulder topped with a smaller one. Viewed from a distance they look not unlike a herd of grazing deer.

These hunting methods are known all over arctic Eurasia.

The Barren Grounds (749) are to be considered rather rich in food in summer, but no place is poorer in winter; so that the band was compelled to retire to the woods in the late fall. There they passed the season hunting caribou, small game and some fur-bearing animals. PRICHARD says that in August the Indians shoot the young Canadian geese, spruce grouse and ptarmigan. Trout also were eagerly sought after by the Indians when the caribou were scarce. This fish was also caught through the ice, a weary work requiring patience and endurance. Nets and fishing lines of their own making, either of babiche or twine, rarely failed to take food enough during that part of the year when caribou flesh was un-

Fig. 281. The Barren Ground Naskaupee's camp near the Tshinntivish Brook at the eastern shore of Indian House Lake, Summer 1905 (119).

obtainable. Especially in early spring, when hunger more or less regularly troubled the Indians, the unfailing wooden hook of ancient days dropped through the ice-hole helped them over the crises. The hooks were made of wood and caribou bones, consisting of two pieces about four inches long tied together in the middle which, when the fish bites and the fisherman strikes, separate and stretch across the jaws of the huge trout.

As has already been stated none of the Barren Ground Naskaupee were specially prominent as trappers at this time; the fur animals they caught were only a few ordinary and arctic foxes. Therefore some energetic trappers from the Whale River (= Mackenzie River) Naskaupee (from Hudson Bay, they trade at Fort George) used to join the group for this purpose.

Tools and implements were largely of bone and antler: arrowheads, fish-barbs, skinning tools of bear tibia, cut obliquely, hair-scrapers of caribou leg bone sharpened at one side, etc. In woodwork there were shallow bowls, spoons with flat wide bowls, needles for sewing hare-skin garments and robes, net needles, canoe mallets, knife- and awl-handles and net-floats. Shelter was provided by the conical, many-poled caribou skin tepee; Figs 281 and 282. The hunters' temporary shelter was an open-topped, head-high windbreak of skin thrown about the poles. Light came from wood fires, protection from cold depended solely upon the immense fires built low within a trench in the snow.

The likness to the material culture of primitive Lapps is very striking.

Fig. 282. Evening effect in the Barren Ground Naskaupee's camp at Indian House Lake in 1905 (119). The construction of the scaffold is precisely the same as the Lapps' 'suonjer' and the tepees as the Lapps' 'lavvo' (watch tent).

Judging from the information I collected a primitive kind of communism existed among these Naskaupees even in times very recent. McLEAN, nearly a century ago, said that nobody could show so little selfishness towards his equals as these Indians. CABOT, who knew their conditions well, gives us some examples of this, and says that it would be difficult to find a people more devoted to their own tribesmen and more ready to make sacrifices for them than the Naskaupee. This mutual helpfulness finds expression not least among their hunting regulations which are very ancient in type.

The Barren Ground Naskaupee have never had individually owned hunting districts. They always hunted in company and the booty belonged to the group as a whole (123). They had usually plenty of caribou flesh and were very wasteful of it, and it was the same with other edibles, fish, birds and eggs. Even if deer decoys were used and a hunter was alone or with a few others he, according to hereditary custom, only took some dainty pieces of the caribou he had killed, e.g. the head and the fat round the entrails, while the rest went to the whole clan. There were many strange regulations for the hunting. One curious rule among the Barren Ground Naskaupee was that if two men were hunting together and the one shot an animal, the skin fell to the man who saw the shot discharged. But in any case the main rule of late years is that game of any kind belongs to the one who sees it first; he lays the flesh in the caribou skin and slings it over his shoulders like a rucksack. McLEAN (610) said in 1849 that if a caribou was only wounded, and could still run a little, it was considered to belong to the hunter who first reached it. — The same writer also noticed a curious example of community of goods in other respects; when an article had been bought in the Company's store it seldom remained in the hands of the purchaser for longer than two or three days. The rapidity of interchanges may be greatly facilitated by the practice of gambling so common amongst Indian tribes, HIND considered.

Superstition had created a complete dogma regulating the relations of the hunter to the guardian spirits of the game. As a sign of reverence for these spirits the skull of the animal killed had to be hung on a pole over the tent. WAUGH found two bear skulls, one facing north and the other south, lashed near the top of a pole with stones piled around at the bottom to keep it upright. The skulls were painted with vermilion, a circle around the snout, circles round the eyes, ears, round the top of the head and the lower edges of the jaws, and a strip from between the eyes down the top of the nose. He also found bundles of bones of various animals suspended on poles set up in the ground on small trees. A trout skull was suspended in one of the meetch-wops.

As regards the socio-religious practices during this period it is known that they seemed to agree with what prevailed among their neighbours in the north, the Chimo Naskaupee. We shall discuss this later.

The unrestrained joy of living of the child of nature is said to have been very distinctive in the primitive community of the Naskaupee at this time; they knew how to live happily even with their small resources. For days after the caribou spearing-battle the hunters would sit by their fires and there enjoy the greatest luxury the wilderness provides: the marrow of the caribou. If the game was plentiful and the hunt succesful the encampment would remain for weeks in the same place and the Indians eat and eat of the common stock of meat till it was all gone. And then they began to hunt again. The Naskaupee have known the art of producing intoxicating drinks from berries and could do this in a few hours. How they made the juice ferment even in winter is a riddle which no outsider has yet solved. Like aborigines in general they loved to become drunk, it is said; it was perhaps for them a kind of magical action by means of which they gave their spirits an opportunity of rising above the earthly life, and this could continue for weeks. One of my informers, who was supposed to be an expert, considered that this drinking had not only harmed their livelihood but also undermined their health. In other ways this desire for liquor has been a misfortune, for some smart traders are said to have taken

advantage of it to secure their skins at very low prices. It is therefore not surprising that it is forbidden by law — at least since 1857 (325) — to sell intoxicating liquors to Indians or Eskimoes, or even to offer them
a drink; the legal fine for the latter is 200 dollars.

It was said that the Barren Ground Naskaupee long remained isolated
from the outer world. McLean (610) describes the Naskaupee of Ungava as
very averse from locomotion;
many of them grew up to man's
estate without having visited a
trading-post. Before the establishment of Fort Chimo they
used to meet in the interior and
give their skins to an elderly man
of the tribe who took them to
the King's Posts (six trading-
posts between Seven Islands,
Lake St. John and Tadousac) or
Eskimo Bay (Hamilton Inlet included Lake Melville) and traded
them for such articles as were
required. Dominique, the chief
of the Montagnais of the Moisie
River, told Hind in 1861 that
at the southern end of Ashuanipi Lake there were some
Naskaupee families who had
never been to the coast before.
All Indians at Ashuanipi who
did not come down from the
Moisie went to Petitsikapau

Fig. 283. Naskaupee squaw with baby visiting
Davis Inlet (536 a).

near the Hudson's Bay Company post or to the coast of Hudson Bay.

In olden times, as at the beginning of this century, Fort Chimo at the
estuary of the Koksoak River seems to have been also this group's chief
trading-place, especially in winter. The very scanty store of civilised luxuries
which found their way to the Naskaupee was obtained here by exchanging
smoked skins, moccasins, and pelts mainly for aliments and munitions. But
they also bought tea, tobacco, matches, large fish hooks, brightly-coloured
cottons and clothes and wools for embroidery; mouth organs were also popular.

They examined the articles before purchasing them, carefully weighing them in their hand, for they must think of the long portages where every extra ounce is important (761).

Sometimes the men of the band used to pay hasty visits to Davis Inlet or Nain to trade,but the road to the Atlantic was troublesome. The road to Davis Inlet followed a deep valley the river of which was filled with rapids but with more than twenty lakes, and finished with the Assiwaban River system; even when water conditions were at their best this was a trying journey because the heavy loads of the canoes had to be carried over the many portages. Yet the journey went quickly, for the Naskaupee are masters in handling their light, open canoes; I was once amazed when I saw small boys paddling

their way Fig. 290. The Barren Ground Indians for this reason generally visited Fort Chimo in summer; then they carried the canoes over to the Whale River (the Naskaupee call this river Manouan = the place for egg collecting) with its many rapids and then by different ways came to the Koksoak River. Cf. Figs 285—287.

Fig. 284. Some future hunters of the Davis Inlet band. Photo the author, July 1939.

The idyll in the wilds was, however, not everlasting. The caribous' migration paths were changed in some years, and sometimes the Indians met only a few caribou and starvation visited their camp. It is therefore no exaggeration to say that the Naskaupee were completely dependent upon the caribou for their existence. When it failed they were forced to go to the coast where they could perhaps supplement their hunting trips with trade and begging. They had never hunted sea mammals, for they had previously always lived in the interior, only visiting the coast in the summer to trade. When the caribou chose definitely to take up new, unknown paths of migration it was a catastrophe for the band.

THE VOICEY'S BAY PERIOD.

In 1927—28 the band numbered 56. They spent much time at the coast, chiefly at Voicey's Bay, and made hunting excursions from there, says STRONG. In 1939 the band numbered about 100 persons.

The Davis Inlet (cf. p. 654) and Barren Ground bands have always, as has been said, been dependent on the migratory caribou. Their ritualistic practices mostly centre round it. The headquarters of the Naskaupee were still in 1910 on Indian House Lake, the shores of which have been for generations the battle-ground where this little tribe has had to fight for its existence (761). Some fifteen years ago, STRONG stated in 1929, the main herd of caribou suddenly stopped coming to Indian House Lake, due, the Indians believed, to their ancestors' neglect of the caribou rites; in this part of the country great

piles of caribou bones were still to be found and they said that the living caribou 'smelled' these bones and turned back to 'Caribou House', a mysterious mountain into which the 'Caribou God' took them (cf. SPECK, 877). From the descriptions it seems as if the life nerve of the Barren Grounds Indian's existence had been broken by the catastrophe of the caribou's changed migrations.

Why did not the band start out to find the new migration paths? Can superstition have so completely blinded them that they did not even try to discover the herd's new paths in the tundra to the north-west? Possibly another circumstance was the cause of their resigning themselves to their fate in this matter. As

Fig. 285. Naskaupee Indian of the Barren Ground band in 1909 (761).

STRONG says, the material culture of this group varies very little from that of the Chimo Naskaupee described by TURNER and depends to a great extent on the presence of forest, and this inability to live in forestless regions apparently prevented the Indians from finding out the whereabouts of the main caribou herd.

CABOT (123, p. 196) who twice visited these Indians at Barren Ground Lake stated that few of them were regular visitors to the eastern coast of the peninsula. However in 1912 just before Christmas the Naskaupee came once to camp near Voicey's Bay, a couple of miles away on the south side of the Bight (to the south of Voicey's), WAUGH (1002) reported. A Hudson Bay man from Davis Inlet came and drove them away or persuaded them to leave and go with him by promising a lot; he thought they had a great many skins but was mistaken. Next year the band came again and has been coming ever since. The Indians camped in the same place until 1918 when they pitched their camp quite close to Voicey's. From there they usually go back into the country in between to trap. In the summer of 1939 they also stayed near Voicey's.

Their hunting-grounds are now always in the forests, but they do not extend to the forest limit. They generally camp near lakes for they still often fish through the ice for trout.

On the hunting-grounds they usually divide into two groups: a larger group going north from Voicey's Bay, while the smaller group has its land to the west of it. Yet the idea of definite hunt-ing areas for each group seems to be largely lacking among these people; e.g. if it is necessary to have many hunters to surround a herd of caribou they often unite for that purpose and separate again afterwards. Each group has thus a huge general area for hunt-ing and fishing, but if one party is specially successful they are visited by the less fortu-nate (909). Fall-traps are made by the Nas-kaupee for wolverine and bear, WAUGH noted, and caribou snares are placed among the trees, or the animal is speared. Spruce part-ridge are captured with a noose on the end of a pole. The chief fur animal is the fox but also some few mink and otter skins are exchanged at Hudson's Bay Company's

Fig. 286. Naskaupee Indian of the Barren Ground band in the sum-mer of 1905 (119).

posts. The idea is justified that the eastern Naskaupee can under present conditions — the plentiful supply of game, fish and fur-animals and the absence of competition — manage quite well with the products of nature in their part of the country.

Regarding the migratory habits of the band nowadays I received the fol-lowing information in 1939 from Mr. BUDGEL, the representative of Hudson's Bay Company in Voicey's Bay:

In the middle of May they are generally to be found with their families in camp at Voicey's Bay; there they remain till the middle of June when they return to the interior. Some of them from their true country to the east make hunting excursions to the barren district in summer as in the olden days of caribou migrations; some of them have taken courage and begun again the

great caribou hunts near Indian House Lake which start in July, but in principle most of them have given up the caribou for the fox. They stay in the
interior till the middle of August when the men come down to trade, but only
stay a few days and then return to the hunting grounds.

In the middle of November the men again come down to Voicey's Bay with
their skins to get food but soon return to the interior; they are back again after
a month with a fresh supply of skins to exchange for food. After that they
remain in the hunting grounds till May.

With the band's move to the forested areas a striking change in their
dwellings has also occurred. At the beginning of the century the movable
tepee was the dwelling of the hunting-people in all weathers. Now they use
the same type of meetchwop as the Montagnais — Fig. 262 —, of calico and
with an iron stove, and, as far as I could discover, never the tepee.

According to WAUGH small tents (*midjua'p* or *mitcuap*) of canvas or caribou
skin are used while hunting. They also make a full-sized tent of poles set closely
together and covered with spruce boughs (a fire inside). This is a hunting shelter
too, — *citagwen'tcuap*. Also big rectangular tents for two or three smokes or
stove-pipes WAUGH found in use; cf. p. 655. It is interesting that STRONG (909)
saw these Indians occasionally using snow houses when caught out in a storm
on barren ground. The relation of this house to the block-built Eskimo structure is uncertain. It is of rather unusual type: a heap of snow is piled up and
allowed to freeze for half an hour, it is then hollowed out with a snowshoe
and deeply bedded with spruce boughs if such are available.

The Naskaupee of olden days have gone, like the caribou they hunted, and
the hunters wander about to-day in the woods near the Atlantic coast. Rotting
tent poles on the George River and Barren Ground Lake tell a sad tale of life
in these now deserted parts and speak as if from the grave of a race that has
passed away (761). The descendants of the Barren Ground Naskaupee are no
longer a tundra people; in the forests they are in their element to-day; the
tundra is the despised land, without wood, without fuel, one can freeze to
death there. Yet the call of the life of freedom seems still strong; they prefer
to live in the forest wilds and seem to have little or no intercourse with the
whites or the Eskimoes and they even seem to be outside the limits of the
Catholic Mission. They are said to retain their old religious ideas — their chief
deity is the Caribou god — whom they believe to be angry with them (p. 678).
As a result even the Davis Inlet people are very careful concerning the taboos
affecting the caribou. On this account for instance the sacred grease prepared

on ceremonial occasions may only be eaten in the lodge, and the plate cleaned at once (909). They are especially concerned with preventing the dogs from touching the head, horns, or long bones of the caribou.

The Ungava or Chimo Band.

The isolation of the northern Labrador enclosure has caused it to become a vast preserve of primitive elements which belong fundamentally to the vast

Fig. 287. Naskaupee of the Barren Ground band in 1909 (761).

circumpolar culture belt (871), and this concerns especially its northernmost section, where the Ungava band or Chimo band now lives. This band is now quite small. According to WAUGH there were about 50 Naskaupee at Chimo (12 men, a few more women and some children). Sometimes Naskaupee from farther west also come to Chimo. Earlier the group was estimated at about 500 souls. The band really inhabits a district outside the country here treated, but the information regarding it obtained by TURNER fifty years ago and by WALLACE in 1905 seems to throw some light also upon the conditions and ideas which must be thought to have prevailed among the Barren Ground Naskaupee during the period to which we have given that name, so they will be considered here to some extent. We shall follow TURNER's report.

The annual cycle.

The Chimo Naskaupee had special activities for each season of the year. In the autumn the old men, the women and the children were left in a fixed camp while the hunters with the older boys spread out in small groups to hunt during the autumn and winter. Now and again the hunters would visit the main camp to bring them game and other foodstuffs there.

In early summer when the rivers broke up the hunters went to Fort Chimo to exchange the result of the winter's trapping. But before they started they sent word to the main camp which was then struck, and the company started by short stages and many halts for rest to the trading-post. There they all met. There they settled their business; their favourite haunt was the shop, and with endless discussions they exchanged their skins for all kinds of goods: guns, ammunition, tobacco, cloth, flour, biscuits, peas, beans, rice and especially sugar and molasses. Usually the manager of the station could get the Indians to bring down the logs and firewood he needed for the winter. Sometimes also a few Indians were engaged to help with the salmon-fishing organised by the Company; but most of them proved unfit for such work, because if any game turned up in the neighbourhood they went after it, leaving the nets at the mercy of the winds and waves which so often meant their loss in the strong tides of the Koksoak River estuary. The whites could always rely more upon the Eskimoes than upon the Indians to do regular work.

When the men had agreed where the summer and autumn camping-place should be the people moved there and the hunters spread out in groups to find the game for the band to live on. In the autumn the tribe again moved up to a common camp in the forest south of Fort Chimo and with that the simple annual cycle of their economic life was concluded.

At that time the Naskaupee knew the habits of the animals in that region so well that they could usually point out with certainty where they would be found. Of course the caribou was still incomparably the most important and gave them all the food and clothing they required. The Chimo Naskaupee, like many other Indians, lived like predaceans on the caribou and it may be said that their whole life was based upon this animal. If they lacked caribou meat they considered they were starving, however much other game and fish was at their disposal. The caribou they procured in different ways.

To understand their hunting methods it is good to make acquaintance with the main lines of the caribou's habits in the Chimo district half a century ago.

In the spring the does leave the herd and go away to the forestless hills and mountains near Cape Chidley. There the wind has blown away the snow from many spots, and there some lichen and last year's grass is always available for the caribou even after the snow surface has hardened in the woods and made grazing difficult or impossible. In the latter part of May or the beginning of June the does seek out some spots protected against the snow storms and early freed from snow, and there give birth to the calves. When the latter have grown somewhat they follow their mothers to other places in the mountains where there is summer grazing in the meadows among the hills where it is cool and the wind blows away the mosquitoes and other insects. Here the bulls also come later in summer. Thus in summer the caribou are to be found now here, now there in the mountains in the north-east, for they always move against the wind.

When autumn comes and the insects are no longer troublesome they collect in small parties each with a bull as leader and go down to lower districts, especially the areas of fine lichen near the Koksoak River where it is moist and tasty at this time. They were so regular in their migrations that they might have had a calendar, it is said in the old descriptions. In August the skin of the young caribou is suitable for the Indians' clothes; that of the female was used for this purpose while that of the male was made into leather and babiche. The fawns' skins were reserved for making children's clothes and underclothing, the fur next the body. The skin of the old animals is best in October and they are then also fattest; the fat on the back can be ½—1 inch thick. It is now time to prepare for the great autumn hunt.

The caribou rutting time is October. This is the best season for the Indians' hunting. Driven by the sexual instinct they wander in small herds to and fro over the rivers. At certain places the animals like to swim over the great rivers, and at such hunting passes occurs the same senseless slaughter as that already related in connection with the Barren Ground Lake (p. 670 sq.). WALLACE reports that an agent of the Hudson's Bay Company told him that he had seen nearly four hundred caribou slaughtered in a few hours at the Koksoak River (995). Then the whole camp jubilates; a magushan is arranged, the drums roll. We may, however, think that the senseless murdering of the caribou is justified by the Indians' outlook, he believes that the great Caribou spirit sent him the game (cf. p. 688).

There were also big hunts in the winter when the snow was deep and prevented the animals from moving or the hardened snow surface could bear the hunter on his snowshoes but the caribou fell into it. The animals were also caught in snares; and they were sometimes driven by a few hunters into deep banks of snow where they were unable to help themselves and could be speared at will (995).

The Fort Chimo Naskaupees also devoted themselves in winter to the trapping of fur-animals and hunting of small game, chiefly hare, and birds were caught by the thousand. In the summer women and children collected eggs in great numbers, and salmon were caught in the great rivers. In the lakes other fish were caught with net or hook, in the winter through a hole in the ice.

After the autumn slaughter the bodies of the caribou were generally skinned and left to dry on the river banks in the sun and wind. If the supply of meat was small it was dried over the fire in the tepee until it was friable and had a slight taste of smoke, and this is still done to-day by all Naskaupees. Then it was ground to meal and placed in a leather bag or a bladder with pure suet, and the mixture became kneaded together into a mass, *wutu't*, which is nothing else than what polar explorers call pemmican; this strong food is thus an Indian invention. Another of their special conserves is called *uinastikai* and is prepared by mixing in the caribou's stomach blood and a mass of partially digested lichen from the entrails of the animals; the mixture is boiled and evaporated and becomes a brownish, granular mass like gunpowder. When in winter the hunter went off for the day he put a handful of uinastikai in a cup of water and drank the mixture; he needed no other food until he returned from the hunt. The Naskaupee also loved marrow. But his greatest delicacy was the caribou foetus.

The Naskaupee liked berries so much that it is said that they made »berry farms» by burning a piece of forest land and growing berries there, chiefly kinds of bilberries and cranberries, which they picked and boiled to a concentrate in the form of cakes. They are also very fond of cloudberries which they use as a prophylactic for scurvy (799, 119).

At the beginning of the century when CABOT was among them, the Naskaupee used neither bread nor salt. The food was always boiled; the drink was lukewarm water; they never drank ice-cold water as do the Eskimo.

— On the basis of his observations in the winter of 1905—06 WALLACE (995) gives some information which supplements TURNER's notes about the Chimo Naskaupee: With open water in the summer the Naskaupi came to the Fort with the pelts of their winter catch. During a short rest, they exchanged them for arms, ammunition, knives, clothing, tea and tobacco, then disappeared again into the fastnesses of the wilderness above, to fish the interior lakes and hunt the forests, and no more was seen of them until the following summer. The Ungava Indians thus never go to the open bay in their canoes: it would bring them bad luck, they believe, for there they say the evil spirits dwell. Of all the Indians that visited Fort Chimo only two or three had ever ventured to look upon the waters of Ungava Bay, and these had their view from a hilltop at a safe distance.

Only a few of the younger men used to emerge from the silent land in the south during Christmas week to barter skins for such necessaries as they were in urgent need of, — and to get drunk on a sort of beer, a concoction of hops, molasses and unknown ingredients that the Post dwellers make and dispense during the holiday festivals.

Thus at that time the Naskaupee were a really primitive hunting people. Their simple

SOCIAL CONDITIONS

were in agreement with all this.

The tribe had a chief who had sole control and determined the arrangements of the hunting and the distribution of the game, etc. The chieftainship seems [?] to have passed from father to son, or another near relative if there was no son. In the Chimo group the chief collected a certain tax; of all game taken he kept one of ten skins for himself. In later times the Canadian Authorities had given him a police star with the word »Chief», which he proudly wore on his chest. No other arrangement tending to social control existed.

The division of labour between the sexes was definitely fixed by ancient custom (606): the only worthy task for the man was then, as now, the hunt; the woman's was to serve him like a slave and blindly obey his orders and whims for good or ill.

PRICHARD reminds us that in civilization a woman's work is never done, and far more is that true of the helpmate of the semi-savage man of the Barren Grounds. Among the Naskaupee the women are the slaves of the men, McLEAN (610) notes. When they move from camp to camp in the winter the women set out first, dragging sledges loaded with their effects and such of the children as are incapable of walking. Meantime the men remain in the abandoned encampment, smoking their pipes, until they suppose the women are sufficiently far advanced on the route to reach the new encampment where they overtake them. The woman has to pitch and break camp, to cook, to cut up and carry her husband's kill, to dress the skin of the deer, to fashion the foot-gear and the greater portion of the clothing. On a journey she often paddles the canoe, and at the portages she carries the heavier load. The squaw's life is not easy, but emancipation will probably never come; in the wilderness the provider of food, Man the Hunter, has reigned, reigns now and ever will reign. The laws of the Northland do not change (cf. also 761).

A hunter took a wife when he was quite young, that is when he could provide for her; sometimes the girl was only fourteen or fifteen years old, or even younger. No betrothal or marriage ceremony was known, except that the girl was asked if she would live with the suitor. If she refused it often happened that she was taken by force. The fetters of a marriage by capture are said to have always been very weak and could be broken for the least cause by the one or the other party. 'Shanish's uncle took his wife away from him and he was obliged to be satisfied with another', reported WAUGH. Intermarriage within the band was the usual thing.

According to McLEAN polygamy was practised and it was not unusual for a man to marry two sisters one after the other, or both at the same time; HIND tells in 1860 of a Naskaupee who left two wives-sisters. All the wives usually lived with the hunter in the same tent, and when jealousy broke loose among them it is said that there were terrible fights and an awful noise until the hunter himself, who at first lay on his back and laughed at them, found it was time to interfere and shouted at them to stop. The Indians told WAUGH that a few decades earlier their forefathers still used to have two or three wives sometimes. Even to-day polygamy is tolerated in principle though difficult in practice because of the lack of women; during this century there has scarcely been more than one man among the eastern Naskaupee who had two wives at one time; though WAUGH reports that 'Adaka had a young girl in addition to his wife' (wife and concubine). Relationship is no hinderance to sexual intercourse out of wedlock; incest is said to occur. Nor were promiscuous relationships rare among these Indians, but sexual intercourse outside the tribe is said to have been unknown.

As paternity was often uncertain the Indians had agreed to regard as the father the man with whom the mother was living when the child was born. While the children were growing up the boys always enjoyed every advantage at the expense of the girls.

The men, as has been said, know only one task in life: to hunt. It falls to the wife's lot to slave in the camp so that before she is thirty years old she begins to look worn out and when she is fifty she is an old woman and ugly as a witch. But a Naskaupee woman knows no other fate; she is brought up to it and cannot think of wishing for anything else. She understands, says CABOT, that the fate of all depends in the last resort on the hunters. And she knows also that it is usually the women and children who survive the famine time, while the men in attempting to procure food for the camp die from the hardships of the wilderness.

SUPERSTITION.

The country inhabited by these Indians is precisely such as would engender and foster superstitious ideas (435). The Chimo Indians' conceptions of the highest powers in creation are attractive; they reveal something of their mental life and explain certain habits and customs. In many respects these ideas, according to TURNER's notes, resemble very much the old ideas of the Labrador Eskimoes.

The whole existence of the Naskaupee has been terribly ruled by superstition. Nearly everything that happened was due to the action of the spirits. Every incident of life was associated with the supernatural, everywhere the Indian sensed the active participation of non-human agencies. According to TURNER there were many spirits, of whom the less obeyed a great spirit who lived in the sky [— can this be a new acquisition they have adopted from the Catholic Mission?]. The spirits were generally regarded as being unfriendly towards humans.

An evil deity, Atshem, was the Terror and bugbear of the Naskaupees. They believed that he assumed the form of one of the most dreaded conjurors of olden times, or, as a frightful giant, wandered through the forest in search of human prey. When the report spread that his tracks had been seen near the camp, the Indians fled in consternation from the neighbourhood, and lived for weeks or months in continual terror. Time and again therefore the spirits had to be propitiated by sacrifice. Such sacrifices were demanded by the spirit of the individual as well as by the spirits of natural objects. Sometimes the spirits appeared in different guises and it was important to know in which of these a certain spirit acted. The only one who could reveal this and knew the right way to deal with them was the shaman, and it was dangerous to offend him by expressing doubt of his art. Shamanistic powers were entirely the result of personal predisposition and were not inherited, though they might be passed from one living man to another (909).

The shaman was the oracle. At least until quite lately the rite demanded that the shaman should be hidden from the eyes of the tribe while he was performing his mystical incantation; cf. Fig. 264. For this purpose he withdrew to a little skin tent — t h e s h a k i n g t e n t —, which was sewn together with great care so that no ray of light could enter it. After installing himself in the tent the shaman began to mumble, but gradually raised his voice till it became shrieks and screams which were heard far away. Finally the drums were used and the tumult increased until the magician fell into a trance. Then he could tell the anxiously waiting suppliants outside that the spirit he wanted to see had appeared, and he was asking his assistance for the desires of his clients. After a while he told them that the spirits had come to his help. Then the shaman's tent was pulled down and the ceremony was over.

Sometimes the shaman could reveal the meaning of a hunter's dream after a few minutes.

According to what I learned the shaman seems always to have had a favourable answer. If this was not so he said it was because some necessary action had not been performed or some circumstance had not been told him by the client; — for instance that someone had broken off a twig, or turned a stone on its side or something similar. This had offended the spirit, who withdrew without saying the whole truth. Then the client had to pacify the spirit with a sacrifice, and the shaman decided what this should be; — he usually chose something which was difficult to obtain so that he himself had time to provide against any possible doubt as to his power, says TURNER.

Simplicity and individualism characterize the institution of shamanism in Labrador, which lacks organisation just as the other non-material cultural elements (871). This shamanism seems to have been a quite harmless kind. It is a long time since the conjurors have induced the Indians to commit outrages or bloodshed and murder. It is known that in 1831 the Indians of Rupert's River and James's Bay attacked the Hudson's Bay Company's post and killed 12 persons (435). This is however the last Indian insurrection against the whites of which I have found any note in what has been written concerning Labrador.

The idea that every animal consisted of a body and a spirit played rather an

important role in the life of this hunting people.   This spirit belonged to a
lower order than the spirit of a human. The animals' spirits are, like all others,
indestructible, and if the body containing it is destroyed it can spontaneously
again take upon itself the material form in which it had previously lived. But
this implied, according to the Naskaupees' idea, that even if an animal was
killed its stock was not decreased, because the spirit only deserted the dead
animal to incarnate itself in the form of another animal of the same kind. Thus
death only meant that the spirit left the corruptible, material form. But it
sometimes happened that the spirit withdrew for some time to its original home,
e.g. to the 'Caribou god' who rules over all the caribou spirits, and this could
have dangerous consequences if for example the incarnation did not take place
very soon, for in the meantime the humans might be obliged to starve. In this
case the conjuror could intervene, for he had the power to attract the spirit and
persuade it to take again the guise of a caribou. Notwithstanding it was well
for every hunter to be careful to be on good terms with the animals' spirits.
Never for instance must the dog gnaw the killed caribou's bones (p. 675). For
this purpose he also carried a piece of the skin concealed on himself as an amulet;
if the booty was a bear the underlip was cut off and its flesh side painted red
with hematite powder mixed with fat, and this was a fine amulet. In 1860
the Naskaupee men were still tattooed on the cheek for the same purpose,
generally from the cheek-bone to the nostril on either side. HIND was told that
the women also were often tattooed. WAUGH reports that totems did not exist.

This is not the place to go into details concerning this attractive but little
studied subject. The reader is referred to TURNER's and SPECK's works. — The
belief in supernatural owners of the game is widespread in the rest of Labrador
also, and according to SPECK it is apparently exclusive to the Montagnais-
Naskaupee among northern Indians. In northern Fennoscandia we know that
the same conception has existed or exists; amongst the Skolt-Laps he is called
Kufedar. It brings up again that perplexing question of eastern North-American
and Palaeasiatic correspondences (871, 915).

The Labrador Indian's activity consists wholly of hunting, setting traps,
and fishing. It is clear from what has been said that if the hunter desires
success he must conscientiously observe certain inherited ritualistic manners
and customs. The trapper was compelled to act *mentu* (cf. SPECK), which
meant in a way that would please the invisible powers. In hunting and trapping these non-material factors were at least as important as the mechanical
exercise of the hunt, which thus became a ritual action. In some way the
hunter dematerialised nature, says SPECK. This explains why it was so important to know the right way to deal with the supernatural powers who
control and guide everything in the life of the Indian hunter. For this reason
he must be careful with the remains of the carcass and avoid anything which
could offend the animals' spirits and disturb their comfort, delighting them
with feasts and the drums before and after a good hunt.

In 1853 PATER DUROCHER made some notes about the southern Naskaupees. These believed that the spirits of particular animals would become hostile to them if they gave the bones to the dogs. At certain feasts they sacrificed the flesh of animals killed in the chase by burning it to cinders. In times of scarcity they sang and danced to the sound of the drum until they fell down with weakness, in order to get a glimpse in their dreams of the places where the wild beasts congregate.

In the difficulties the Naskaupee has in procuring food and clothing in this extremely poor country he believes that his dreams guide him in his actions. He therefore follows his d r e a m s with the utmost precision, wholly regardless of any consequences other than those to which the fulfilment of the dream may lead him.

When anyone was sick the Naskaupee sang until they were overcome by sleep, in the hope of seeing in their dreams the enemy who had cast a spell over the invalid, or that they might discover the herbs which were capable of effecting a cure (435).

In the time of the Jesuit missionaries also the Montagnais were noted for their reliance upon the power of their dreams as explained by their conjurors. They believed and still believe that in dreams they were in direct contact with the spirits who watch over their daily life. When the Indian dreamed of firelight on the ground (?), it meant that the caribou were on their way to his land. But it was not given to everyone to correctly explain the dream, so that the interpreter of dreams held a very important position in the little company. Such work belonged in the first place to the conjuror or some other Indian who had been declared an expert.

A ceremonial feature connected with dreams diffused over the whole north is s c a p u l i m a n c y, divination by means of scorching an animal's shoulderblade and observing the cracks and burnt spots. This practice of divination is general and fundamental among the Montagnais-Naskaupee hunters. As SPECK points out we encounter in this idea one of the most significant indications of circumpolar culture affinities that can be cited (871); (cf. Fig. 255). The individual hunting ceremonialism is well illustrated by the following passage which we find in SPECK (877): »You speak about singing for game», said Petàbana of the Ungava Naskaupees; »about using the rattle and drum before and after dreaming about the caribou. Yes! I have a song and I dream songs, and read the shoulderblade divination. They all do, he» (he pointed to Mictaben, who sat near), »he» (he pointed to Nabe's) »in fact all: That shows we kill the animal and live».

These Indians conceived of the dematerialised objects of nature, the spirits, as real beings, in about the same way as seems secretly to have been the case — and perhaps still is — among the half-nomadic Lapps in Finland. The spirits of the animals are to a Montagnais and a Naskaupee as real as their visual observations: no one among them would dare to doubt the existence of the cannibal giant Windigo, the under-water people, the owner of the caribou (N.B. the Skolts' Tshats'ile and Kufedar), for their ancestors had depended on them and the spirits had always given them success with the game so that they could live. The witness of the shoulder-blade and the bear ceremony (see SPECK 877) call to their minds things as actual as canoes and snow-shoes. Thus for both the territorial Indian clans in Newfoundland-Labrador an unbelievable number of invisible beings move about the mountains and the wooded valleys. They are especially active when darkness falls and that explains why every sensible savage in Labrador — like in general everywhere in the Old and the New World — does not like to be out after dark, for one cannot know whether the unseen powers are out on friendly or unfriendly errands. In this way the ideas of the Labrador Indians resemble somewhat those of the Eskimoes and the Palaeasiatics, and all of them are clearly branches of one and the same stock. There is much in the superstition of the Labrador Indians which can be found practically similar among, for example, the Skolts in Petsamo.

Knowing all this one has more practical thoughts of a bizarre action which has been in use among the Naskaupee until quite late, namely that the old and incapable, and troublesome cripples, who could not contribute to procuring food for the tribe were simply got rid of: they were often at their own request abandoned at the camp when the clan moved on, sometimes tied together like a bundle with babiche. The condemned wretches soon starved or froze to death, after which beasts of prey took care to dispose of the corpses. It was considered right that a son, a brother, or a close friend should carry out this execution, and one must do them the justice to say that the parent himself expresses a wish to depart; otherwise the unnatural deed would probably never have been committed, for they in general treat the old people with much care and tenderness.

This horrid practice shocked the white population. If, however, we consider that according to the Indians the spirit is indestructible, death only means that it has the possibility of exchanging a worse for a better material guise. Thus this action was really considered one of love by the Indians themselves. The son generally performed this kind office for his father; and, it having been a practice among them, they wonder at our considering it an act of inhumanity, CURTIS says (196). Perhaps, too, this was a happier death for the suffering old incapables who placed the honour of the hunter higher than the shame of sitting with the women in the tent while the others followed the caribou. — The Naskaupee considered that lighting flashed when an old man's or woman's spirit left its material covering; *tci'bän'is kwrte'u* = death-fire (WAUGH). The Northern Lights signify that the dancing spirits of the dead are holding their nightly revels.

BRYANT was informed at the end of last century by one who had spent two years at Fort Chimo that a custom of killing the aged and helpless still prevailed among the Naskaupee. It is now considered improbable that the Indians, as little as the Eskimoes, get rid of their old people.

— For most of these ideas and customs I have found confirmation in the writings of SPECK, to-day the best scientific expert of the Labrador Indians' non-material culture, and the reader is referred to them (see bibliography) for more detailed information.

Reasons of the primitiveness.

This curiously primitive society of the Naskaupee presents a positive uniformity in the conviction of the nearness and the potency of supernatural powers (606). HIND stated that it has become quite an axiom for everybody who knows them that it may be impossible to wean them from their belief and the wild excitement of a life in the woods. They care for no other pursuit than fishing and hunting, they live and die on their hunting grounds, with no inclination to rise in the scale of civilization (522, 1049). They feel no need of any other life, and nature favours it, and they will certainly continue their old-time ways and their migrations if no outward cause prevents them (cf. 1049).

As far as I know few positive measures have been taken to assist the northern Naskaupee in assuming modern civilization; — cf. e.g. 1082 a. I have no knowledge that a single Indian of Newfoundland-Labrador has been given an opportunity to learn to read, write and count. In social and religious respects the Indians have been but little raised above their state of development at the time when the Catholic Mission first discovered them. Animism is still said to control their view of life. But when their world is so poor as the deserted woods and barren grounds of Labrador it is only natural that they should seek to reinsure themselves where they may find sympathy and protection, so they do not despise what the priest may be able to give them and their families by reading the Latin mass and giving them alms at the trading-post of Davis Inlet. Though really their mental life still belongs to the Stone Age.

The recent metamorphosis among the Naskaupee.

It has been said that the Naskaupee are probably the most isolated Indians in North America. Yet certain changes have taken place during the last half century both in their material and non-material culture. At the beginning of the century they seem to have lived in a curious age of mixed metal, bone and stone. They were still using on their spears and arrows points of hornstonelike calcedon, or of caribou bone or antler.

Stone implements are nowadays restricted to hand mauls for breaking caribou bones and pounding meat, net sinkers and whetstones and the stone tobacco-pipes. Even before the end of last century new tools, new clothing for the bodies, new weapons, new foods, new drinks and new ideas and ideals began to make their way among them. The metal tools, axe, and pot were infinitely superior to those of stone. However, their specific Indian tools and equipment are, of course, unchanged and still in use. There has been

Fig. 288.   Dancing Naskaupee Indian.   Photograph of a
model in natural size presented by Prof. FRANK G. SPECK
to the National Museum in Copenhagen.   The garments are
original and made of white chamois leather painted with
patterns in red and blue.   The ornaments are so skilfully
elaborated that it is difficult to imagine they were originated
in the Naskaupee's present environment.

no change in the canoe, the snow-shoes and the toboggan, simply because these means of transport cannot be surpassed in those natural surroundings. The man-high bow and arrows are still used when ammunition is lacking; the Naskaupees are considered splendid bow-men. The crooked knife is still in common use; also the gill-nets both of twine and habiche, and the bone-pointed fish-hooks, like the snares and fall-traps. The shaking-tent (*gocaba'tcige'nc*) and the drum (*deven-igan*) are still used in conjuring for luck in hunting and fishing. The sweat lodge, crescent in shape, is much used.

Yet if a cultural loan has been economically advantageous they have not hesitated to adopt it. The winter dress consisted a century ago (610) of a jacket of deer-skin, worn with the hair next to the body, and a coat of the same material reaching to the knees, with the hair outside. Leather breeches, leggings and moccasins protected the lower extremities and the hands and arms were shielded from the intense cold by gloves and gauntlets reaching as far as the elbows. When in full dress, they wore a cap richly ornamented with the claws of the bear and the eagle. The garments of the women consisted of a square piece of dressed deer-skin fastened round the body with a belt and suspended from the shoulders by means of straps, a leather jacket, leggings and moccasins; cf. Figs 288 and 289.

Fig. 289. Costume of a Naskaupee woman in olden days in the collection of the National Museum in Copenhagen. Dress and cap are of white chamois painted in red and blue.

Clothing among the Naskaupee has nowadays quite a large percentage of European styles (coats, trousers, shirts, underclothing, socks and stockings, skirts, blouses and sweaters). Foods have evidently changed considerably, too, since CABOT and WALLACE visited the Naskaupees. The Indians whom WAUGH visited considered flour a necessity and sugar came close to the same category; also baking-powder and salt. Coffee is used, also molasses and canned goods of various kinds.

Owing to geographical conditions these Indians have meanwhile remained untouched by outside influence to a degree that can hardly be paralleled elsewhere in the American Arctic (871). For them there has been but one source of contact borrowing open for consideration and that is with the Es-

kimoes of Labrador and those of Hudson Bay. It is therefore most interesting
to find how much the Naskaupee have borrowed from the Eskimo. Some of
them have procured draught dogs. STRONG says that at that time the Davis
Inlet and Barren Ground people used a few Eskimo dogs, which were harnessed
to the sled in fan-shaped formations, but this art, like the *komatik* with runners,
was only acquired from the coastal Eskimo some fifty odd years ago. Prior
to that the men pulled their own toboggans, and the small white Indian dog
was only used for hunting small game. These dogs are now almost extinct.

Fig. 290. Naskaupee youngsters paddling their canoe with the speed of the ordinary
motor-boat. Photo the author, July 1937.

The Eskimo 'dickie' is very largely used by the Naskaupee, particularly the
outer calico wind-breaker variety, which is worn over the caribou-skin coat,
although the latter hangs some distance below it. The Eskimo skin boots are
also worn, especially in the wet season, WAUGH stated in the 1920's.

Some of the bone articles among Montagnais-Naskaupee are also Eskimo
tools, imitated by Indian workmen (871), and the Naskaupee women's knives,
cagwigan, are just the *ulo* of the Eskimo women which they seem to have
borrowed quite lately (p. 529). These facts illustrate the conservatism or
receptiveness of the Naskaupee community to new stimuli.

After the death of *Ostanitsu* — the clan chief of the Barren Ground band
in the beginning of the current century, known and esteemed by all — the

communistic order of former days was broken up. He must have been an extraordinarily estimable, affectionate and clever old man from what the Rev. PAUL HETTASCH told me; he could hold the people together and arrange things for their best. Since then no new man has arisen with developed intelligence and the gift of leadership; his son Charlie, who succeeded him, died young. The band then could not agree as to where each should hunt, and split into small parties of five or six families; it was easy to differentiate between them at the trading station, the Rev. HETTASCH told me, for the tents disappeared in groups even as they had made their appearance. There was, of course, a leader in each party who called himself chief but his authority and influence did not at all correspond with that of the old-time chiefs. STRONG (909) explains that the chieftainship became of little importance and was usually vested in the oldest and wisest elderly man of the party, but his authority was rather nominal.

Communal hunting had to be given up when the great caribou herds took up new, unknown migratory routes. Finally even the smaller herds of 75—150 beasts, who some decades ago used to keep to certain areas near the coast, disappeared; the people then spread out over the district in small parties. It was better to hunt up groups of two or three caribous in small parties. Trapping, of course, profited by this development, but whether it was fortunate for the Indians, as I judge, time alone can prove. The material culture of the Indians is sufficiently elastic to make the adjustments necessary to prevent chaos resulting, I think; cf. p. 653.

These Naskaupee have finally become christianized to a greater or less degree according to the meaning attached to the term by members of different denominations (435). It is said that Roman Catholicism has made little headway; they are so off the beaten track that they seldom or never hear or see anything of a priest. Yet in their spiritual anguish an apostle arose: old CHIEN, a Montagnais who had married into the band, began to perform the different ceremonies — of marriage and burial etc. — according to the Roman Catholic ritual. Chien seems to have been a blessing, and according to what I heard from a white man who had had much contact with these Naskaupee, this century has seen great improvement among them. Moral precepts have become stricter; all miscegenation with the whites has ceased; cross-cousin marriage is considered most correct and is urged on the younger people by their elders; no illegitimate children are said to be born. My informer said that also the health conditions of the clan were perfect. They seldom go hungry nowadays for there is plenty of meat and trout for the people in the small groups, he said. They manage, however, better without the white man's food, which is clear from the fact that the traders at Voicey's Bay notice the Indians are in better condition when they arrive there in May than when they depart in June; and nowadays they buy only a two-weeks supply of flour as a reserve. Another striking circumstance is that the younger men are more moderate in their consumption of beer than the older.

Yet it cannot be said, as it is by some observers, that the old rites and customs have been given up. In support of this the poor development of the symbols in their applied art has been cited; further that art is considered (1002, 909) as almost entirely for purposes of decoration, not for magical purposes. From my conversations with the local representatives of the Hudson's Bay Company I have, however, got a somewhat different impression of this matter; the decorations of the hunting gear for instance, were interpreted for me as having a decidedly magical purpose.

It thus appears that the Naskaupees' way of living is gradually coming to resemble that which we have learned to know among the present-day Montagnais; some experts even hold that there is practically no difference between the two. It is undeniably surprising to find that tundra Indians compelled to live in the forests have spontaneously adopted habits suitable for the new environment (cf. 1049). From this it is clear that under primitive conditions the demand of the environment is a stronger incentive than tradition. This ethnological convergence is an interesting circumstance which the administration can build upon.

The recent development amongst the descendants of the Barren Ground Naskaupee should thus give a more optimistic aspect of the future of this little Indian group than that suggested by CABOT in the beginning of our century. He then believed he had found some symptoms of a décline in social matters amongst them. They were becoming increasingly dependent on the goods of the whites, and in the absence of trade competition this might gradually bring them under the control of a more or less monopoly business and reduce them to a kind of economic serfdom by a gradual shifting to the trader's advantage of the relation between the purchasing-price of the Indian's goods and the sellingprice of the store goods; — which would have struck a blow at the heart of the problem of the Indians' conservation (cf. however 811).

The Indians of Newfoundland-Labrador represent a very primitive and perhaps deculturised stage of Algonkian culture.

Simplicity and primitiveness are the most salient features of the whole Indian culture of Newfoundland-Labrador. Moreover the Algonkian of the far north-east compared with other Algonkian groups show an early phase of the culture of their own stock, a fundamental, archaic, interior culture, very old, SPECK has pointed out (871). By force of economic circumstances due to climatic and geographical conditions it seems safe to say that no other type of culture can have existed on the far north shore than that possessed by the Eskimo along the coasts and the Indians who occupy the adjacent inland region today.

Now, there arises the question, is the hunting culture of the Indians an endemic form of culture stabilized by convergence, that is an autochthonous culture, or must one assume that it came into existence by a process of disaggregation and deculturisation from a more elaborated cultural stage? Let us review these alternatives.

As I became better acquainted with the cultural conditions, both present and former, of the Labrador aborigines, it surprised me very much to find so many fundamental resemblances to or corresponding features of, for example, the culture of the Fennoscandian Lapps. In the old (Iron Age) culture of the coastal Lapps on the one hand and of the Eskimo on the other, as well as in the inland culture of the half-nomadic forest Lapps and for instance the Naskaupee Indians, certain elements and not least some ideas in the sphere of non-material culture clearly indicate that a common root or roots must have existed between the primitive cultures of Fennoscandia and Labrador. Is that agreement due to culture inheritance or to convergence?

Everywhere in surveying Indian culture in the far north, SPECK has pointed out, one encounters the adaptation to internal winter conditions: the dependence upon large game animals, the hunting on snowshoes or skis, the game-drive, trapping for fur clothing, fishing with a spear and by nets, portable shelters, the necessity for complete body-clothing, the dependence upon wood-fires for heat and cooking, the necessity for adequate means of transportation in summer and winter, canoe, hauling sledge, snowshoe and reindeer or dog transport. The simple family group, the absence of political cohesion, the fundamentality of the family hunting district, the weakness of the power of the chiefs, and the unpracticability of large and prolonged religious festivities together with the corresponding development of individualism, all seem to set the peoples of northern Eurasia and America apart from other groups with a clearness which justly entitles them to rank as one of the typical world-marginal culture types emphasized recently by anthropologists.

If the huge areas where these resemblances can be traced are considered, the first inclination is to urge in explanation the process of convergence: to adapt oneself to the living conditions of the milieu is more and more imperative the farther north a hunting people lives.

As a possible explanation of conditions in the far north-east SPECK (871) figures that the simplicity of the culture exponents are together the ear-marks of a long period of existence in the country of an archaic, sedentary culture of a relatively unaffected old Algonkian type. This reasoning seems to pre-suppose, as there is some reason to do, that the migration into the eastern area must be pushed back to a very early period.

We have seen that in historical times the Naskaupee have been forced farther towards the north (p. 581) and the Montagnais have followed in their tracks. It is probable that also other Indians have earlier moved towards north-east Labrador in diffuse waves. It seems to SPECK that the drift in older times has been always eastward, and there has been no outlet for a back-flow of either culture or people toward the west. Meanwhile, the natives did not in-crease in numbers; because of the rigours of the environment a balance may have been reached, the natives' ingenuity could not overcome the unresponsive-ness of barren nature by such inventions as would have enabled them to exploit the resources more fully. The rather marked uniformity in many respects among the Montagnais and the Naskaupee may be taken as evidence of a more recent dispersion from earlier centres. The many references to south-ern animals — mere names to these people — and other contents of the myths in which reference is made to such things as palisaded villages suggest a rather recent northerly movement from a happier country. When the country was inhabited by a new folk wave the invaders perhaps met a more elementary type of culture, normal for those parts, as a function of the poorer geographical environment. They were compelled to adapt their ways of living to it. When the displacement of a group of people with a certain type of culture takes place to a poorer environment, I believe that we must assume a process of decul-turisation to be normal. In the present case it progresses gradually with the migration north-eastward.

At least in historical times the changes in inherited culture led to a degene-ration in the new areas. In the new surroundings it lacked both justification and possibilities, and many of its elements had to be given up as the people adapted themselves to the new limited possibilities and demands. Perhaps these Indians have to some extent degenerated mentally too in their new milieu. Certainly their ideas are not able to rise above the primitive interests of the hunter. The beautiful national dress, the long white chamois coat with its artistically painted decorations in red and blue, the white skin leggings and fine beaded white moccasins as well as the head-band have all gone the same way as the artistic stone tools. Symbolism in the old ornamentation is mostly forgotten, but its artistic expression bears witness perhaps to a native cultural development of a long period in a more hospitable area than their present one, where it will be completely abandoned. The mythology of the easterly Naskaupee, says STRONG (909), forms an interesting link between that collected by TURNER in the north and by SPECK in the south (869, 877). It is therefore probable that the deculturisation advanced gradually towards the north. The fact that Indian place names, especially towards the coast, seem to be comparatively scarce, while they use Eskimo or white names for rivers etc., also favours the suggestion that the deculturisation is of fairly recent date. I suppose that every invasion was followed by a deculturisation process in the new area. This being the case the primitiveness of the Naskaupee must be a secondary one.

I therefore think there is good reason to agree with STRONG's conception (909) of these Naskaupee, which at the same time seems to hold good for the Montagnais, that considering their lack of, or poor symbolism, their simple material culture, and their unique religion based on the caribou god, it would appear that the north-eastern Naskaupee represent a very simple and presumably old type of Algonkian culture. Pushed into the barren region where they now live, they abandoned their earlier higher customs and adhered to the old primitive cultural survivals; only slightly have they been influenced by the coastal Eskimo and the whites in recent times.

The thought was not strange to SPECK that the primitiveness of culture can have come about through recession from a more complex phase of culture under which these tribes had previously lived. But after more mature consideration he concluded that the likelihood of the eastern migration of the Algonkian is not a strong objection in this case, because at an early time it would have served to remove part of the primary Algonkian group from the reign of outside cultural influence, leaving this division as the conservator of the original simple culture by reason of its self-isolation. The north-east can therefore be considered as an archaic, marginal, culture zone, where human groups have resided for a long time apart from cultural changes and inno-vations in a cold and inhospitable country removed from outside contact by reason of its continental terminal position and distance. The natives of the north-east may have followed the same mode of existence for a very lengthy period, adjusted to the requirements of a far northern existence and clinging to the habitat of the caribou, the beaver, the seal and the bear, the canoe, and the long semi-arctic winter (871). The newcomers became assimilated into that culture.

The development of culture consists of a series of complex processes. It moves over the world quite independent of distance and the ethnic composition of the peoples. Some people die out and every trace of them disappears, but their culture spreads and the development continues hand in hand with it. It can be difficult to prove the connections in the broken links of the cultural chain if only the ethnological characteristics are considered. But here we come upon questions which not only fall outside the scope of this work, but also can only be answered when a systematic, archæological exploration of Labrador has been carried out. —

The white people.

The settled population of today: Subdivisions.

There was hardly any noticeable white population in Newfoundland-Labrador before the end of the 18th century (920 c). The first efforts to establish permanently inhabited trading- posts here were made by the British captain CARTWRIGHT (134, 135, 137, 949). His start was rather successful; his establishments, e.g. the Caribou Castle at Sandwich Bay, seem to have been able to export a lot of fish and furs to England. Unfortunately the piracy during the Anglo-American conflict harmed CARTWRIGHT and he went home to England. When conditions became peaceful again white men began to settle on the ruins of CARTWRIGHT's enterprise in the southern part of the coast. During the 19th century the white population gradually spread their habitat all along the coast up to Hebron in the north but not beyond it, if the Hudson's Bay Company's posts are excepted, the personnel of which do not live directly upon the products of the land and sea. Yet their increase at the coast has been very slow, as appears from the round figures given on p. 439; these include also Eskimoes, Indians, passing traders, missionaries, doctors and their families, etc., but shew approximately the main variations in the numbers of the settled inhabitants. In Fig. 210 the settled population has been plotted according to the census of 1935 (222).

The earliest white pioneers were fur-traders whose interests were the opposite of those of a settled population. They had no permanent stake in the
country. Few of them, if any, had brought wives with them, and none of
them thought of cultivating the soil. They were a floating population (510)
and probably for the most part disappeared by assimilation amongst the
Indians.

It may be said that the varying supplies of cod and salmon, fur-bearing
animals, seal and whale are still the main support of the permanent white
population of Labrador.

The semi-settled fishing population soon found it economically advantageous to combine with their own the occupations of the aborigines — the caribou-hunting and fur-trapping of the Indians and the sealing and hunting of
the Eskimoes. This combination required however a semi-nomadic life: the
Eskimo life in summer at the coast and the Indian occupations in winter in
the hinterland forests. In that way they laid a secure foundation for their
permanent existence. We shall later see that quite early a certain adaptation
to their milieu too can be traced; thus those who settled on the outer edge
of the sea obtained their chief livelihood by sealing and fishing, those who
settled in the wooded bays turned chiefly to hunting and trapping.

I have only been able to obtain a vague picture of the life of these people
in former times. At present there is all through the district a tendency to
form two specialised types: T r a p p e r s, whose subsistence depends almost
exclusively on the catching of fur-bearing animals and who have only one
fixed home, and L i v e y e r e s, the chief sources of whose livelihood are
fishing and seal-hunting, but who also engage in trapping, and who yearly
migrate between two dwellings. Yet no clear-cut line of demarcation is
possible between the two categories; there are scarcely any who are purely
fishers or purely trappers, for the most important occupation is determined
by local resources and the opportunities to exploit them. The energetic
trappers of Dove Brook in Sandwich Bay (p. 721) could just as well be
termed liveyeres. In winter they are purely trappers, but from the end of
the trapping season till its beginning in autumn they follow the typical occupation of a liveyere. Yet I have retained this classification in order to present
more clearly the variations between the two types.

Trappers.

It is perhaps surprising that in the limitless forest lands, which for example
between Indian Harbour and Lake Ashuanipi extend for some 700 km, it is

scarcely possible to find any large area which is not visited and exploited in winter by trappers, white or coloured. Moreover, it is not the fortuitous hunting of wandering people, but a planned and stabilised trapping. For centuries the fur-bearing animals of the boreal forest belt have attracted men to set their traps there: the marten, mink, otter, beaver, weasel, ermine, musk-rat, lynx, some bears and foxes, both the valuable silver and black as well as the ordinary. There too the furs of Labrador are considered amongst the finest in the world, their closeness and softness being due to the low winter temperature. The trapper's income varies naturally with the quality of the skins, but is usually good, and therefore a great number of them have established themselves in Labrador. The skins are most beautiful at the end of October and that is when energetic trappers ought to be on their land. According to its distance from their summer homes the end of September or beginning of October sees them moving along the watercourses to the wilderness.

At the Straits of Belle Isle the trappers start from the river mouths in autumn towards the river sources, in the same way as the dwellers by the rivers St. Mary, St. Lewis, Alexis, and George, and those who inhabit the small gulfs towards the north up to Cartwright. The trapping-lands of the Dove Brook trappers stretch from Sandwich Bay up to some 80 miles inland along the valleys of the White Bear and the Eagle Rivers, and in the valley of the Paradise River they reach as far as the water divide towards the Alexis River. The men who live at the upper end of Lake Melville, again, have several hundreds of miles to go; while those in the Makkovik and Hopedale district have no more than 70 or 80 miles. These variations in distance of course modify somewhat the customs and habits of the different groups (70, 645, 717, 910, 993).

As an extreme example we shall first take the trappers who travel the greatest distance and have their lands within the Lake Melville water system; they generally go under the name:

NORTH WEST RIVER OR GRAND RIVER TRAPPERS.

A tacit agreement has developed with regard to individual rights to use certain f u r-p a t h s, i.e. certain stretches in the forests where a trapper may lay out his traps to the exclusion of all others.

Principles of custom for trapping.

When a man goes up the river beyond an established fur-path he builds a 'tilt' (blockhouse) in a suitable spot and sets out his traps, and from henceforth the ground is his and his descendants'. Trapping-paths are thus handed down from father to son. Here custom raised an invincible barrier to theft of property or indiscriminate trapping (1123). In some cases paths are rented by the season to trappers outside the family by tacit consent; the owner then generally supples the 'tenant' with a canoe, a tent and a stove and in exchange receives a third of the catch obtained. This custom fits into the interesting conditions of inheritance among the whites of the area. The youngest son, at least among the North West River trappers, is the general heir; the elder sons usually marry early, have already their own households and have laid out their own fur-paths, while the youngest still lives with his parents. When they grow old it is he who looks after them and does the heavy work, fetches wood, runs the nets and so on. Even after he marries the youngest son usually remains at home and continues to care for the old people. In these conditions it is quite natural that he in return inherits their house, traps, nets, motor-boat, guns, and trapping-ground. — It is interesting to find here the same inheritance arrangements, in essentials, as among the semi-nomadic forest Lapps in Petsamo.

Individual trapping-grounds and their tenants in 1937.

On the map, Fig. 253, I have marked the different trapping-grounds according to information given me by the trappers themselves who even helped to draw them on my excursion map; my gratitude for this is chiefly due to HENRY BLAKE, MURDOCK McLEAN, GRAHAM BLAKE and JORDEN GOUDIE.

The numbers of trappers in the different areas are given below; cf. the map:

I. Grand River district:

1. Robert Michelin — Traverspine.
2. Carl Hope — Muskrat Falls.
3. Stuart Michelin — Sandbanks.
4. Charlie Graves — Porcupine.
5. Joed Blake — Gull Island.
6. Harvey Montague — Horseshoe.
7. Joseph Blake — Blake Waters.
8. Dan Michelin — Mounis.
9. Henry Blake — Mounis Rapids.
10. Gordon Goudie — Lake Winnikapau.
11. James Baikie — Elisabeth River.
12. Dan Campbell — Metchin River.
12. Donald Baikie   »   »
13. Wilfred Baikie — Goose Cove.
14. Edward Michelin — Big Hill.
15. Arch Goudie — Grand Falls.

16. Jorden Goudie — Flour Lake.
16. Victor Goudie
17. Air Best — Gabbro Lake.
18. Cecil Blake ⎱ Valley River.
19. Victor Goudie ⎰
20. Ray Blake — Osokmanuan Lake.
21. Francis Tiffani ⎫
22. Walter Blake ⎪
23. Herbert Michelin ⎬ Lobstik Lake.
24. Fred Blake ⎪
25. Dick Blake ⎭
26. Henry Baikie ⎱ Sandgirt Lake.
27. Fred Goudie ⎰
28. Will Bird ⎱ Michikamau Lake.
29. Henry Mesher ⎬
30. Omen Mesher ⎰
31. Donald Baikie ⎱ Michin River.
32. Dan Campbell ⎰

II. **Kenamu River district:**
  a. Ross McLean.
  з. Wallas McLean.
  b. John Whity.
  c. Irum White.
  d. Donald Michelin.
  e. Mark Best.
  f. Henry Best.
  g. Douglas Best.
  h. Philip Blake.
  i. Alvin Blake.
  j. Jim Goudie.

III. **Goose Bay district:**
  k. John Grove.

IV. **Goose Bay River district:**
  1. Robert Best.

V. **Grand Lake district:**
  1. Sid Blake — Caribou River.
  2. Willie Baikie — Berry Head.
  3. Robert Baikie — White Cockhead.
  4. Friman Baikie — Beaver River.
  5. John Montague — Red River Monikyn.
  6. Colvin Bromfield — Mable Island.
  7. Bent Blake — Seal Island.
  8. Gordon McLean — Crooked River.
  9. Murdoch McLean — Wabustan River.
  10. Bent Blake ⎫
  11. Duncan Blake ⎬ Seal Lake.
  12. Herbert Blake ⎪
  13. Tom Blake ⎭

VI. **District on north shore of Lake Melville**
  14. Fred Ritchie — Butter & Snow.
  15. John Michelin ⎫
  16. Edward Michelin ⎪
  17. Charlie Michelin ⎬ Sebashkau.
  18. Gushne Michelin ⎪
  19. Peter Michelin ⎪
  20. Hildred Michelin ⎭
  21. Robert Baikie ⎫
  22. Tom Baikie ⎪
  23. Bert Baikie ⎪
  24. Hugh Campbell ⎬ Mulligan River.
  25. Cyril Campbell ⎪
  26. Glyn Campbell ⎪
  27. Ramen Campbell ⎭
  28. Russell Chaulk ⎫
  29. Byron Chaulk ⎪
  30. Pierce Chaulk ⎬ Pearl River.
  31. Holam Chaulk ⎭
  32. Peter Shepard ⎫
  33. Rennel Shepard ⎬ Valley Bight.
  34. Bill Shepard ⎭

VII. **District on south shore of Lake Melville:**
  35. Sam Bromfield
  36. Irum White
  37. Jim Gear
  38. Math White

VIII. **Lake district on the southern water-divide.**
  39. Reginald Blake ⎫
  40. Graham Blake ⎬ Minipi Lake.
  41. Isaac Rich ⎪
  42. Solomon Seward. ⎭

Perhaps the divisions on the map ought to have been drawn with two lines: one — the t r a p-l i n e — stretching to the one side, the other on the opposite side of the watercourse and both mainly at right angles to it with deviations due to the terrain. Yet it was impossible to draw these fur-paths exactly with the material available. As no white trapper with any hope of great success would set his traps between the traditional fur-paths, schematic boundaries had to suffice.

The map also shows that the trapping-grounds in the Hamilton and Naskaupee Rivers' system are rather closely covered. The number of traps here is estimated (in 1939) at about 15,000. But within the whole of the Lake Melville system it is considerably greater. To the former must be added the Indians' traps and those of the trappers on the eastern part of Lake Melville, chiefly east of St John Is., in the district of Rigolet and at Backway, both white men's and Eskimoes'. — The Rigolet people set their traps chiefly around Double Mer.

In consequence of the inheritance arrangements already described new land is constantly being taken up for trapping as the population increases. The men of Lake Melville system — when excluded from the grounds in the east by the Hopedale and Rigolet trappers — tried to extend northward, towards Davis Inlet for example. But this has not been successful; the country is mostly treeless, only along the watercourses are there densely wooded tracts, some miles in breadth, and in them the most valuable fur animals — the marten and the beaver — are rare or whole absent; so these trappers returned to the Lake Melville drainage area (187). Since then the chief expansion has been westward.

Thus the trapping-grounds are branching ever farther inland from the coast (cf. p. 649 sq.) to areas formerly exclusively Indians'. The grounds are becoming more and more covered with traps. From place names it appears that the Indians in bygone days spread their hunting as far as to the coast.

Valley trappers and Height of Landers.

The Hamilton River trappers usually divide themselves into two groups: valley trappers and Height of Landers (= the trappers working near the water-divide in the west). The boundary between the two lies at the Big Hill Portage where the Grand Falls must be avoided. This portage rises steeply 700 feet in a quarter of a mile, and here the deep part of the valley is left behind in reaching the Lake Plateau. According to information I received the trappers had scarcely any grounds on the Lake Plateau at the turn of last century. Attempts were made to take up land there for trapping, but the Indians destroyed the traps and burned up the blockhuts (644). Since then the white population has rapidly increased and it has been absolutely necessary to extend their trapping-grounds also on the Lake Plateau. Only some small remnants of the great-family areas of the Montagnais Indians (p. 637) are still used by the Indians. But not even those are recognised by the other Indians or the whites as inherited Indian family grounds from which new comers are excluded (cf. p. 634 sq.).

As is seen from the map the Indians are practically excluded from the largest valleys and are now compelled to seek their livelihood on trapping-grounds of second quality. It could be presumed that many conflicts, fights, and arson would result from this, as has happened in other places where aborigines and invaders have met in competition for the same districts. Yet here the Indians demonstrated their disapproval and indignation by setting their traps alongside of the trappers' fur-paths, as Mathieu André did beside Fred Goudie's at the beginning of November 1930 (644), to the disadvantage of both. I was also told that sometimes outraged trappers have marched, loaded rifle in hand, to reclaim their »stolen furs» (731). But no continued, serious conflicts have occurred. Many among the trappers I talked with about this matter could speak the Indian language as well as the Indians them-

45

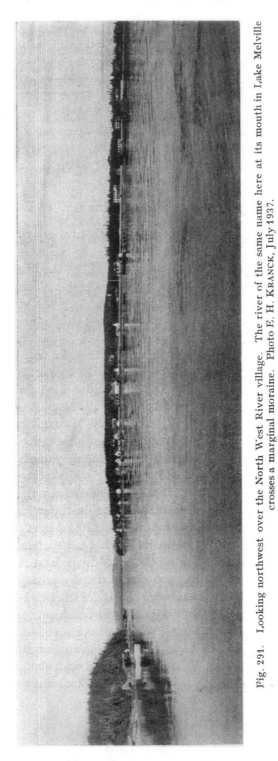

Fig. 291. Looking northwest over the North West River village. The river of the same name here at its mouth in Lake Melville crosses a marginal moraine. Photo E. H. KRANCK, July 1937.

selves and seemed to have had the most friendly relations with them since childhood. It is indicative that when the Indians visit the white communities they generally settle down beside the house of a good white friend, whom they call an »old chief», and usually ask his advice on all possible questions concerning the mysteries of the new civilisation. There is friendship between them.

As already indicated our map illustrates a cultural-historical event which appears to have general validity for the primeval forest areas of the subarctic: the districts of the small groups of aborigines become gradually surrounded by the white men as they spread out, and then by mass effect are either assimilated or exterminated (cf. p. 649 sqqq.).

Fixed dwellings.

All the Grand River trappers have only one fixed dwelling and around this the different phases of their life circle. Most of the trappers live in small, neat villages. At the eastern end of Lake Melville there are three: North West River — Figs 291 and 292 — at the outflow of Grand Lake, M u d L a k e half-a-mile from the mouth of Hamilton River with 10—12 weather-worn houses, a little white church and a cemetery, and the third little community has arisen in M u l- l i g a n on the north-west shore of Lake Melville. However, some trappers still prefer to live alone in the wilds. For instance Fig. 293 shows J o h n Grove's fine house surrounded by primeval spruce woods on the outer edge of

Fig. 292. Looking from the east over a part of the North West River village. To the right from the wharf is the Hudson's Bay Company's complex, to the left Int. Grenfell Association's hospital and school and private houses. In the lower corner to the right the football ground, from here obliquely to the forest the doctor's house and from thence obliquely to the right in the forest the Church of England church. Most of the private houses are outside the picture. The number of inhabitants is about 250.
Aerophoto of the author, Aug. 3rd. 1937.

the delta which the Goose River has laid down in Goose Bay. His nearest neighbour is his brother, whose house lies some miles farther into the delta on an arm of the river, but his other neighbours are about ten miles away. Carl Hope's house lies twenty-five miles up the Hamilton River, at the tide limit like an outpost towards the wilderness. These small centres of European civilisation present many surprises, both great and small.

North West River offers a delightful view from a distance, its white painted houses with red (= Hudson Bay Company's buildings) or black roofs (= trapper dwellings) standing out so beautifully against the background of the spruce forest. The village makes one think of a neat little Danish bathing resort, though the spontaneous orderliness of the village planning, Fig. 292, is different. In North West River the population live in the most intimate contact with two important disseminators of culture in Labrador: Hudson's Bay Company and the International Grenfell Association, and the community clearly reflects their good advice and instruction. The walk along the shore at North West River is dominated in the centre by the Hudson's Bay Company's complex (the managers' houses, the houses of the staff, the shop, workshop, storehouse etc.), not far off to the east are the Grenfell Association's buildings (hospital, school, doctor's house, workshops, etc.). Beside them lie the trappers' well-built, white-painted homes. I have had the pleasure of being a guest in some of these last, both old and new, and the interiors made the same impression on me as beautiful miniatures; I shall always remember the atmosphere of warm, dignified intimacy which prevailed there.

Journey up the river.

Let us now devote some attention to the occupation of the trappers. Thanks to the accounts of my trapper friends and MERRICK's captivating descriptions I have a rather good idea of the life of the trapper within the otherwise locked doors of the wilderness. Here fur-skins are the only source of income. It may therefore be considered that the cycle of their livelihood begins when in autumn they travel up to their trapping-grounds. We shall now accompany the North West River trappers on their journey.

Those whose fur paths lie farthest in the interior start first. It has become the custom for these so-called Height-of-Landers to begin to move to the Lake Plateau on 10th September for the w i n t e r   h u n t. The lazy days of summer are as if blown away, energy has returned, will and muscles are on the stretch. Laughing and joking a little flotilla of about thirty canoes make their way from the North West River village. Most of the trappers have a canoe each, but it sometimes happens that two men find room for their equipment and food in the same one. After the flotilla has started those left behind generally let off shots into the air as long as the boats remain in sight; — just as was the custom in Lapland only some decades ago. The departure is an important event, for the separation lasts for months.

The journey up the river, with alternate paddling, poling, and tracking from dawn till dusk, is made in all haste, for some of the men must travel more than three hundred miles. But the speed varies with the weather; in the large slack waters the flotilla may be obliged to lie by the wind and it is therefore necessary to hurry in between if one is to arrive at the destination before the trapping has to begin. If a canoe gets a little hole from grazing a stone, a little butter or lard is smeared over it, and this will harden in the cold river and keep the water out till evening when it can be properly repaired. The only long rests are taken at the portages, for there depots of food are set up for the return journey. Day after day, week after week they

Fig. 293. The trapper J o h n G r o v e's house in Goose Bay, Lake Melville. In the house there is in splendid order; the men have to work in the porch with their furs, dog harness, carpentry etc. etc. and there they hang up their wet clothes to dry. The primeval forest reaches to the house; no garden plots exist. Photo the author, 1937.

continue alternately paddling, poling, and carrying their canoes until finally after a month the trapping-grounds are reached; J orden Goudie (cf. p. 703) reckons that on the average it takes a month to get to his fur- path at Flour Lake from North West River.

Trapping and life on the trapping-grounds.

On the grounds the trapper has built himself a tilt; there he instals himself with all his stores while the canoe is laid up on a scaffold somewhere near and covered with spruce branches. Those trappers who stay on their grounds all the winter and only return by canoe when the rivers melt again have considerable stores in their little houses. The tilts are of different kinds, some with a floor, some without, windows are rather uncertain, the main thing is a good iron stove. Thin and draughty walls are improved very simply by pouring some pails of water over them, this will freeze up every crack. It is said that a tilt here will last fifty years. Some are nearly as small as a kennel.

Some days after the arrival the trapping begins: the fur-skins are darkest in the middle and end of October. Little do the future wearers of the furs in 'Vanity Fair' dream of the hard work that lies behind this industry in Labra-

Fig. 294. Trappers from North West River. To the left W i l f r e d
B a i k i e, to the right M u r d o c h M c L e a n.   Photo the
author, 1937.

dor, how much wet and cold, hunger and hardship a trapper must endure in the wilds so little friendly to mankind, before he succeeds in his efforts! Now the trapper hastens to lay out the traps on his fur-path, work which takes at least a week, for he may have to lay out 200—300 traps on one path, sometimes at every quarter of a mile, sometimes at greater distances (286, 579, 599). As has been said, this is not a path in the ordinary meaning, but a zig-zagging, uncleared line, with a row of traps, and for that reason it is called a trap-line. This line must be blazed with an axe at such distances that it is possible to see from one mark to the next; a triangular mark is made in the tree beside each trap in order to find it again after the snow falls. Besides that it is usual to change the positions of the traps at times (cf. 70, 268, 599).

The trapper is no daily wage-earner like the lumber-jack. Even to choose the position of a fur-path in those hundreds of miles of wilderness is a complicated science; it must run over open ground with deep thick woods where the marten thrives, and the darker the woods the darker the skin of the 'Labrador sable'; there must be little streams liked by the beaver, the mink, and the otter, and narrow lakes with open shores and bare land near where the fox wanders. It is said that in the forests the trappers read and understand the tracks in the snow as we read an open book, and he 'lives' the habits of the animals' life. There in the wilds he feels himself the highest of all creation, he knows that he surpasses even the Indians; his catch of fur-skins is always greater than theirs. To lay out traps correctly (579, 644) demands experience and instinct; it is some secret understanding with the spirits of the animals which helps him, some say. It is an art to lay the trap so that the red fox will place his paws on it. It is cold, hard work, especially when traps are to be set for the otter and the beaver, for they must be laid under water and baited with rotten fish. It is said to be good to smear a little beaver castor on the traps, for its smell exercises a magic attraction on all fur animals. — The poisoner-trapper is unknown here.

Fig. 295. Grandmother M o n-t a g u e with her pets. Photo the author, 1939.

A trap-line is usually so long that it takes three days on snowshoes to examine the traps. For this reason the trapper often sets up six or eight small tilts, like big dog kennels, usually built up on four layers of logs, where he can rest after the day's exertions, melt the game and skin it. Thus on a trap-line there is much more to be done than an uninitiated person would think. About four o'clock in the morning he begins his day and can only pitch his camp when twilight has fallen. All day he is seeking and arranging his traps, carrying a sack with the game and the bait. When in the evening he reaches his hut he must make a fire, melt and skin his booty, eat, and bake bread for the next day (for the white man in Labrador does not eat stale bread). When everything is in order for the next day it is often past eleven or midnight. The cold often wakes him in the night, and once or twice he must get up and put wood on the stove, the fire is his guard. If he neglects to chop wood in the evening for his stove he may freeze to death in his sleep, as did John Pardy's brother. On the fur-paths the trappers carry neither blanket nor sleeping-bag, for the sack with his game is heavy enough. 'Tender in the body as a boiled fish' he gets up in the morning many a time conscious that he must 'get going' — 'just for badness' as he says —, he can rest on Sunday. With the swish, swish of the snowshoes he repeats 'I am not tired, I'm not hungry, I'm steel' (645). A lazy worm would soon be below the mark here, but the Labrador trapper can live this life; he is so trained and hardened that an endurance test is said to be a daily event for him. The wilderness is grim and heartless in its desolation but it obeys him, it rewards him, he has become its undisputed ruler, says MERRICK.

Fig. 296.   The trapper family H e n r y   B a i k i e from Mulligan, Lake Melville. Photo the author, 1937.

Trapping is both an occupation and an art, and to learn it the Labrador trapper must begin school early. And fine speciments of manhood do we find here! cf. Figs. 294—296. Ewert Michelin already as a ten-year-old boy kept his family supplied with fresh hare and ptarmigan all through the winter; he had snowshoes, his own gun and axe, a big dog and a little sledge with which he fixed and visited his traps and snares. At thirteen years of age Lawrence Hope was a good hunter. Bert Blake began his trapping in the wilds when he was twelve, Harvey and Cecil Goudie were ten and thirteen when they made their first journey up the Grand River to the trapping-grounds, and my guide, Henry Baikie began his journeys on the Grand River just about as early; already as a boy he knew every dangerous rock and eddy in that river, more than 500 km long. John Michelin was fourteen when he started out alone to his fur-path in the wilds. John Pottle was trapping alone at the age of seventeen. Russel Grove was nineteen when he started for himself in a previously unknown part a new fur-path 300 miles from home (644).

It is possible any day to see little children five or six years old shooting from canoes on the water with their own express rifles WATKINS wrote (1000). At first father and sons generally go trapping together, boys from eleven to fifteen, so that under good direction they shall be initiated into all the mysteries of the calling. When a boy is fifteen he is considered a *man* in Labrador. The result of this training is that an accident or a death occurs only extremely rarely in the wilderness (see further MERRICK, passim). 'Reading books is no good if you've got nothing to eat' the father teaches his son. But the same father himself teaches his boy to read, write and draw. There are no illiterates among the trappers. But there are indeed some calligraphists whom I think a lithographic institute would at any time welcome to their service.

This life in the wilds compels of course early selection, and most of the Labrador trappers who survive are prodigiously hardened and powerful. On his seventy-fifth birthday Joe Michelin went for a little walk of 36.2 kilometres from Traverspine to North West River. At the age of 93 years it is said that he also trapped a little, keeping more or less close to home. The fever of the wilds continued to burn in his veins, snowshoes, guns, canoes and boats played a more important role in his life than did the comforts of home. Unfortunately he was susceptible to the seeds of epidemic diseases.

It is said that there are many kinds of animals in our Lord's paddock; there are also many kind of trappers. What I say here gives essentially my own impressions of those with whom I had the pleasure of associating, and what they so convincingly related of their lives. It is scarcely possible that many Indians know the life in the wilderness better than these trappers, its charm, attraction, and dangers, and the ways of controlling it. In the modern sport of orientation the trapper would carry off the prize both on sea and land; his habits of dealing with the unknown, his ability to find his way in the dark, the fog, and the snowstorm is absolutely miraculous, he always finds the way and it is usually the shortest to the goal. He knows full well that to lose his way would in many cases mean death, one must be resolute and endure. The writer of these words began his own incursions into the wilds before he was ten years old, and for half a century he has met there great hunters and fishers. But never has he had so unbounded a respect for the white man's capability in the wilderness as when he learned to know the white trappers of Labrador. Life among the dangers and difficulties of such a country educates one to courage and resolution and the trappers stood out as both superhuman in ability and artists in life. A trapper could tell like a child of the dragon-fly's dance in hunting the mosquitoes on the sandy beaches and the dance of the hare in a moonlit glade in the spruce forest. But immediately afterwards he could, if necessary, strain nerve and sinew, and his face became as if hewn in stone. He loves his wilderness, I think it was John Michelin who said that he felt more his own master on his fur-path: 'in North West River there are too many people, one can seldom go out without meeting somebody and being obliged to chatter'. When he wanders about the forest quite alone, or lies month after month in his tilt, with no other company than the stars in the sky, the soughing of the trees and the sparks from the stove it is no wonder if he becomes reflective, enquiring and quite otherwise introspective than the people of inhabited districts.

In this way month after month passes on the fur-path. Each man's rights are clearly defined and there is no encroachment. If in the course of his travels a man comes upon another man's trap and there is a fox between the jaws, he will take it out, hang it to a tree close by and then re-set the trap. This practice is strictly followed, not from force or fear but by mutual regard for traditional behaviour. 'Good Custom is better than Law' is the trappers' motto. This is the unwritten law that prevails in the land of the trappers. Woe unto the man who dares to break away from the 'good custom'! The whole body of stern Puritans would turn upon him in righteous indignation, and his trapping days in that district would be cut short. With this state of affairs existing small wonder is it that the trappers instinctively dislike legal injunctions (1123).

To go round his traps usually takes a week it is said, but on the seventh day he rests wherever he is. The North West River trappers are Scottish Presbyterians and therefore always keep the Sabbath. Sometimes visitors come, and charm and intimacy may steal into the soul, and it is as if a closed book opens. The experiences of life unroll as in a film, tales and ghost stories are told and they play cards. As a matter of fact the rules as to what one may or may not do on the Sabbath form a quite complicated dogma among the trappers; see MERRICK.

Journey home.

As Christmas approaches restlessness comes over many of them: the longing for home breaks out: Going home! It sounds as if it was just a little walk, whereas for many it is a tramp of hundreds of kilometres. Some yield to the call about the 20th December, others not until about the 20th January; the store of food seems here to be the deciding factor, but the longing to get back to a young wife can also be overpowering. Yet if his store is sufficient and he can overcome the longing for his cosy house and the family the trapper stays until the ice has left the rivers, for then he can go down by canoe. Yet he is usually forced by lack of food to begin the troublesome journey home on foot as soon as the trapping is at an end, in which case he must trudge back later to fetch his canoe. The cause of his lack of food is often that he has too generously shared his store with needy Indians or other trappers.

The journey home on snowshoes in midwinter is no walk on roses. The trapper packs his skins and his tackle on to a tabanask, and this he himself must drag in all weathers; in a thaw, which is not rare if a cyclone comes and softens the snow, the tabanask can be unbearably heavy on account of its large friction surface. The nearer he gets to his home, the more he hastens;

he can attain even 6.5 kilometres an hour; there have been cases when trappers dragging tabanasks have in 48 hours covered 130 km in completely untrodden land. What I have already said about the confused growth of Labrador's primeval forests should be remembered here (p. 383 sq.).

Journeys in connection with the winter hunting comprise for 'Height-of-Landers' about 600—700 miles in the worst case. As all trappers own a span of husky dogs one is somewhat surprised to hear that they prefer dragging their own heavily laden sledge to taking the dogs to the fur-path. But the dogs would be more trouble than use, they say; they would eat at least one pound of food per day and if one must go hunting to procure them food trapping would become suffering for them as it is for the Indians (p. 651 sq.). Besides this husky dogs, Fig. 297, are born thieves; in an unguarded moment they will swallow all they can get at, 'from mitts and woollen jackets to blood-stained penknives', they soon learn how to get loose, and wander about and steal from the traps. Under these circumstances it is easily understood that the trappers do not care to take the dogs with them. The only animal that I have heard can be trained to be

Fig. 297. Sledge-dogs on summer vacation (536 a).

an extraordinarily good help in this work is a cross between the husky dog and the Newfoundland.

There is joy in the village when the trappers return. When most of them are back, usually in the middle of January, there is a real midwinter celebration in North West River, usually lasting three days. People stream thither with their dogs and their children even from Sebashkau, Mulligan and Pearl River, north of Lake Melville, as well as from Goose Bay and Mud Lake south of it. There the old dances — the quadrilles, lancers etc. — are danced in the schoolhouse by girls in fine dresses, and healthy, strong, elegant trappers; Fig. 294. And no one ever gets drunk.

Spring hunt.

In the middle of February, however, the wilderness again calls its faithful children. Muscles swell, snowshoes are strapped on and the trapper starts out on his so-called s p r i n g   h u n t, which usually lasts till the middle of April. Now it is good to have a dog to draw the sledge, and usually a little Indian dog also which sometimes even helps to draw: clothes and other light things are packed into a seal's skin which the little dog can drag. The days now begin to be longer and the small party trudges from 5 in the morning till 8 in the evening, covering often 35 miles in the time!   The going is easier since the snow is more closely packed, and in the middle of March, after the midday thaw, the surface hardens. Now the track is fine and one might travel all day. Yet the fur of the animals has become discoloured in the sunshine and the skins are of little value. Therefore also trapping is forbidden by law from the 15th of March; skins taken then are forfeit to the authorities.

At the beginning of April many take advantage of the hardened track to carry special stores up to their main tilts by dog sledge, because if the worst comes to the worst the trapper must rely on his cache of food. When all has been put in order and the work with the tilts and traps is finished the trapper has something better to do than to lie up there and wait for the summer. The canoe is therefore placed on a *catamaran* — a low sledge with broad runners — which the dogs drag home. Not until late in June, when the freshet is over, can one travel down the rivers by canoe.

As can be seen from the above the trapper's life is not an easy one in our eyes; if on the one hand it offers much of real romance and gives him a surprisingly good income in some years, it is on the other hand full of privation and hardship. Many of the white fur hunters in the Hamilton River valley, however, know no other life. And just as hunting attracts them, so does the free, strong life in the wilds where so much happens.

Home duties and home life.

S e a l - h u n t i n g calls the trapper to the traditional spots between Sandy Run and Green Island where the wealth of seals is great. One haul of the net has been known to give up to 30 of them. North West River trappers have also caught seal at the blow-holes near the ice edge outside Hamilton Inlet, but this is difficult and dangerous, for an old Greenland seal could easily drag the hunter into the depths if he were not on the alert. Like the Eskimo the trappers consider the seal's liver, heart and tongue as delicacies, the flesh of the young ones is wholesome food; the carcase goes to the dogs. The skin is used for all kinds of clothes, dog's harness, etc. as among the Eskimoes.

I c e-f i s h i n g, too, attracts people in April. In May the ice begins to break up, lanes are formed. Then the migratory birds come back: geese and ducks. Men begin to h u n t f r o m c a n o e s; the hunter generally hides behind a little sail like the seal hunters use in Fennoscandia —, and with this camouflage he resembles so closely one of the ice blocks drifting about in the mirror of blue water that he can easily get very near to his prey.

But also work is waiting that never comes to an end: tree-felling and cutting, which for one family, including its dragging home by dog-sledge or floating in summer, its chopping and stacking, takes at least a month. Without sufficient fuel, they may freeze to death in the coming winter.

We have now come to the s u m m e r. Potatoes, cabbage, and lettuce are planted. Barrels are made and other carpentry is done in the home, salmon is caught for one's own needs, and when the cod begin to approach the coast some families — now however only a few — make their way out to them in Indian Harbour. A decade or two ago it was quite general for the families to move out, bag and baggage, twice or three times in the year, to different parts of the Hamilton Inlet to fish and catch seal. But it was a tiring journey, both men and women had to row their utmost to move against the tide. Now they have decked motorboats when the family moves, which shows that their standard of living is secured. Yet in spite of this cod-fishing has practically ceased.

One possible form of livelihood, on the other hand, is almost wholly neglected by the trappers: cattle-breeding. And yet conditions for this are good in many parts. It is true that a few individual cows have found their way into the community, as in North West River and Kenamish, but there is no question of any real cattle-breeding. We shall return to this matter in another connection.

In the summer many of the h o m e d u t i e s occupy the men.

In daily life the trapper must be able to turn his hand to many things. And as a matter of fact he does them more or less well. He is a carpenter and a wood-chopper, he must saw, do wood-work and cabinet-work; he is a plater, a mason, a boat-builder, a repairer of engines; he sails, he hunts, he fishes, he grows potatoes and other vegetables (lettuce, cabbage, carrots and beetroots). But above all he is a hunter: 'In the woods I'm like boss of the world' a trapper said to MERRICK. Many of them are the equals of the Eskimo in seal-hunting and dog-driving, and many surpass the Indians in skilful trapping and manœuvring the canoe in the seething waters. I myself have been convinced that the trappers are first-class seamen in navigating modern vessels out on the open ocean. 'Sometimes it seems as though they had taken for their own the best qualities of the three races, the Eskimo's laughter-loving happiness, the Indian's endurance and uncanny instinct for living off the country, the Scotsman's strength of character and will' says ELLIOT MERRICK after living in North West River for two years, 'I began to feel that these were happier, finer people than we had ever known before. The primitive hard life seems to produce character as civilized existence cannot', he adds. 'There are few finer trappers, probably, in the world', Dr PADDON emphatically exclaims (731). And I heartily endorse his expression of admiration.

Only in a few exceptional cases does a trapper's wife move with her husband up to his trapping-grounds, and then no farther than to Winnikapau Lake in Hamilton River. It is her lot to be the good genius of the home during the trapper's absence and she is a bundle of energy just as much as her husband. She has no one to wait on her when he is away. She brings water from the river and wood from the pile, warms the house, keeps it clean, and polishes the lamps. She looks after the children and washes the family clothes which she then hangs out to dry with snowshoes on her feet. She cleans the fish and plucks the birds, skins the hare and the porcupine; she prepares the food, she bakes, she feeds the dogs. Then she also procures food for the family by shooting ptarmigan and catching hare or ice-fish. She picks berries and makes pies. In the spring she prepares the seals' skins and makes shoes and clothes for the family: sweaters, parkas, moccasins, underclothes, overalls, dickies, etc. It falls to her lot to mend the clothes, but besides that she must make flour-bags, tents, gun-cases and sails for the trapper, and also knot salmon and trout nets (645).

Here life educates the personality; husband and wife are good comrades, happy, healthy people, who make life joyful for each other and their children. In the evening all gather round the crackling fire on the hearth and exchange their thoughts and experiences. The trapper's home is a little heaven on earth, says MERRICK.

In 1915 the young men were called upon to enlist and many made the supreme sacrifice. But many who returned had seen and learnt to know in Europe the most refined life which civilisation there can produce. But all trappers are so attached to their home district that none is willing to desert it when they have reached a mature age. I know only one exception: that is John Grove's son, who is now professor of mathematics at a Technical High School in the States. His father has been called a 'Titan in his own land'. By trapping and some trade he has amassed a fortune considerable even according to American standards, but he continues his trapping. John Grove once took a fancy to move to the United States, but after some years the longing for home was so strong that he could not withstand it. 'Here we are again, and here we shall stay' said his wife to me. There are brains, as well as brawn, to be found north of Belle Isle Straits, Dr PADDON stated with pride (731), and the world will be the poorer for the passing of the self-reliant and resourceful, long-distance trappers of the Newfoundland Dependency.

I happened to enjoy the hospitality of John Grove and his wife quite unexpectedly. During an excursion in Goose Bay we were surprised by the low tide and our motor-boat was sinking deeper and deeper into the mud while the lashing rain wet us through. John Grove, who heard the ineffective hum of our motor, understood our situation and semaphored to us in the blackness with a flashlight. When we finally got out he welcomed us as guests to his beautiful home which is to be seen in Fig. 293.

The house comprises a porch, kitchen, dining- and drawing-rooms on the lower floor and on the upper floor a sitting-room, four bedrooms, dressing-rooms etc.; one almost waded in the bearskins with which the floors were covered. My host gave me dry underclothes of such fine wool as I had never seen the equal. I spent my night in a splendid bed of which the linen was beautifully embroidered.

All the interior bore witness not only to wealth but also to a cultured home, and I believe that, for example in the circles of the officials of the Old World, one would have to go rather high up in the scale to find a corresponding standard of life. The meals gave the same impression. Breakfast at 6.30 in the morning began with fresh oranges and porridge, then bacon and eggs were served and fried salmon, and the meal finished with coffee, tarts, and orange marmalade.

This was of course a specially wealthy house I had the pleasure of visiting. The normal trapper family's yearly income is said to average about $ 1000 or a little more, and in such a family one can scarcely fill the mouths of the children with oranges brought by airplane, for they are, like everything else in Labrador, very expensive.

Incomes.

The trapper's life can thus be very profitable for those who understand it. In the winter of 1936—37 John Grove earned from his skins $ 7,000, and Wilfred Baikie $ 4,000. It seems to be generally reckoned that on good fur-paths an average income can be obtained of $ 1,000—3,000 in good years. Incomes have risen somewhat since the close time for the beaver stopped in 1936, and trappers may again catch up to eight beavers each every winter. Nowadays it is scarcely possible to count on getting more than one silver fox per winter.

Origin of the trapper communities and how they maintain their standard of life.

The North West River trappers may be thought to represent a cultural paradox. The trappers carry on a purely predatory life. Upon the interior of the country they have set no mark and no crops grow upon their land; they harvest the wild (761). One would think that in the prolonged isolation in a virgin country, far away from the motherland, the cultural range must have diminished for lack of vivifying contact (cf. 522, 743, 891, 1049).

We know that these people's forefathers were of varied origin.

Montague emigrated from the Orkney Islands when a boy; he had for years followed the life of a' planter'; that is, had engaged in the fisheries during the summer and trapped in the winter, trading with the Hudson's Bay Company and drawing his supplies from them (105).

The older families are, with few exceptions, the descendants of Scottish and English ancestors. Many of these restless spirits came here in the service of the Hudson's Bay Company when it began to set up stations on the coast (Fort Chimo 1831, Rigolet 1833, North West River somewhat later, and others). The Company brought carefully selected people to manage these trading-posts, chiefly from Europe, from England, Scotland and even from Norway. Many of the descendants of the company's managers and clerks as well as others of the staff were so charmed with the free life in the wildernes and found trapping so profitable that after their contract had run out they settled down as independent hunters of fur animals. Some of them married Eskimo women. In some families there is even a mixture of Indian blood. Thus Fred Goudie's paternal grandmother was a full-blooded Cree-Indian. One of the trappers is a half Scot, one quarter Eskimo and one quarter Indian, another is 7/8 English, the rest Eskimo. It is said that the oldest family in North West River is now in its fifth generation.

Now, it might be thought that when married to the natives the immigrants must soon have given up most of the culture to which they had been accustomed, because of transplantation and the inheritance of native patterns and deculturisation (cf. 510). Some *coureurs-des-bois* and the trappers who came from France at the beginning of the European expansion left their women behind. They intermarried with Indians and became used to Indian ways, and the purveyors of the intrusive culture disappeared in the second generation, leaving no survivals. This is, however, not at all the case in North West River. If you go to visit people, for instance on Sunday afternoon, you will find good tone and dignity prevailing in the company, which will transport you into a little centre of old English culture (170). You will find that the people have understood splendidly how to assimilate the different material cultures around them without allowing the Scottish-Presbyterian faith, the old English speech and customs of their white forefathers to deteriorate; just as little have they lost their innate *instinct of white people*. Evidently these traits of old culture were brought out by the trappers who came later and their wives preserved them with dignity. It is just the woman who intuitively takes responsibility for fidelity to and maintenance of certain traditions (139 a). She tries to make the new life as much as possible like the old. She trains her daughter in loyalty to her wishes and to be able to carry on this life after she is gone, to love her duties, to be conscientious over them and feel pride in them. Without the traditional order, 'o u r w a y', the trapper's wife fears the life would become disgusting like that of the poor savages. Thus the woman as the pivot of family life makes possible the continuance of the old home culture (510).

But of course many impulses for the retaining or the raising of the standard of living have reached this community also from prominent men who have been in the country for some special task. In North West River Lord Strathcona, the Old Grand Man of Canada, and the one who took the initiative for the Canadian Pacific Railway, is one of those who made part of his career there.

DOVE BROOK TRAPPERS

represent a still more complex type of occupation. We find in their cultural framework two basic patterns amalgamated: the trapping and the hunting of the Indian culture with (superimposed upon?) the seal-hunting and fishing of the Eskimo culture. A good deal of Indian nomenclature has also been retained in the trappers' language. The fact that the way of living here has not developed into a more specialised type, as amongst the North West River trappers, is perhaps due to the geographical conditions in Sandwich Bay which favour the more complex means of livelihood.

Trapping.

During the season, October-March, the Dove Brook trappers are out on their fur-paths. The families remain in Dove Brook and now and then their menfolk come to see them, to leave their furs and supplement their equipment and provisions. These repeated visits are possible because the distance to the trapping-grounds in the valley of the White Bear River is not so great.

When the coats of the animals begin to be bleached by the spring sun the men definitely go home.

Sealing.

Soon after their return the trappers go out to the ice edge along the ocean coast and hunt seal which are necessary for dogs' food among other things;

46

Fig. 298. In April they also carry supplies by dog teams to the places where they will fish cod in the summer: Paks Harbour or Gready, etc.; Fig. 299. They also catch some fish through the ice of the lakes. In May, when long lanes open up, the water-birds are shot when they return from the south.

Salmon-fishing.

At the beginning of June the families, with their most necessary utensils, move to their fixed salmon-fishing places in the fiord. The salmon pass up in the middle of June or somewhat earlier on their way to the rivers; they are then caught in nets and are a very important source of income. Salmon-fishing in Sandwich Bay is rationalised. The fish is taken fresh from the catch by the Hudson's Bay Company's collecting boat, and in this way is quickly exchanged for cash. As an example of what salmon means for the economic life of Labrador it can be stated that the Hudson's Bay Company in 1937 handled about 1.25 million pounds of it in the south-east part of Labrador. The fish is frozen in Sandwich Bay on a

Fig. 298.   Trappers on the way to seal-hunting in the spring.   Photo Mrs K. M. KEDDIE.

specially equipped 7,000 tons steamer, *Blue Peter*, and then sent to England, where it tastes good in London a month after it was caught in Labrador.

Cod-fishing.

Salmon-fishing comes to an end about the 20th of July, and in the first days of August the families move out to the fishing places near the sea, mostly to Paks Harbour, there to collect with trawl and jiggar the cod needed for the winter. They also sometimes catch herring in September and October. Usually, however, the families move back to Dove Brook in the middle of September.

Home duties and home life.

At the beginning of the autumn they are in the village looking after their homes and getting their tools and tackle in order before the trapping again calls the men to the woods at the beginning of October.

In between these periods the trappers do much work about their homes; they look after their potato plots; other vegetables are not cultivated in Dove Brook because of the absence of the people from their chief place in summer.

At the beginning of June the snow has melted even in the woods, but the earth is still hard and night frosts are common, the drift-ice still lies out at sea. For this reason potatoes are only planted late in June and taken up in October; in 1937 no potato-tops were showing on the 28th of June. In Sandwich Bay again there are no cows; the only domestic animal is the half-wild draught dog. The people said they do not yet keep cows or goats because they fear that the dogs, if they broke loose, would at once put an end to them: 'It has always happened like that before when people here tried to take up cattle-breeding.' Now the children must be satisfied with vitamin-free condensed milk. Salmon, bacon, salt beef and tinned food with Nestle's condensed milk (the original from Vevey) are prominent in the home food. Factory products are more and more replacing local, home-

Fig. 299. Harnessed dog sledge. Photo Mrs K. M. KEDDIE.

made products. The acceptance of the urban pattern is evidently due to their usefulness for self-maintenance.

Thus the life of the Dove Brook people is still conditioned by the country's own resources and the yearly rhythm of nature, an unchanged, hunting and fishing half-nomadism. It is thus in its essence the same kind of life as we know among the Skolts of northernmost Finland a century ago (Fennia 49, No. 4) who lived in nearly the same natural conditions. In this way one understands that in Labrador, too, the livelihood of the people of Dove Brook is quite secure, because the risks of bad years can be distributed over different occupations. The present high standard of living seems thus to be secured for the future so long as these nomad-hunters follow this persistent, self-sufficient complex of livelihood and the trapping-grounds are big enough for all.

Dove Brook with its white, green-edged houses with black roofs looks charmingly cultured against the background of the dark, unfriendly spruce forests to the north; cf. Fig. 300. This village has a saw-mill of its own (to the extreme left of the picture) and a church-schoolhouse round which the family houses have grouped themselves. Each family has its own house with storehouses and sheds; to the left of the picture George Bird's two-storeyed house can be seen, also Tom Bird's house, the church and Robbot Bird's house. In the middle of the village there is a curious building, a wooden erection with a platform on which the dogs' food — chiefly seal carcases — is kept, so that the animals shall not be able to get at it if they manage to break out of their stockaded kennels; these lie inside the wood and cannot be seen on the picture. The houses are very roomy for the children are numerous; Fig. 301.

When I visited Dove Brook it was empty of people as they had moved out to the salmon-fishing places. But one had only to enter; no house there is ever locked when the owner goes away. Robbery is unknown; a trapper never touches anyone else's things and he seems to have a very high conception of the obligations of a visitor and of the demands of cleanliness. The houses were models of order and spotlessness. — In all the many thousand kilometres that I travelled in Labrador I never saw vermin; cleanliness of home and person were generally exemplary measured by the standard of the Old World. — The living-rooms were comfortable; beds and chairs had down cushions, the tables were large. Every house had a radio, which gives them the news from England and America. No newspapers reach the coast; the people neither subscribe to nor read newspapers, for in Labrador no sensible person thinks of wasting time in this way. Though they like books and magazines very much, most of the books seen were of a religious nature, prayer books, etc.

A stranger wonders how a church was built for so small a community as Dove Brook. In reply to my enquiry the clergyman of Sandwich Bay, the Rev. Spershoot, told me that the church building serves both the spiritual and the social well-being of the people. The village folk helped in its erection and its maintenance lies directly in their hands. The white population all over Labrador is predominantly very religious; when beginning a meal the head of the house never omits to ask a blessing aloud. Twice during the winter they are visited by a priest, in the summer they travel to the main church in Cartwright. Here the spirit of social and individual ownership also comes into play as vitally as the religious feeling (cf. 510). Like people of the wilds in general they are inspired by a strong feeling of common sense and solidarity; voluntary, generous collections for different purposes are quite spontaneous among them, and co-operation is usual. Also in Mulligan a school-chapel was built in 1938 by free labour, the men of the place sharing the task. The small communities in the North West River district have had community radios for several years.

The church in Dove Brook also serves as a lecture hall; — the people love lectures, and visiting strangers are at once invited to speak on different subjects. It is no dull half-civilisation which meets us here. School education has had fine results, as one could notice after quite short conversations with the young people.

On the whole the inhabitants of Dove Brook are a healthy, happy and good-looking folk; cf. Fig. 301. Intimacy and loyalty of the most agreeable kind stamp the daily social intercourse. The people seem to have no idea of nationalistic politics. Nothing is known of universal suffrage. I met no one on Labrador who had any idea of what is meant in Europe by nazism, communism and other such -isms. All religious forms and inherited customs are strictly observed. This is precisely what JUNEK found among the people of Blanc Sablon.

CONCLUSIONS.

An interesting conclusion may be drawn from a study of these two trapper communities. If we compare the favourable economic conditions of those who live on Lake Melville and Sand-

Fig. 300. The winter dwellings of the Dove Brook people in Sandwich Bay. On both sides of the church-school are the families' homes ranged along the brook whereas the warehouses, dogs' kennels etc. are placed at the edge of the wood behind the houses; near the houses small potato plots. At the left margin is seen part of the people's sawmill for household use. The scaffold in the foreground is the cache for the dogs' food. In the foreground masses of trees which the freshet brings every spring. Photo the author, June 27th 1937.

wich Bay with those of their neighbours the Indians we get an indication
of an interesting psychological circumstance: the Indians live for the day
only and rather despise the trouble that the whites take for the future,
while the whites arrange their life with foresight and prudence.   In Labrador
the root of civilisation lies in thought for the future. The Indian has remained
at the fluctuating stage of childhood, he exists on the remnants of the higher
cultural conditions in which his forefathers lived farther south. But he is so
undeveloped that his brain cannot conceive the usefulness of uniting the

Fig. 301. Members of two families Bird from Dove Brook at their salmon-fishing place
in Sandwich Bay. In the background are seen the southern spurs of the Mealy Mountains.   Photo the author, June 26th 1937.

coastal culture of the Eskimo with his own inland culture. The trappers
have made this amalgamation and in this way the white immigrants have
raised themselves to be the undisputed rulers of the country. Their descendants may unconditionally and permanently conquer the land by means of
economic peaceful penetration.

   But the question must be asked: Will this condition of affairs continue
even after industrialisation has forced its way hither? In North West River it
can already be noticed that the desire for ready money has caused the trappers
to work for wages and neglect sea-fishing completely; so much time is lost on
the long journeys the latter demands that they find it more advantageous to

buy their supplies of winter fish than to catch them themselves. The few (five or six) families who of late years have still travelled out to Indian Harbour with their fast motor boats to fish cod seem to do so more as sport than to fill a need. In 1937 it was said that only six trapper families still fished for salmon in Lake Melville. On the other hand nearly every family catches seal for domestic use (boots, dogs' food) on Green Island. Here we meet an indication of an evolution towards a more specialised economic life and for this development the natural conditions are the best (cf. p. 746 sq.).

Circumstances are the same as in Dove Brook in the main for the white population north from Lake Melville, as well as the half-breeds in the country round Makkovik, Hopedale, Davis Inlet, Voicey's Bay, Nain, Nutak, and Hebron. The farther north one goes along the coast the shorter are the water-courses falling into the Atlantic and the shorter therefore are the stretches of trapping-ground into the interior. I have never heard that the whites there lay their trap-lines very far inland. Out from the first four districts on the other hand the islands are usually the haunts of many foxes, and there they place the traps. In the northern districts the smaller income from furs is compensated from other sources.

Here we come to another type of population, the so-called

Liveyeres.

TERMINOLOGY AND ORIGIN.

The resident white fishermen of the outer coast are mostly known under the title of l i v e y e r e s, a West of England word supposed to be a corruption of 'live here' (325). On the southern coast a distinction is also made between *liveyeres*, English-speaking, semi-settled fishermen and *habitants*, semi-settled fishermen who speak French. The former are also termed 'p l a n t e r s' or 's e t t l e r s'. The name 'planter' is erroneous; the local inhabitants when talking of planters mean fishermen from Newfoundland, who return to their homes when the season in Labrador is over. FERLAND (253) wrote: *planteurs*, je dois employer ce nom de planteur que se donnent les habitants de la côte, quoiqu'il n'y en ait que deux ou trois parmi eux qui plantent des pommes de terre. On the other hand the term 'settler' should be eliminated because it may give rise to confusion; in the coastal district north of Makkovik this is the term given to some semi-settled half-breeds of white and Eskimo; it is better therefore to avoid the word, or to use the term 'white settler' to avoid misunderstanding. North of Hopedale there are said to be practically no pure white residents except the personnel of the Moravian Mission and the Hudson's Bay Company. North of the Kiglapait Mountains there are no hamlets but only some houses on the lonely posts of Nutak, Hebron, and Port Burwell (275). Thus the resident population becomes fewer and fewer as the northern limit is approached and the harshness of the climate and the lack of food increase; cf. Fig. 210.

The liveyeres all dwell beside salt water in summer. Meanwhile there are few signs of man's presence on the shore, only some few small clusters of fishermen's dwellings fringe the coast. Some lie in rocky convolutions on the open shore at the extreme edge of the mainland, other, in some few cases, on islands or inlets or bays or river estuaries presenting an aspect forbidding in the extreme. Geographical factors have naturally largely determined the distribution of the summer settlements, but they have also restricted their contact with the outside world. If it is only possible the people settle near some lines of communication; in this respect Gready is typical of southern Labrador, having grown up at a narrow tickle separating two islands, along the shores of which small houses are scattered (275). At the northern part of the coast in the long bays some lonely families may be found here and there. At the head of such a bay will be the house of an elderly settler, whereas his sons and daughters, having married, have built dwellings of their own on either side of the bay right down to the sea (761). Most of the bays, however, are still untenanted. The map, Fig. 210, shows approximately the distribution of this coastal population. There is a visible relation between the size of the fishing village on the one hand and the calmness and strategic position of its harbour on the other.

The importance of these fishing villages must therefore not be over-estimated; one can speak of only one town on the whole Newfoundland-Labrador coast, namely Battle Harbour; Fig. 340.

The clusters of liveyeres' dwellings present a picture which differs much from those of the trappers.

Duncan (233) wrote: As a permanent abode of civilized man Labrador is, on the whole, one of the most uninviting spots on the face of the earth. This impression of the liveyeres' home country we find repeated in the litera-ture. Tucker (961) in 1839 stated regarding the liveyeres: I am not placing them very high in the scale of civilization. There are none among them who can read or write, save the few traders... utterly ignorant... no very high value is placed upon the virtues of chastity and morality... the vice of intemperance prevails.   The small fishing villages which dot the shore fringe between the misty ocean and the forest wilderness in the rear are still (1940) at the folk level.

They have arisen in different ways.

McLEAN (610) a century ago described the [planters] liveyeres on the Atlantic coast as consisting for the most part of British sailors who preferred the freedom of a semi-barbarous life to the restraints of civilisation. They passed the beginning of the summer catching salmon, which they bartered on the spot with the traders for such commodities as they required. After that they proceeded to the coast for the purpose of fishing for cod for their own consumption. Late in autumn they went to the interior where they passed the winter in trapping fur-bearing animals.

The liveyeres of to day are evidently to some extent descendants of those pioneer fur-animal hunters and seal catchers who married Eskimo or Indian women. The features and the brown skin, e.g. in Batteau and Battle Harbour, indicate in many cases that either Indian or Eskimo admixture on the mother's side must be present. In the primitive conditions of these wilds natural selection was a hard and merciless reality. Certainly many of mixed blood perished or have been assimilated with the natives. Only those more favoured by nature have issued victorious from the struggle.

But especially south from Sandwich Bay a good proportion of the liveyeres are descendants of Newfoundlanders. These stayed on the coast to take care of the fishing-rooms and property left there over the winter by the people on the schooners (p. 748) and remained from lack of initiative or ability to get away (325). Other people finding living too expensive elsewhere and dissatisfied with their settlements in Newfoundland and hearing of the plentiful fishing and sealing in Labrador came in parties from the eastern coast of Newfoundland to establish themselves here and try their luck as fishermen-trappers; in this way the south shore seems to have been settled (435). Many too began to know Labrador from passing schooners and finally settled here because of the fishing. Occasionally women came from Newfoundland and were incorporated in the society through marriage. Many of the white settlers from Newfoundland claim an ancestry originating in Devonshire, Cornwall and the Channel Islands (510). Many of the liveyeres are still Newfoundlanders by birth but Labradoreans by adoption (7).

Some of the liveyeres are however the descendants of people who immigrated direct from Scotland and England. Unfortunately there is no register which makes it possible to reconstruct the genealogical tree of the liveyeres.

The English tongue is universal amongst the eastern Labradorean liveyeres but many of those living on the Straits of Belle Isle also speak French, the language common along the Gulf Coast, especially westward from St. Augustine. There are many English dialects among the liveyeres which are very difficult for a foreigner to understand and may give some indication of the original race from which their ancestors sprang (127). A particularity of some of the Labradorean liveyeres is to give sex to all the ordinary objects of daily life; I left he [the rifle]; put he [the kettle] on; he's [the wind] come right across (761). On the other hand the local people have some difficulty in guessing the home country of an English-speaking foreigner; a lady who had wondered very much from where the writer came finally exclaimed: »You are a Scotsman.» The French spoken in Labrador has many picturesque archaisms but was not at all difficult for me to understand.

The semi nomadic life and the two dwellings.

The aspect of the dwelling-houses of the liveyeres vary from place to place. At the Straits the wooden houses are neat and roomy, ordinarily built in two floors: on the ground-floor is the kitchen, a sitting-room and the sleeping apartment of the parents, whilst the dormitories tenanted by the youngsters are upstairs. Cleanliness and comfort are common. In olden times the houses were brought ready made from elsewhere. Now to-day, since saw-mills have been built, they are constructed in Labrador itself.

But when one comes to the Atlantic coast the habitations are mostly very humble, and at lonely places the famous 'tilts' which Hind reported are said to exist still. They were generally formed of stakes driven into the ground and covered with moss and bark.

Semi-nomadic life and two dwellings west of Red Bay.

Most of the liveyeres still possess two homes as in bygone days (961, 235). The one for s u m m e r r e s i d e n c e — *la maison de large* (253) — is close to the sea at the fishing-station; cf. Figs 302 and 303. Wilder homes can scarcely be imagined than, for instance, the summer dwellings on Ragged Island, where the people can walk literally only some meters on land, all movement being on the sea; cf. Fig. 303. The w i n t e r d w e l l i n g — *la maison de terre* — occupies a most sheltered position, and is situated some distance up a wooded valley. The people from Chateau Bay, for instance, spend the winter some five miles up the valley, and the Ragged Island liveyeres move up into the

Stag Bay Brook's forests. Some of the northern liveyeres prefer to pass the winter in the Mission settlements where they have educational advantages for their children.

Similar conditions seem to have prevailed in the 1860's; PACKARD says that at the Straits several settlers had different dwellings for summer and winter, often at a distance of two to three miles from each other. About the 20th of October they would leave their summer homes and travel up to seventy miles,

Fig. 302. A liveyere's summer residence at Battle Harbour. Photo the author, Aug. 1937.

for example, along the Eskimo River for trapping and hunting caribou, fox and grouse. Early in the spring they would move down to their summer homes — 'tilts' — at the coast, for from the 25th of March to the 4th of June they devoted themselves to seal-hunting. During the fishing season — June till October — they lived in their rude summer houses.

HIND and many other travellers found it difficult to understand how white people from the downs and coombs in England can live happily with their families in these wild, barren and desolate bays, in a country of privation and hard labour. However, in spite of its hardships and precariousness the life seems to have attractions, and many instances are given of families coming to Newfoundland and then emigrating to the States or Canada with a view to bettering themselves, but returning again after a few years' trial to their old homes on bleak and barren Labrador, to poverty and starvation. On one side they have here the cold, foam-flecked swell of the sea, on the other naked rocks, swamps and bogs, terminating at the foot of hills or ridges which rise many hundred feet. The natural factors have combined to make the dwelling-places comparatively inaccessible, to isolate the inhabitants from the rest of

Fig. 303. Summer residence at Ragged Island near Makkovik, used by a liveyere when cod-fishing. Photo the author, July 1937.

the world and even to a surprising degree from one another. They live in what may be called a back-yard (510, 743).

In this elongated culture area the small communities, highly self-contained, are very alike in many of their ways (510). It is no wonder that many observers have found that the lack of progressiveness is the most conspicuous trait of these liveyeres; they are proud of their adherence to the ancient customs of ancestral founders. The imitation from generation to generation imposes a marginality upon them, JUNEK says. And yet they are ingenious and honest, hard-working people, Fig. 304. It is worth while to read the life history of the two Labrador coast dwellers, written around their assumed names, Bill Bayman and Henry Salt (761, p. 184). Here one finds a perfect instance of the restricting influence of natural conditions upon cultural development (522, 891, 1049); GRENFELL gives masses of illuminating examples (see bibliography).

Development into one fixed dwelling west of Red Bay.

The first attempts to settle on the coast of Newfoundland-Labrador goes back to the eighteenth century (cf. p. 480 sq.). In 1763 there were thirteen 'settlers' on the Labrador coast (325). Letters in the Moravian archives state in 1797 that two Englishmen had come to settle near Hopedale. Seven other Europeans were reported in the neighbourhood, two of them having married Eskimo women. In the beginning of the nineteenth century the whites began to advance upon the Moravian Mission's districts. The resident population from Blanc Sablon to Sandwich Bay in 1856 was computed at 1553 persons. In 1868—69 the liveyeres numbered 2479 between Blanc Sablon and Cape Harrison, including about three hundred Eskimoes and Montagnais Indians (325).

From all descriptions one must conclude that this isolated population was much neglected and undeveloped. Yet time has not stood still here.

At the Canadian part of the coast there is a settled population as far eastwards as Natashkwan, and between it and Blanc Sablon the population is predominantly permanent and cattle-keeping. In the 1860:s HIND wrote of the Gulf Coast people: They have some pigs and sheep and are preparing to bring cows into the southern villages. Nowadays there are fixed habitations on Newfoundland-Labrador right up to Red Bay; according to my skipper in 1939 people here have also procured cows and the families no longer move inland in the autumn for trapping.

Fig. 304. The liveyere J o h n B l a k e. He is one of the best representatives of the Labradoreans. He lost his right arm years ago but supports his large family and is free from debt. Photo the author, July 1937.

On the Straits of Belle Isle, east of Red Bay, there are some few liveyeres' dwellings. But the people here live 'like grandfather did'. In the spring they move to the sea to their cosy, pretty summer house which brightens the landscape so well to hunt seal. In the beginning of the summer they fish for cod, in August and September for herring. This is true, for instance, of the liveyeres in Henley's Harbour. To the east and north from Chateau Bay during the summer, Labrador trappers and people from Newfoundland come and fish for cod. But Chateau Bay is used by these liveyeres-trappers only in summer; during the fishing time they live there in their neat, white-painted houses. In the fall they move to their homes some three miles farther inland to the storm-protected valleys' while the Newfoundlanders sail home. The liveyeres at the coast from Cape Charles to the north move farther inland, e.g. from Battle Harbour — Fig. 302 — the nine miles to Mary's River, while the liveyeres who live in summer on Seal Island, Bolter's Rock, etc., move about twenty miles into Hawke Bay; Indian Tickle people move about thirty miles to Table Bay and Sandy Hill Bay, and so on. Thus the summer places at the coast are quite empty in the winter, for even the traders go away in October to Newfoundland with the mail steamer; Fig. 311. I think this kind of hunting and fishing semi-nomadism will continue amongst these liveyeres, because there is scarcely any pasturage for cattle on the islands and headlands where they live in summer.

In the most southerly part of Newfoundland-Labrador, therefore, an important cultural-geographical boundary runs somewhat east of the political one. With the exception of occasional visits from seal- hunters all the harbours east of Red Bay are empty of people in the winter. From Red Bay along the Straits of Belle Isle and the Gulf of St. Lawrence, on the other hand, a settled population is living in the bays on the northern shore. Half-nomadism has been abandoned. This settled population has thus grown beyond the liveyeres' stage.

Yet this population does not seem to resemble that of Newfoundland-Labrador in their life otherwise, whereas it does resemble the Canadian culture of the Gulf; not only do they keep cows, at Point Amour I found tracks of a motor car. This gives in a way a new characteristic to Newfoundland-Labrador. JUNEK (509, 510) has given some attractive descriptions of it, so I shall add nothing here.

We continue further north.

GEOGRAPHICAL GROUPS.

The geographical distribution of the liveyere stock in Newfoundland-Labrador according to PADDON falls into well-marked sub-communities, chiefly distributed round the larger bays, each of which is almost a little world in itself (cf. 731).

Battle Harbour is the centre of a considerable group living round St. Lewis' Bay; cod and salmon afford their main means of livelihood since the market for seal products has fallen away so greatly. It is a straggling group for the most part.

Another group reaches from about thirty miles north of Battle Harbour to about thirty miles south of Cartwright, with the Hudson's Bay Company post at Frenchman's Island roughly at its centre. There is little prosperity in this group.

Cartwright is the centre of the most outstanding salmon-fishery along the whole coast. There is fairly good trapping and some successful agriculture too.

The eastern basin of Lake Melville is occupied by liveyeres and is one of the worst areas in the country for living on the dole.

North of Hamilton Inlet a few enterprising and prosperous liveyere families are to be found.

Let us now take a look at the ways of living of the semi-nomadic liveyeres.

SOCIO-ECONOMIC CYCLE.

Among the liveyeres almost all means of earning are bound up with the seasonal variations. Fish is the liveyeres' quarry almost from one year's end to another (761).

The socio-economic cycle can be considered as beginning with the fishing.

Salmon-fishing

goes back to the first time the white man appeared on these coasts; already early in the 1500's there are reports of salmon-fishing.

In the early summer, from about the middle of June to the end of July, when the salmon has returned to Labrador's tickles and bays, it is eagerly caught by the liveyeres. For this purpose they use so-called salmon nets which the Hudson's Bay Company usually lend them. The amount of the salmon catch varies but is in general rather large. I was told that at some of the larger fiords some liveyere families manage the salmon-fishing so successfully that they never need to move to a special trapping-house, but this may be exceptional.

Experience of the salmon-fishing places north of Makkovik is said to indicate that the catch is steadily decreasing; Jim Lane's Bay was famous a few decades ago for its supply of salmon; now nothing worth mentioning is caught either there or further up towards Cape Chidley. When one compares the information given by CARTWRIGHT in the 1770's concerning the salmon caught in Labrador and its yield to-day it is evident that the diminution is astounding.

As the importance of salmon-fishing declines northward that of the s e a
t r o u t in July and August from Hopedale to Hebron increases. The price
paid for it is 5 cents per lb. and it is exported salted to England.

Cod-fishing.

Fishing for cod is the outstanding and primary occupation of the liveyeres.
The sea is considered common property. However, Newfoundland fishing
schooners and other outsiders are not permitted to enter the liveyeres'
fishing domain, except when the weather is so rough that the trespassers

Fig. 305. 'Jigging' for codfish (536 a).

cannot stay outside the harbour, but even then the vessels are only allowed
to cast anchor. Cod-fishing is carried on independently by each family unit,
though the success of a family depends considerably upon the indirect co-
operation it receives from other such units.

Some notes left by JUNEK concerning the work are given here.

In the southernmost part of Newfoundland-Labrador the fishing season begins late in June, when the caplin and cod make their appearance in ever-in-
creasing shoals. JUNEK found in Blanc Sablon in 1934 that during the week of
June 17—24 the caplin were plentiful and vast shoals of cod had already started
to come in. The caplin turn shoreward to 'breach' and become the prey both
of the fishermen who use them for food and of the cod. It is perhaps the best
bait used, with clams and 'squid', a species of cephalopod.

North of Hamilton Inlet the cod seldom come shoreward before the first
week of July, and the farther north the later (cf. table p. 755).

The cod are caught nearer the Gulf-coast in t r a p-n e t s. These are however very expensive and most men are unable to procure them. They therefore, like all the liveyeres of the northern coast, earn their livelihood by j i g g i n g; Fig. 305. A 'j i g g a r' is a leaden, troll-like decoy cut in the form of a caplin, and having two sharp hooks. A line is attached to it. It is thrown overboard and the fish are caught by suddenly jerking the jiggar up and letting it sink down. This is a laborious process, especially when the water is deep and not many fish may be taken in this manner. Probably every one looks forward to the day when he too will be able to afford trap-nets.

In the autumn on the other hand, when the cod goes out again toward the open sea, to deep water, the 'h o o k a n d l i n e' or 't r a w l', also called

Fig. 306.   Receiving on the stage the codfish catch from the 'jigging' boat (536 a).

b u l t o w or l o n g l i n e, with over one thousand hooks on it, is used instead of the jiggar.

The fish is brought home, still alive, to the wharf — Fig. 306 — and later conveyed to a box attached to one side of a large table in the shed. The cleaning of the fish begins. The group of fishermen arrange themselves round the table. The 'cut-throat' reaches into the box, grasps a fish by its eyes and makes a lunar incision on its ventral side across the throat from gill to gill. Then he makes a longitudinal incision through the belly of the fish thus allowing the entrails to fall out. In this condition the fish is pushed to the neighbour, the 'header'. He rips out the liver of the cod, dropping it into a special barrel under the table, tears off the head and tosses it, together with the entrails, over the side of the wharf into the shallows below, where the husky dogs are waiting to gorge themselves. He then pushes the headless fish across the table to the 'splitter'. He runs a square knife under the vertebral column of the fish lengthwise, cutting out the column on a return motion and dropping the now shapeless mass into another barrel filled with sea- water.

The cleaned fish is transported by wheelbarrow to the 'stage', a small shack where the fish are salted and stored, arranged in rows belly-up on the floor, and sprinkled with crude salt; Fig. 307. In this state the fish are left until they are ready to be dried upon the rocks or 'flakes', the last step of the cod-fishing industry; Figs 308 and 309.

This process is repeated over and over until finally the morning's entire catch has been cleaned. Very little motion is wasted in these processes until the handling of from twenty to twenty- five quintals of fish is finished (1 quintal = 112 lbs Avoirdupois; 1 'green' quintal = 224 lbs). — Often quantities of salmon

Fig. 307. Piles of cleaned codfish, 'green fish', in the shade (536 a).

get caught in the trap-nets together with the cod; the pink salmon is reserved for human consumption, the white salmon — 'shinks', — is kept as winter food for the dogs.

During August the fish left in the 'stage' is taken out and the 'making' of fish begins. The washing of the salted cod and its drying on the rocks or 'flakes' take a good long time. The flakes are a framework raised a short distance above the ground and covered with old nets or spruce bark. Lightly-salted cod if not washed and dried is soon invaded by maggots. If August is rainy and foggy and the fish cannot be placed out to dry the catch of the summer can go to waste when drying. One must be perpetually on the alert to spread the fish to be dessicated as soon as the sun begins to shine, to pile them into heaps and cover them with sailcloth or spruce bark, Fig. 309, at the slightest indication of rain or fog. This process goes on for about a week. More fish is taken out of the 'stage', washed and dried until the whole season's catch has been disposed of (510).

An important by-product of the cod catch is cod-liver oil. The liver is ripped out of the cod by the 'header' and dropped into a special barrel. When this is full, its contents are dumped into a larger barrel. Left exposed to the sun and air the liver decomposes and the oil rises to the surface. At the end of the fishing season, about the fifteenth of August, the oil is drawn off, poured into barrels and sold.

Cod-fishing goes steadily on in the northern parts of the coast the whole summer and autumn, in the southernmost parts on the other hand only until the end of July. By then at least two hundred quintals of fish per family must be caught, salted and dried in the southernmost part of our district, or economic disaster follows, according to JUNEK. About this time the dogfish, a variety of shark, comes and not only drives the cod away but also tears the meshes of the cod-traps.

Fig. 308. Codfish spread out on the rocks to dry (stock-fish). Photo the author, Indian Harbour, Aug. 1937.

Some liveyeres are employed directly by the Hudson's Bay Company in cod-fishing in the southern part of the coast. They are furnished with fishing-gear and other utensils, wharf and stage. In return they receive twenty of every hundred quintals of fish they salt and cure. They have to pay the company a rent for the boats used and the current price for gasoline (510).

The h e r r i n g generally come shoreward at the southern part of the coast late in August and in September and are caught in nets that have meshes one and one quarter inches square. The herring is cleaned, the entrails are removed, and the fish salted down in barrels. But the profit is most meagre, and herring-fishing is looked upon as a minor occupation only. It is considered more advantageous to keep the herring as winter food for the dogs than to dispose of it by trade (510).

The liveyere works exceedingly hard during the fishing season and must be difficult to beat on his own ground. When the weather prevents him from going to sea nets and lines are repaired, and in this work the women also take part. From about the middle of August the women and children are engaged in berry-picking, especially bake-apples (*Rubus chamaemorus*) and plumboy (*Rubus acaulis*).

In the late autumn wood-gathering is the second of the primary activities. It is gruelling work, from year to year the distance increases to the places where firewood is to be had. Before the snow falls the wood is cut and the resinous roots of stunted conifers dug out and piled in heaps until it can all be transported to the home by dog-sledge. At home the wood is piled in conelike heaps, which give a peculiar aspect to the villages.

Fig. 309. Piles of dried codfish covered with spruce bark ready for export. Photo the author, 1937.

Transference to the winter-dwellings and trapping.

In September or October the fishing has been brought to an end and the family transfers to their winter-dwelling.

When the snow begins to fall hunting and trapping become important activities. In the south the hunters now go many miles inland. But in the north, PRICHARD wrote, the liveyeres are said to be unwilling to penetrate into the interior: They seldom or never go over the threshold of the unknown, 'it is a place where men starve', they say. In fact, few things have struck me more, PRICHARD continues, than the almost universal fear which lay upon the fisher population of this barren coast with regard to the rear hinterland upon whose eastern lip they live. The reputation of many difficulties for travelling has attached itself to the country: 'he was compelled by the want of provisions to relinquish his project and return home', we hear again and again; they fear the sterility of the inland, difficult navigating of rivers, scarcity of food, not a track or the glimpse of an animal, and the severity of the eternal loneliness of the forests. Their belief has naturally been strengthened by the calamitous results of different expeditions, for instance LEONIDAS HUBBARD in 1903, KOHLER, CONNEL and MARTIN in 1931, two American engineers in 1930 (?) etc.

For many liveyeres there are therefore only the fox and the weasel to trap through the winter. The musk-rat and others are not encountered by them between November and May; they live around freshwater accumulations which freeze over, forcing them to hibernate. Marten, mink and lynx are very rare in the forests near the outer coast (510). The trapping and hunting goes on until the end of March or beginning of April (cf. p. 716).

Return to the summer-dwellings. Sealing.

The seal-fishery was one of the principal inducements to the first settlers on the coast, but it has since the beginning of our century ceased to be commercially pursued by the liveyeres. However, the liveyeres still take up sealing on a smaller scale, JUNEK reports. The enormous masses of ice which the Labrador Current brings from the north during March bring also thousands and thousands of seal which have sought refuge on the ice floes for the purpose of rearing their young; these are the so-called 'seal meadows'. Now thousands of men from Newfoundland are engaged in sealing. Fantastic are the catches reported by HIND; in 1827 in one single week forty-one vessels laden with 69,814 seals arrived at St. John's; one vessel took upwards of about 3,500 seals in six days (45 a, 1085).

During and after the second half of March seal are hunted openly on ice floes also by the liveyeres; they shoot them when they are sleeping on the ice floes, because the flesh of the young seal is always welcomed. This is difficult hunting for people who do not know the easily manoeuvred Eskimo kajak. Throughout the latter part of April and almost all May, which is the main season, sealing is carried on also by means of trap-nets more than a hundred fathoms long by ten fathoms wide with meshes eight inches square, which are placed under the ice, sometimes several of them together. At the Straits the nets are placed so that the entrance is to the westward because the animals are going northward. The seal fat taken in the spring is softer and more mellow and forms the pale seal oil of commerce.

*

Thus in the organisation of the liveyeres' material culture pattern traits are encountered of Eskimo and Indian ways of living. Only this could permanently guarantee their self-sufficing means of livelihood. Traits of modern Euro-American pattern which are appropriate and expedient are, however, increasingly assimilated and the new mechanical means and methods bring the liveyeres nearer and nearer to the universal culture (510). They thus

represent a rather complex social product. Though there is still an important
gap in their culture accretion: cattle-breeding. The natural conditions in
many bays seem suitable, and I was not a little astonished the first time a
liveyere told me about the impossibility to keep cattle in a village with husky
dogs: the dogs will soon kill and finish a cow, he said. Later I heard of many
instances confirming this opinion. FERLAND (252) says that up to that time
[1858] with two or three exceptions no settler had succeeded in raising any
domesticated animal because of the dogs; cats, cows, pigs and sheep have

Fig. 310. Indian Harbour, a much frequented fishing harbour with two merchants. To
the right Jerrett's trading-post. To the left S/S *'Strathcona'*. Photo the author, July
1939.

all been destroyed by them; only one pig and one goat had escaped the
general massacre when FERLAND was on the coast. GRENFELL told of a goat
which was landed from the steamer when he came to see a patient. He went
with the settler's wife up to the house, but turned before entering to look at
the landscape, and what did he see! the goat killed and partly torn to pieces
by the dogs.

Yet I cannot believe that this explanation is anything else than an in-
curable superstition. The development at the Straits of Belle Isle challenges
the old opinion, for there both dogs and cattle are to be found and still no such
accidents are reported.

In this people, however, whose food is almost exclusively fish, dietary deficiencies often result in avitaminoses of varying nature and severity (510). The flesh of the fur-bearing quadrupeds and seals as well as berries is of a certain value in this respect, but for fighting the consequences of the avitaminoses and fortifying the health conditions of these weak people the most practical means is probably a supply of fresh milk.

A closer scrutiny however reveals that edible plants grow among the hills, most of these are berry species: bake-Apple (*Rubus chamaemorus*) plumboy (*R. acaulis*), crowberry (*Empetrum nigrum*), blueberry (*Vaccinium pennsylvanicum*), strawberry (*Fragaria virginiana*) and cranberry (*Vaccinium vitis idaea*) There are three varieties of currants. The raspberry is frequently found in sheltered places in the southernmost part. The gooseberry also grows here. From high bush cranberries (*Viburnum pauciflorum*) a jelly can be made. Edible plants of the green-leaf class are alexander (*Ligusticum Scoticum*), in taste like parsley, and dock (*Rumex occidentalis*) (985, 510).

On the shores of Northern Labrador two common plants grow which could be relied on as a source of certain vitamins and foodstuffs: *Atriplex glabriuscula* Edmonst. and *Elymus arenarius* L. var. *villosus* E. Mey. (275).

SOCIAL PROBLEM OF THE LIVEYERES.

The liveyeres present a very difficult social problem for the government of Newfoundland (392 a, 97). In the 1860's PACKARD wrote as follows of those south of Sandwich Bay: Dirty, forlorn tilts, smoked and begrimed with dirt, the occupants in some cases thoroughly harmonise with their surroundings. Their features and hands are smoked dark and their rough life is more or less demoralising, but certainly law and order are well maintained on the coast, and no cases of immorality came to our ears. The pictures of these people's lives as found in descriptions by GRENFELL and others are terrible. Involuntarily a sentence of WHITBECK and THOMAS flashed to my mind: Adjustment to the subarctic regions means not only adjustment to a very cold climate, but it means adjustment to poverty of materials with which to build and work, poverty of vegetation and foods, poverty of opportunity, poverty of association, poverty of mental stimuli, and, normally, an almost complete absence of all that makes for civilization (1023).

On the whole the liveyere stock must be graded as unsuccessful, Dr PADDON writes about those on the Atlantic coast (731). Why?

First, these bay sub-communities have sprung from small groups of settlers aggregated around trading-posts, and it is well known that paucity of bloodstock is prejudicial alike to physical and mental welfare in any community. Moreover the fusion of white with native blood was bound to result in a very difficult problem of adjustment for the half-breed. Perhaps the Indian or

Eskimo instincts predominated whereas the family life and surroundings tended towards the European. Such a man needed understanding, encouragement, stimulation and firmness and too often met with the opposite. To Dr PADDON's personal knowledge, many of those stigmatised as hopeless when left to themselves can be both docile and reasonably industrious men when supervised and handled with common sense.

Secondly, choice of occupation is at a minimum. Trappers, especially, are born rather than made; and many men are fooling with a few traps rather than aiming at self-maintenance. It is to no trader's interest to fit them out for trapping; it is almost certainly a losing game. At the same time they crowd out the trappers who might succeed in some districts at least.

Thirdly, the problem of settlement life versus isolation is one that hampers not a few. Attracted by church and school facilities or the chance of a job there is a tendency to congregate unduly at times. Fuel soon gets cut out for miles round a settlement of any size, there is not enough game and fish to go round, and there is danger of a contaminated water supply. Moreover some really best fitted for settlement life persist in isolation, and drag down their nearer neighbours who are within sight of success if unencumbered, while others who might succeed in a humble way if living and working on their own account fail as members of a group.

Fourthly, there are groups of human limpets clinging to unproductive rocks without hope of any success and with no higher ideal than the dole, while there are rather favourable tracts elsewhere either totally uninhabited or under-manned.

My skipper — WILL SIMMS — who knew very well several of these despised liveyeres south of Hamilton Inlet, gave me a greatly modified picture of their conditions. He said that at the beginning of this century they had in the main worked themselves up into an energetic and relatively wealthy population, who lived in good houses, were well dressed and well fed. The complication which followed in the tracks of the first world war had, however, altered the picture: Poverty seemed to have come to them again. Constant undernourishment is showing its influence on the people's physique, and the standard of living is falling every year. They see that the yield of that exhausting work, fishing, is only losing in value, so it is not surprising that many of them have lost hope of a brighter future, they have resigned themselves to the worst, suffer in silence and lose interest in life generally; they get the dole of six cents per day and with that they can of course keep alive, but the physical condition of the people is slowly deteriorating; the new generation is born and brought up in extreme poverty and hunger. For purely physical reasons many of them are obliged to give up trapping in the wooded valleys; they lack the strength and the food, and then set their traps in the neighbourhood of their winter homes where the game caught is scanty. The majority earn their livelihood by fishing.

From another quarter I heard an opinion which surprised me at the first moment: that these people's poverty was due to the district being 'over-populated', that is to say in respect of the trapping. The woods in the river valleys in the south-eastern end of the Atlantic coastal strip of Labrador are thin and considered comparatively poor in fur-bearing animals, and in support of this it was said that not even the Indians, who have of course an unusual ability to move all over the land, carry on trapping there. These liveyeres therefore only set traps on a small scale for the foxes around their homes.

This, however, cannot be the reason for their poverty, it was said from other quarters presumably in a position to judge, while referring to the fact that the sea offers an enormous quantity of fish and seal, and if the people were not indolent they would not need to be in want because of nature's forbidding harshness. He considered that a contributory cause of the depressed situation was that the standard of living had been raised far too high during the prosperous period of the world war, after which the decline was catastrophic in its effects.

Of course the author cannot take up any personal standpoint on this question. A specialist among the traders told to me that he reckoned that a little family of these 'longshoremen' south of Cartwright required an annual average income of $ 160. According to the same informer the yearly income of a normal family in the Dove Brook community (p. 721) was about $ 600, whereas the trappers of North West River (p. 719) have a yearly average income of $ 1000. These figures place the economic position of the liveyeres in strong relief. What can be the reason for the striking difference which distinguishes socially and economically the southern liveyeres from, for example, the Dove Brook people, who could strictly speaking also be called liveyeres (p. 747)? In our days when one looks from the south to the north it is obvious that the modern influence in a general way diminishes in this direction. But no interrelation seems to exist with the degree of prosperity and the increase of progress in modern ways; JUNEK also found that the cause is not one of distribution only. There must be others, e.g. racial, or historical causes or both. Observers who have been able to watch their development for years have told me that the different origins of the ancestors of these people are to a great extent responsible. The immigrants from Newfoundland have seldom been able physically and mentally to measure themselves against the strong and healthy, enterprising immigrants from Scotland and England. Nor have the latter had to drag with them the old sour dough of prejudice and outlived custom like the former. In the struggle for existence under the hard conditions on the coast these basic differences between the two population elements have become more and more distinct and divergent; the former have mostly been unable to preserve the physical vigour and mental buoyancy without which man is subdued, gives up hope and finally sinks into the hard, subarctic milieu, while most of the latter have worked themselves up to a condition of being well off and retaining their mental elasticity and good traditions. It must, however, also be mentioned that even among them certain symptoms of degeneration can be observed, e.g. imbecility and diseases traceable to avitaminosis and other scourges of the wilderness.

It is to be deplored if I am making a mistake. But I have been unable to discover any one who has a thorough, objective interest for the temporal welfare of these people since Sir WILFRED GRENFELL's death put an end to his unselfish activity for them. And it seems that nobody has yet been appointed to look after their temporal interests.

Supplementary information regarding the liveyeres is to be found in GRENFELL's work and the fairly numerous articles written by more or less occasional travellers to the Labrador coasts, and the reader is referred to them (6, 7, 13, 14, 15, 88, 97, 233, 248, 268, 321, 337, 343, 364, 380, 384, 392 a, 485, 509, 510, 522, 709, 718, 820, 833, 920 c, 961, 1095, 1123).

A retrospect. Differentiation of the white population with permanent abodes in Labrador.

The white people of Labrador are of very motley origin. While the new comers were adapting themselves to the harsh natural conditions they became welded, *in statu nascendi* so-to-say, into a new people at first quite homogeneous. At the time this began to happen contacts with the outer world were only occasional, and in order to procure the necessaries of life they were obliged to manage in some way with what the land and the sea could supply. To do this they adopted a half-nomadic way of living which combined the hunting and fishing of the Eskimoes at the coast with that of the Indians in the interior (cf. 522, 743, 891, 1049).

The variations from tract to tract of the nature of the country slowly brought about an internal differentiation of the people. In those parts where the hinterland is rather narrow, the valleys are occupied by comparatively thin woods where there are few fur-bearing animals; in such cases the white inhabitants' means of livelihood became centred near the coast like that of the liveyeres. If on the other hand the hinterland was of considerable depth and the distance great from the bay heads to the sea their means of livelihood became like that of the trappers centred on the fur-bearing animals and the forests that were rich in game.

As the coast became gradually included in the network of Canadian means of communication, the cultural environment also became changed, chiefly by the application of numerous loans. The liveyeres slowly gave up their semi-nomadism and adapted themselves to the new conditions where cattle-rearing was possible, and fixed dwellings grew up as in Red Bay and districts west of it. In other parts, this half-nomadism with seal-hunting and fishing as its integral part still survives, but is accompanied with a languishing existence, with decreasing income, increasing hardships, degeneration of the people, and increasing, secondary primitiveness.

The trapper communities have arisen by the same amalgamation process as the liveyeres', but the trappers' ancestors seem to have been of other stuff than the latter's. The strong, energetic and enterprising trappers made their way hundreds of miles into the dense woods of the interior where the precious fur-bearing animals live, and they developed such skill in the capturing of these and other game that even the Indians could not compete with them. Their income from the furs increased every year and was infinitely greater than that of the liveyeres and the Indians. They could therefore omit from their annual cycle the troublesome stage of being at the coast, the semi-nomadism was given up and the trappers became settled. Prosperity will increase still further when cattle-rearing becomes included in their economy. An invaluable advantage of settled dwellings was the possibility of having schools for the children and of linking up the adults with Euro-American civilization.

*

Let us now consider the second division of Labrador fishers, comprising the men of the fishing-fleets, who are only summer visitors, coming down through the ice from Newfoundland in the early days of June and July and leaving in October.

The floating population.

Already in the first years of modern times that part of Labrador bordering the Straits of Belle Isle was visited every summer by fishing vessels from Europe. Since DAVIS in 1586 discovered the fishing banks off the Atlantic coast of Labrador it seems probable that more and more such vessels have also paid longer visits to the coast every year. At the present time Newfound-landers are persistent, annual visitors. They come in boats of all sizes, fish

Fig. 311.   The Labrador mail-steamer *'Kyle'* (536 a).

here during the summer and return to Newfoundland with their catch at the end of the season.

To get an insight into the nature and working of the fishery in Labrador it is appropriate to make acquaintance with the terminology used there for the different kinds of fishermen. A distinction is made between three types:

R e s i d e n t.	T r a n s i e n t.	
Liveyeres or	*Planters* or *Stationers*	*Floaters*
Long-shore men	(white shore fishers	*Labradormen. Bankers*
(white, or Eskimo	mostly from Newfoundland)	(white schooner fishers
half-breeds)		mostly from Newfoundland)

'Shore-men' are thus either resident or transient. The floaters and planters never penetrate the interior or even pass out of sight of sea-water as do the liveyeres.

The cod-fishery must be studied in connection with the fishery of New-foundland. It falls into three divisions: the first in point of time is the deep sea 'b a n k   f i s h e r y' (996), which begins in the early spring on the banks off Newfoundland. The second, and by far the most important, is the 's h o r e   f i s h e r y', which begins in June and may last till October. The third division is the 'L a b r a d o r   f i s h e r y' beginning in July and lasting also till October (6, 13, 14, 15, 102, 234, 332, 364, 341, 342, 343, 329 a, 644, 883, 1028, 1083 b). It is the last division which we shall study here.

Fig. 312. Fishing schooners, 'floaters', 'going down to the Labrador'. Labrador schooners are splendid vessels, they are called the world's best sailers, they can do their 11—12 knots in a strong wind. Unfortunately their skippers are not always correspondingly skilled in navigation and wrecks are not unusual. (536 a.)

Planters or 'stationers',

also called 'l a n d s m e n' (1051), are as already pointed out seasonal visitors to Labrador.

The planters are mostly recruited from eastern Newfoundland, e.g. from Conception Bay, Bay of Islands, Bonne Bay, Trinity Bay and Bonavista Bay.

A r r i v i n g   a t   t h e   c o a s t   f r o m   t h e i r   h o m e s t e a d s.

At least five hundred planters travel as passengers on the earliest schooners or mail steamers going to Labrador; Fig. 312. One must come early: 'First come, first served', is the rule here. Many times it has happened that the stationary crew on their arrival at their customary posts have found the best trap berths in the vicinity, which they have been in the habit of using, taken possession of by a 'floater'.

In the 1860's PACKARD wrote a vivid description of this floating company of people: The decks of the schooners were crowded with men, women and

Fig. 313. View of a fishing harbour in south-eastern Labrador. Probably Black Tickle (52° 27.5′ N. lat., 55° 45′ W. long.) just off Square Harbour. Aerophoto the FORBES-GRENFELL EXPEDITION, 1931.

children, with all their belongings, a number of small fishing boats and dogs on the deck, as well as huts for pigs, goats and hens; the people had, like the old Norsemen, brought their families and stock with them for the summer's stay on the coast; 70 or 80 persons are said to have been crowded into a little schooner without any convenience of any kind (325). Half a hundred years ago there may have been a total employed of about one thousand men and girls with servants, besides the squatters, who fished on the land in stages (862).

On the eighteenth of June, 1937, I had on *'The Kyle'* — Fig. 311 — a
company of about four hundred planters with their stocks of food, their tackle
and even some boats. They spread themselves over the extreme edge of the
coastal strip from Cape Charles to Holton Harbour and Fanny's Harbour near
Cape Harrison and even as far as Ironbound Island in Makkovik's Archipelago.
They were successively dropped at the outposts, Figs 313 and 310, in the
neighbourhood of which they had small cabins of planks, so-called f i s h i n g-
r o o m s, Fig. 314, at some places forming small hamlets; Fig. 17. The sum-
mer-houses, also called 'tilts', look quite nice, usually painted white, red or

Fig 314. Cape St. Michael (upper picture). 'Stationers' houses at Francis Harbour
(lower picture). Photo the author Aug. 1937.

yellow; the landing-jetty, called 's t a g e' Fig. 306, gives a special character
to the whole. It is connected with the 'f i s h-h o u s e'; Fig. 307. The fish-
houses are rude structures like low sheds, roofed with turf and built on piles
because of the tide, reminding one of the prehistoric pile-dwellings. In some
places the beach is lined with stores and 'flakes' (p. 738) on which cod-fish
is set to be dried.

Fishing for cod.

Here the planters remain all the summer months to fish cod. This shore
fishery is conducted in small boats by parties of two or three men — Figs
124 b and 305 — who used to rely for propulsion on sail and oar, but nowadays
they mostly have motor boats. The fishing is done either by hand-lines or by

hook-and-line, baited from the small boats. Some southern planters are said
to have fish traps also. The introduction of the gasoline motor, if it has
added somewhat to the gross cost of production, has greatly extended the
mobility of the planter-fishermen.

The fishing stations present a busy and animated scene in the season. The
fish is brought to the stage — Fig. 306 — for curing. Before salting it is split
and washed in tubs filled with salt water until it is clean, then put in the
waterhorse, packed up with a light sprinkling of salt. It is then brought out
on the rocks or on the bawn and spread 'heads and tails'; Fig. 308. After the
first day's sun it is made up in small faggots, not more than a half-quintal in
each. On the second day it is spread out again all the one way and made up
in the evening in larger faggots of two quintals or more; Fig. 309. The next
time it is spread out it is made up in piles of twenty quintals and well covered
with rinds and tarpaulins round the heaps to keep them thoroughly dry. It
remains there for a few days to press and work in the pile (cf. 862). The opportunity to dry and cure fish on shore is thus of the greatest importance, for
in this way it is much easier to increase the catch. The curing was said to be
rather careless and primitive on the Labrador coast.

It is surprising that the majority of these planters stop fishing as early as
August, when the cod fish begins to become most valuable (p. 762). Meanwhile, after this, usually at the end of August, the planters generally return
to their homes with their catch. The boats are left permanently on the coast,
for the fishermen find their way from and to Newfoundland on the coastal
steamers or schooners. Cod-fishing is moreover not a regular pursuit of the
shore-men; there is the chance that a lucky day's fishing will give a return
out of all proportion to the average (585).

Their fish, 'shore-cured Labrador' (325), constitutes one of the largest
items in the cod-fishing industry of the New World.

This traffic continues year after year, and is of ancient origin. An old
report was given in 1801 by ANSPACH (15); he says that this so-called northern
fishery was carried on by planters from Conception Bay and Trinity Bay.
In 1826 before the Supreme Court at St. John's a description was given of the
method pursued in shipping crews and fishermen to the Labrador (325). It
does not differ essentially from the present custom.

Yet this method of fishing must be rather expensive and seems quite
incompatible with good management. To the time lost on the journeys, the
arrangement of their baggage etc. is added the great risk factor due to the

unreliability of the presence of the fish, for the planters' radius of action is rather small and his income is dependent on whether the fish comes to the land or not. Morever, in the few summer months they must make provision for their homes' subsistence throughout the whole year as they have no other means of livelihood, it is said. It therefore is most surprising that the schooner-men and planters go to Labrador at all when there is cod-fish near their homes, where for instance the bank-fishery is probably the richest in the world. The reason given by some planters themselves was the lack of profit from their labour in their home districts. I think that for generations this cod-fishing has been a custom, and the power of custom perhaps yearly drives the schooner people up to the bare, barren Labrador coast at the usual time, even when there are enormous quantities of cod near their own homes.

There thus appears to be an enormous difference between the ways of living of the liveyere and the planter. The former can supplement his income in winter by trapping, and perhaps also by some lumbering.

Transitional types.

Many planters have for this reason chosen to set up homes in Labrador between the Canadian border and Batteau and thus gone over into the liveyere group, combining early summer salmon-fishing with late summer and autumn cod-fishing and winter trapping. I have also been unable to discover any essential difference between such longshore men and the liveyeres called sett-lers, to the south from Fanny's Harbour (Cape Harrison) towards Gready.

There are however some planters who come to the coast on their own schooners as we shall soon see.

Floaters.

The schooner fishers are termed f l o a t e r s, meaning that they are not generally attached to fishing establishments on the Labrador, but catch their fish wherever they can get it, and take it direct to Newfoundland ports, where it is cured.

The floaters differentiate between the so-called L a b r a d o r m e n, who fish exclusively on the coast of Labrador, and B a n k e r s, who fish chiefly on the great fishing banks outside Newfoundland, and in the autumn make a trip to Labrador to increase their catch.

Among the f l o a t e r s the

Labradormen

are the most numerous. They work on fore-and-aft rigged schooners of 30—70 tons in teams of 8—10 men. The total number of fishing schooners visiting the Newfoundland-Labrador coast amounts to 2,000 or 2,500 per summer. In the schooners' cod-fishing, as in the seal-fishery, about 90 per cent of the fishing crews are hired for the season on the share system by the captains of the vessels or the fishing stations. Each individual on board, from master to boy, draws, instead of pay, a certain defined share in the profits of the venture. Usually the schooners belong to business men from Newfoundland, and the owner, like the skipper and the crew, gets his share of the catch at the end of the season. Half the catch usually goes to the skipper for finding and feeding the crew. When the fish fails it means that months of labour and hardship give no return, and the personnel of some unlucky vessels must depend for their winter's provision on an advance from one of the merchants, who looks to recoup himself from the result of the next or a future successful season. For this reason the Labradormen are very anxious to begin fishing early in the summer.

The yearly income of these fishermen is, as pointed out above, wholly dependent on the yield of fish in Labrador, and if they get to their places early they may even catch two boatloads of cod during the summer. It may be said that generally the Labradormen find their way to the fishing-ground as soon as the pack-ice belt begins to disappear, generally about 15th June, but some try to force the ice earlier.

Cod-fishing and its organization.

In the end of May or beginning of June or as soon as the ice is gone in the south the Labradormen begin their fishery in the Straits of Belle Isle. They wait here some time desiring to continue their way to their Labrador fishing-grounds as soon as the ice breaks up on the Atlantic coast. As first come is ever first served along the Labrador coast, he chooses the best berths for his trap; Fig. 312.

The following extracts from shipper NICHOLAS SMITH's fishing reports may give some idea regarding the departure for Labrador:

Extracts from skipper NICHOLAS SMITH's (862) fishing reports from Hamilton Inlet (Bluff Head, Smokey, Holton):

Arrival at the fishing station	Cod appeared at the coast	Cod traps set out	Departure for Newfoundland
1874 June 15	—	—	October 11
1875 » 17[1]	July 7	July 12	—
1876 » c. 20	June 25	» —	» 10
1877 » 30	July 12	» 30	» 15
1888 July 12	» 15	» —	» 10
1890 June 27	» 7	» —	—
1896 » 22	» 4	» —	—
1897 » —	» 3	» —	—
1900 » 18[1]	» 7	» —	September 25
1901 » 17[1]	» 7	» —	» 28
1902 » 23[2]	» 7	» —	» —
1903 » 20[3]	» 15 [6]	» —	October 7
1905 » 20	» 9	» —	» —
1906 » 16[4]	June 20[6]	» —	» —
1907 » 19	» —	» —	» —
1908 » 19	July 11	August 20	» —
1912 » —	» 9	» —	» 25
1914 » 17	» 7	» —	» —
1919 » —[5]	» —[7]	» —	» 16
1922 » 20	» 7[8]	» —	» —
1923 » 17[1]	» 7[9]	» —	» —
1925 » 21	» 9	» —	» —

Some of the floaters live and work all the summer on their schooners and trap-boats.

Others have procured fishing-rooms for themselves, to which belong besides the dwelling a fishing-stage, a small hold for about 300 quintals fish, a bulley or fishing-boat to carry about 16 quintals, and a good bawn for spreading about 300 quintals fish. Bunk-houses are built when the dwelling-house cannot accomodate the whole crew. In the 1870's the dwellings were mostly sodden tilts.

The latter group of fishermen thus forms a transitional type resembling the planters.

---

[1] No ice. [2] June 17 stop owing to a jam of ice, later no ice. [3] Much ice. [4] Ice jam in Hamilton Inlet. [5] Much ice remained in the bay until 20th of July. [6] Fish scarce. [7] Very bad fishing. [8] Very plentiful. [9] Not very plentiful.

After the arrival the first job is to build out stage-heads, repair boats, land the supplies and gear and fit up the houses after the winter's ice and snow. Then it is necessary to go up the bay to secure a load of firewood and timber, longers and shores. During this time others of the crew secure trap berths and try by »jigging» to find out if there are any fish. For all this work part

Fig. 315. The trap-boat goes out (536 a).

of the schooner or steamer crew were also used so that one would not be behind time when the fishing had to begin.

With all the ice gone the outlook for fishing becomes favourable. Good and bad fishing season alternate irregularly; when caplin and cod come in together, it is considered by the fishermen as a good sign. The cod was formerly taken as in the Stone Age, by 'jigging' or by hook-and-line or bultows with up to 1400 hooks on them. The men in the fishing- boats caught cod-

fish by 'jigging' as fast as two men could pull up the long lines. The method
of taking fish always by hook-and-line — Fig. 124 b — was only amongst the
English; the American ships were generally furnished with large seines.

Some seventy years ago cod traps began to be used also among the English
fishermen on this coast. It is a huge box of nets the meshes of which are two
inches square securely moored in the water and with 'leaders' stretching out
in several directions to conduct the shoaling fish into the trap, from which
they cannot find their way out. It is about three hundred and twenty-four
feet long, fifty feet wide and forty-five feet deep; the edges are kept afloat by
cork disks, the bottom edges are weighted down by lumps of lead. A leader-

net two hundred and
forty feet long finishes
with an opening at one
end into the trap. The
price of a trap is be-
tween three and four
hundred dollars (510).

The trap nets are set
at places where the fish
are thought most likely
to be running and it
is astonishing in what
rough and open places
the cod-traps are some-
times put in spite of the

Fig. 316.   The trap boat on Queen's Lakes returning
overloaded with codfish. Photo the author, July 1937.

sea and the fog, as I saw in the berths at Salmon Island. The crew have
in that case also to remain with a barometer on the shore to watch. But
there they often get the huge fish. When there is no ice to interfere with the
cod-trap operations prospects are good.

The trap system may give large catches; it may give none. Large catches
are sometimes made in a very short time with the trap; the fishermen only
sit upon the rocks and wait for the fish to run into it. But it is a hazardous
method, the fish may or may not run in, whereas a catch is generally ensured
with the hook-and-line method.

Trawl-fishing (p. 737) on the off-shore grounds is practised in the autumn.

Weather permitting the men go out to the traps at sunrise, usually in motor-
driven fishing-smacks — trap-boats; Fig. 315. When a good catch — Fig.
316 — is anticipated a motorless boat is attached by rope to the former. The
fish is taken out of the trap by hand and flung into both boats; if good the

morning's catch constitutes from twenty to thirty quintals. The fish caught
are gutted, washed, split, — Fig. 317 — heavily salted and stowed in the hold
until the catch is complete; Fig. 318. At same stations it is also dried on flakes
or the rocks; Fig. 320.

The fishing just described is hard work: dressing down the fish, splitting
and salting it goes on all day until dark, and even then often until one or two
o'clock in the morning in the faint light of an oil lamp. Sometimes the men
have only had two or three hours' sleep when they are called again at daylight
to go out with the trap-boat. Many times there are even conflicts about
breaking the sabbath. SMITH (862) writes: We wanted to finish then as it
was Saturday night, but we were forced to stay in the stage till 2.30 Sunday
morning; this was against every man's wish, but it would be a greater sin to let our beautiful fish spoil for want of being salted. So we took our chance, and I think that the Man above will agree with us. — However, there are not many such unbiassed fishermen as SMITH in Labrador.

Fig. 317.   Curing the codfish on board a floater on
Queen's Lakes. Photo the author, July 1937.

If possible the split and washed cod-fish is spread already in Labrador,
on 'flakes', on branches of spruce boughs strewn on stages so as to allow the
maximum circulation of air. If spruce is not available the fish have to be dried
on the rocks, Fig. 308, but the result is not so good. Artificial driers are not
known on Labrador and yet they would be of the greatest value in this region
where unpropitious weather for fish-curing is very common in some years.
The best cure is obtained from fish dried on the flakes when the weather has
been fine enough to dry them thoroughly but not so fine as to burn them.
The produce is known as 'shore-cured Labrador fish' and
will be sold to the Newfoundland merchants for export. The quality of the
product depends on the size, and the cleanness of the fish, the weather and
the amount of salt used, etc. and the whole is divided into a dozen categories
according to the market for which it is most suitable (585).

The floaters for the most part carry their cargoes of fish to Newfound-land in a 'green', i.e. salted, state. In this condition the fish is packed down in the hold. If the fish is not plentiful in the sea the whole summer may pass before the schooner is fully loaded and ready to return to Newfoundland. When the worst comes to the worst the fish goes bad; it cannot be appetising to see the men who, when unloading such a schooner, go about up to their knees in decayed, stinking codfish quite unfit for human consumption.

It used to be reckoned that a schooner crew of 8 Labradormen would have an average catch of 1500—1600 quintals salted or rock-dried cod (1 quintal = 112 lbs; 1 »green» quintal = 224 lbs). In the summer of 1937 the yield was

Fig. 318.   Schooners in Webeck Harbour heavily loaded with 'green-fish', i.e. cured and salted codfish. Photo the author, July 1937.

fairly moderate and the price then was about $ 2.50 per quintal. — I was surprised to find the prices so low compared with Fennoscandia. Does this result from the smaller size of the fish? Or perhaps it is due to the methods of preparation?

About the middle of September the fish is ready on the heavily loaded schooner, Fig. 318. Earlier it was loaded on board the 'dry fish vessels', where the supercargo inspected it and gave his verdict. Also steamers were chartered to load fish. Each tried to get the first steamer to market and the agents took fish that was not half ready. This was the beginning of bad cargoes, says SMITH (862). The fishermen took advantage of these opportunities to ship fish after one day's sun; it saved them a lot of time and work.

Nowadays the schooners mostly return with the catch to Newfoundland, where it is either sold 'green' to local merchants or cured by the families of the fishermen, Fig. 320. Then the schooner is dismantled for the winter and the crew disbanded. But they meet again the next spring when the traps have to be repaired; by 20th May everything must be in readiness. It is usual to take the same old crew year after year.

Risk is a large factor connected with most of the Labrador fishing; one would say: Their return depends upon the vagaries of the cod and on the skill and assiduity of the fishermen (cf. 585). The true floaters follow the fish. In our days the fishing schooners are found sometimes near Makkovik, sometimes at Mugford Tickle, sometimes at Ramah or even at Cape Chidley, searching for fish. One of the skippers is as good a pilot as ever sailed the coast; he

Fig. 319.   Newfoundland floater going home from Labrador. Photo the author, September 1937.

knows every rock over and above the water from Belle Isle to Nain, and all the others follow in his track. It is said with some exaggeration that Labrador is the best nursery for seamen the world ever saw. Thus many of these schooners have to-day no fixed fishing-station but seek their spoil anywhere from the Straits of Belle Isle to Hudson Strait. Prior to the middle of last century these vessels did not often go north of Hamilton Inlet, and not until about 1890 did they generally go north of Hopedale, writes PACKARD.

Most of the schooners in these days however go direct to the spot of their choice, and most of the men, though living on their vessels, will always be found in the same places. During my journey with Dr PADDON the schooners at anchor marked on his chart were almost regularly found at the spots indicated.

It is quite surprising that the floaters restrict their fishing to the cod species, seeing that the off-shore fishing-grounds are also frequented by the halibut.

American vessels earlier travelled 1400 miles to and fro solely for this fish (325). Nowadays it does not seem to interest the floaters at all.

The number of herring which frequent the waters of south-eastern Labrador is beyond belief. The Labrador herring is perhaps the largest and fattest specimen of that fish I have ever seen. It can no doubt be marketed at good prices. During the period 1860—1880 the herring fishery was also very important, but to-day it seems to arouse little interest. One great disadvantage is that this fish seems to desert the coast in some years or only comes in such small numbers that to catch it yields no profit. — For this work the planters used nets, setting them out in the afternoon and visiting them to take out the fish next morning. The floaters used large seines and it is reported that they have taken in a single haul enough herrings to fill 2000—3000 barrels. The fish must be salted immediately owing to its thick coating of fat to prevent it turning yellow and spoiling.

Fig. 320. Drying the washed 'green-fish' on the wharf at Battle Harbour (536 a).

Bankers.

These schooner-fishermen mostly come from the south-westerly coasts of Newfoundland and are anxious to have the cod-fishing on Labrador rationalised. They fish for the most part in spring and summer on the great banks; hence their name. But, when this fishing is at an end in the latter part of August and the beginning of September and the cod have gone, their schooners generally sail away to Labrador to increase their catch. At that time, however, the schools of fish have also there generally left the shores for deeper water and then the schooners spread their trawls on the off-shore grounds. Their success has often been marvellous.

— It is easy to distinguish the Labradormen from the Bankers when one sees them lying waiting for the wind in a harbour; the former have on the side of the vessel a white-painted, so-called t r a p-b o a t with a motor, whereas the salient feature of the bankers, who use long lines or jiggars, is a d o r y, the light American fishing-boat, for the bankers never use trap nets.

Long use has shown that the shoals of cod on the Labrador coast are apparently inexhaustible. Year in year out the fishery has gone on, and no diminution can be observed, even if one must take into consideration that it is subject to all the vicissitudes to which other fisheries are liable, periods of plenty varying with periods of scarcity. Fish, however, is always to be had somewhere.

Poor years are said to be really rare. The Labrador archipelago and the untouched grounds off that coast allow room for a practically unlimited number of fishing vessels.

CRISIS OF THE LABRADOR COD FISHERY

Under the present regime the income from fishing is very limited in Newfoundland-Labrador and it is questionable whether it is worth while in the long run (585, 1033).

The yield in Labrador is mostly of low quality and does not always keep well, and if more salt is used it reduces the quality still more. The quality of the cod yield on Labrador is not comparable with e.g. that of Finmark; it is for the most part very small and exceedingly thin when caught before the middle of August in the middle and the northern part of the Atlantic coast. To this is added the fact that the preparation is primitive and in places careless. The heads and the entrails are simply thrown away at once into the sea. PRICHARD had good reason to say: the folk of Labrador are entirely predatory people, and the floaters hold the record for predacity!

Perhaps the fish would be better caught later, as for instance nowadays by the bankers, and dried directly as in the Finmark without salt on flakes or 'hjells'. This would perhaps be possible in September and October when the dry land winds prevail. Yet, I have never heard whether such a method has been practised and whether the fish under these conditions also dries before freezing, because in the latter case it would be spoiled.

Possibly the situation could be in some respect improved by using modern methods and bulk handling in a smaller number of centres or floating factories, which should mean a reduction in the unit cost of production. A drastic change in methods of either fishing or curing seems however not probable following LODGE.

The whole catch is rushed off to market as soon as possible with the result that there is always a glut and the returns are small (325).

But here comes another serious inconvenience. There is a formidable difference between the fishing industry in European waters and in Labrador.

The Labrador fishermen, liveyeres and floaters alike, prosecute their calling in much the same way as their ancestors did 300 years ago, whereas the European fishermen utilize an elaborate capital structure in the form of steam-trawlers and drifters based on a very costly share system. LODGE has calculated that therefore the Newfoundlander for his year's work received something less than one-fifth of what his brother on the other side of the ocean received.

Moreover there is one circumstance discouraging for the old time fishery: the chronic overproduction of the fisheries in Europe, resulting in a slump of the prices and especially also on the market where the Labrador fish is sold. The handicap of the European producers is the greater since some Latin nations also send very considerable fishing fleets out to the banks off the colony of Newfoundland.

In a period when the deep sea fishery is gradually being more and more industrialized, and over-production for a market is, and possibly will be, a permanent condition all over the world, with a fall in price since 1929 of 50 per cent, it seems that the kind of fishing practised by the stationers and the Labradormen cannot be made a rational occupation. The Labrador cod-fishery thus appears to be definitely greatly handicapped by the Scots, English, Norwegian and Icelandic fishermen. In one district where the costs of production are unusually great because of the high custom duties, the balance will be decidedly negative. The Labrador fisherman cannot live to-day on the returns he obtains from the fishing alone. Competition with the Norwegian-Icelandic and Scottish-English commodity is not even possible by the application of export premiums and the reduction of some duties to the exclusive advantage of the fishermen.

It may therefore be presumed that this form of fishing will cease of its own accord when it is no longer supported by reduction of import duty and guaranteed minimum export prices ($ 3.25 per quintal in 1939), which will more or less directly favour only the fishermen, says LODGE, who in his character as a former Commissioner of Public Utility in the Newfoundland Commission of Government, has studied the question more deeply than most people (cf. 585, pp. 101 sqq.). Mr. LODGE after his inquiries unsentimentally expresses extreme scepticism as to whether the cod-fishery will ever again be an effective economic mainstay of Newfoundland including its part of Labrador. The government at St John's, however, has at present no other occupation to put in place of the fishing, which therefore continues as a kind of provisorium with great expense to the British Exchequer. The present state of affairs is explained by saying that it is more advantageous to give the relief necessary to save the fishermen and their families from starvation in the form of industrial subsidies than to take over all the direct expenses for their living (585). The fishermen themselves are of course satisfied, have full confidence in their government's actions and are convinced that providence will once again smile on them, the chosen people.

Social welfare work in Newfoundland-Labrador

is, for a stranger from over-organised Europe, one of the most interesting phenomena of Labrador. Wide views are followed and some highly educated private personalities work for the people's welfare in freedom but with responsibility.

Historical sources from olden times (see bibliography) lead one to conclude that for a long time the people were left to themselves and had to arrange their affairs with each other as well as they could. During the period of French control the governmental powers seem scarcely to have extended their protection to this end of the country. Certainly at times concessions were given to different people in return for exploring the country (p. 480 sq.), but the concessionaires did not enjoy much support or protection from the local authorities. It is not known to the author whether any missionaries were at that time in the country now called Newfoundland-Labrador.

When Labrador was transferred to the British Crown in 1763 the energetic governor, Sir HUGH PALLISER, introduced new methods into the administrative and social spheres, of which we have already seen one example (p. 554; cf. 325). According to the custom of the time conditions in Labrador were still arranged by decree or letter, but with a few exceptions (196) governmental supervision seems to have been almost non-existent on the coast for a very long time. One fact unfavourable to the continuity of the administrative arrangements was the repeated transference of jurisdiction over Labrador from Canada to Newfoundland and back again.

In 1826 the first c i r c u i t   c o u r t for civil jurisdiction on the coast of Labrador was instituted. It continued till 1834 when it was found that there was not sufficient business to warrant the great expense, but regular visits were made up the coast. In 1862 the Labrador court was re-established and continued to make annual visitations until 1874, inter alia collecting customs duties. Later it was again reinstituted (325). We may conclude from this that the need for a court did not seem very great, a fact which speaks eloquently for the peaceful and law-abiding character of the fishing and hunting population, as already HIND observed.

Until quite lately both the white and the coloured populations of Labrador have in practice been allowed to look after themselves. They have followed quite freely their own customs and habits, and as a rule have settled their mutual relations according to their own unwritten customary laws and their inherited ideas of justice. Judges and police have concerned themselves extremely seldom with the affairs of these people. There has been no local representative for the governmental authority, at least during the last few decades, and tradition has succeeded in regulating their lives quite satisfactorily, while they have followed their own principles without compulsion from outside. The only judicial authority to which the natives as well as the white people have sometimes appealed has been the J u s t i c e o f t h e P e a c e, and as such the authorities in St. John's used to appoint one of the Moravian missionaries-in-chief and later Sir WILFRED GRENFELL and, since 1913, Dr HARRY L. PADDON. Dr PADDON told me that this appointment meant very little work, and during 25 years he had only been obliged to interfere seriously in two cases. Since 1880 no serious crime has been reported (1937).

J UNEK gives an interesting description of the people's idea of administration and jurisdiction on Labrador's south-eastern coast, and most recently in Blanc Sablon (510). The folk on the southern coast, he says, entertain a vague impression that somewhere there is a government or some controlling body which, when laws are broken or disobeyed, sends out as its representative a punitive ranger. Most of the legalism in this region enters the minds of the people in the form of prohibitions. Judicial and legislative powers cannot be said to be fully conceived by the liveyere; it is a knowledge of officials rather than of laws in the true sense of the word. Civil law comes into play with its violation, criminal law is said to be unknown.

It was not until Newfoundland fell into financial difficulties and was obliged to give up its Dominion rights that the administration customary in the British Crown Colonies was introduced; in 1935 some ten R a n g e r s were appointed in Newfoundland-Labrador also. It is indicative of the people's view of these things that the new officials were received rather coldly; from many quarters they were reminded that formerly the people had managed quite well without police authorities and, in consultation with their Justice of the Peace, had arranged all their difficulties to the general satisfaction. The rangers were personally the kindest people and had the best will to do their duties satisfactorily. In these remarks I therefore like to see a spontaneous expression of the sound social spirit which we meet in these small communities of the wilds. As far as I heard neither the Rangers nor the Magistrate nor the Justice of the Peace had any important business to do after the new arrangement was introduced and up to the time of writing (1940).

For this happy condition of things, however, Newfoundland-Labrador has also to thank two philanthropic undertakings without which the people would certainly not be what they now are: North of Hamilton Inlet T h e M o r a v i a n M i s s i o n, and south of it T h e I n t e r n a t i o n a l G r e n- f e l l A s s o c i a t i o n. These have taken a great deal of responsibility for the social education and the welfare of the people. Their altruistic activities are among the most attractive phenomena a stranger finds in Labrador, an outflowing of unequalled human idealism in the midst of the biological primitiveness of a brutal wilderness.

The Moravian Brethren's work.

With some exaggeration it may be said that the *de facto* government of the Eskimoes and settlers (p. 727) north of Cape Harrison has been the Moravian Mission. This »flock of sheep in the snow» as one has called the Eskimoes, would probably long ago have been exterminated in the same way as their kinsmen on the southern part of the Atlantic coast if the humble Moravian pastors had not here taken care of every one's soul and body. The work carried on by the Moravians is not only one of the most remarkable gains of civilization on this coast, but a monument to the Christian mission history in general, and merits close study. The reports of the Moravian Mission in Lab- rador form moreover a consecutive history of that country since 1752 (652— 667). They give the fundamental traits of the country's natural history too, ethnography, natural resources and their utilization, etc. In the annual reports of the Moravians are to be found invaluable records regarding the varying supply of seal, whale, cod-fish, caribou and fur-bearing animals and the consequent effect on the Eskimoes' life. These reports are very extensive; some comprehensive works have been mentioned in the annexed bibliography (5, 128, 187, 188, 189, 205, 224, 396, 449, 471, 533, 554, 555, 559, 567, 647, 695, 758, 761, 786, 787, 787 a, 788—792, 794, 834, 999, 1078, 1078 b, 1081, 1088, 1089, 1091, 1114). The space being limited I can here give only some general indications of the Mission's civilizing work.

Little was known in the civilized world about this remote corner of the earth and its inhabitants before the Moravian Church arrived in Labrador. The few reports brought by returning fisher crews described the Eskimoes as a totally savage people. Indeed the men who landed from the first mission's ship in 1752 were killed. However, by their quiet work and endurance the

Moravians have gradually changed the savages into a humble Christian people, and the life and atmosphere in their villages to-day give a true picture of life in a native Christian community.

HISTORICAL NOTES AND FOUNDING OF THE MORAVIAN STATIONS IN LABRADOR.

On July 31st 1752 a small vessel which bore the name »*Hope*», arrived on the Labrador coast. It carried a trading expedition accompanied by four Moravian Brethren for whom JOHN CHRISTIAN ERHARDT acted as interpreter and super-cargo. It is believed that they landed near the present Makkovik (cf. 205), took possession of the place in the name of the English King, and thought this to be a suitable place for a settlement. They met there some Eskimoes who exhibited the most evident pleasure at meeting a white man who could speak their language. ERHARDT carried on a brisk barter trade with them in the most amicable manner. On September 5th the house in which they planned to stay for the winter was ready and the ship went further north to find new opportunities for trade. Ten days later it returned with the news that ERHARDT, the captain and five of the crew who had left the vessel in a boat to trade with a tribe of Eskimoes, had not been seen again. What had happened must probably for ever remain a mystery (205).

In the following year an American whaler visited the place (Ford's Bight in 55° 10″?), found the house and the remains of seven murdered men, which they buried.

Ten years later JENS HAVEN, who knew the Greenland Eskimoes, made proposals to the newly appointed governor of Newfoundland, Commodore SIR HUGH PALLISER with the purpose of obtaining permission to go to Labrador and devote himself to the work of converting the Eskimoes, for which work he had prepared himself. SIR HUGH at once issued the following proclamation:

'Hitherto the Eskimaux have been considered in no other light than as thieves and murderers, but as Mr. HAVEN has formed his laudable plan, not only of uniting these people with the English nation, but of instructing them in the Christian religion, I require, by virtue of the powers delegated to me, that all men, whomsoever it may concern, lend him all the assistance in their power', . . .

It may be said this is the Brethren's Labrador Mission's certificate of baptism.

HAVEN was very successful in his first meeting with the Eskimoes (p. 554) and SIR HUGH was greatly pleased with the beginning of his mission. HAVEN went to England and returned in 1765 accompanied by three of the Moravian Brethren. In Chateau Bay they met some four hundred Eskimoes who had as usual travelled south on trading and marauding expeditions. Shortly after SIR HUGH arrived and through the agency of the Brethren the famous »treaty of peace» between the aborigines (probably the Eskimoes south of Hamilton Inlet) and the English was concluded here.

At the very first and for the successful conduct of their mission the Brethren claimed a grant of 100,000 acres land for each settlement they planned to establish on the Labrador coast, and some other privileges too. In these reservations which should be the Mission's absolute property, the Brethren should not tolerate any quarrelling or violent competitors trading with spirituous liquors and using brutal lusts. On the Mission's property none should be allowed to stay except during good behaviour. In 1769 the Brethren obtained their grant on their own terms. Furthermore the Missionaries were vested with the authority of a Justice of the Peace.

In 1770 HAVEN arrived with his companions off Amitok Island near Nain. They soon obtained the consent of the Eskimoes to their appropriation of the locality chosen. This locality is to-day called Nain. The work was to begin (cf. for details 205,325 etc.). All plans were truly laid, the settlements were prepared with the greatest care, and consideration was given to every detail. Their houses were set up and the little colony was very comfortably settled for the winter. In the following year Lieutenant CURTIS (196) was sent to visit the settlement, and in his report he gives a vivid and enthusiastic description of the life at the station: a herd of barbarous savages are in a fair way to become useful subjects, he stated.

The Eskimoes were nomads and moved from place to place on the coast in pursuit of seal and caribou. It was impossible for the missionaries always to follow them, and when the nomads got out of touch, they used to return to their old habits and superstitions. For these reasons some other centers had to be established, where the missionaries and the Eskimoes could live together in winter. In this way stations were successively set up in Nain (1770), Okkak (1775), Hopedale (1781), Hebron (1829), Zoar (1865), Ramah (1871), Makkovik (1900) and Killinek (1904). Zoar, Ramah and Killinek have since been abandoned. Everyone who understands something about the industrial life of the Eskimoes must express his admiration of the excellent choice of the places for the settlements.

Each station has a church, a store, a dwelling-house for the missionaries and workshops for the natives. Around this fixed settlement the nomadic Eskimoes built their small wooden winter huts.

Each station has a missionary, the »House-father», as he is called, with his wife and children. If the number of the Eskimoes make it necessary there is an assistant too. There was earlier also a store-keeper, who was a layman, and attended to the business of the store.

For a man from a better abode the natural surroundings of the stations are poor and not inviting. The storms and cold are much the same at all these villages, differing only in degree, and with the presence of protective trees and sunshine.

STATIONS.

The most northerly of the Mission stations, K i l l i n e k (Port Burwell), on the north-easterly tip of the Labrador, was a good place for walrus, white whale, and seal. It was however abandoned a few decades ago. As appears from Figs 80, and 342 it lay in the most godforsaken wilderness imaginable; it is merely rock and water, not a bush is to

Fig. 321. Looking south-west over Hebron. The harbour and settlement somewhat to the right from the centre. The long house is the Mission station (cf. Fig. 323). Beyond is the Hebron Fiord, its mouth to the left. Aerophoto the FORBES-GRENFELL EXPEDITION, 1931.

Fig. 322.   The harbour and wharf at Hebron.   In the foreground is seen the banded gneissic rock, i.e. 'roots' of the Archaean Labrador mountain chain.   Photo ALEXANDER FORBES, 1935.

be seen, not even a blade of grass which would reach to the knees. Driftwood and train-oil are the fuel. It was bleak and dismal, raining and snowing by turns in the height of summer, misty and raw even in August, with few blades of grass to relieve the sullen greyness of rock. It was surely the most forbidding village of Labrador. Communications were precarious because of the drift ice and strong tidal currents of which the amplitude sometimes reaches a height of 3 5 f e e t (cf. 989, 478, 400, 403).

The station R a m a h was a pretty little village set among the high hills along the shore of a small and rocky bay, Ramah Bay. A few score yards

Fig. 323. The combined church and mission house at Hebron. Photo HARALD LINDOW, 1918.

behind the buildings the cliff rose steeply, and the roar of avalanches was not unknown. It was but a cold and stormy spot with fierce winds and heavy snowstorms. The bitter experiences of famine caused the people to desert this station. Famine in Labrador? This seems extraordinary; but it often happened that in spring when the Eskimo went out in their kayaks to catch seal the hunting failed because of the heavy drift-ice (472, 476).

The station at H e b r o n is like the two former situated outside the forest limit in a desolate area, Figs. 321, 322, 323, but favoured by the Eskimo because it is only some miles from the ocean and the catch is therefore very welcome. For a long time in the 19th century it was the chief point of contact with the natives.

The station O k k a k was the prettiest of the villages. It lay at the head of a sweeping inlet surrounded by high hills of which the grim slopes have patches of green where the coarse grass and the stunted bushes struggle for life. Below, its neat church tower, its white-painted Mission house and store set in a cluster of small wooden huts where the Eskimoes lived formed a rare sight for eyes weary of the sea, says Dr. HUTTON; Fig. 324. To the mind of the Eskimo Okkak was a fine place because it was a good centre for hunting and fishing. Scenery for its own sake does not appeal to them very much; Ah! they say, bears and foxes in that forest: fine place for hunting! In the bay the

Fig. 324. The abandoned Eskimo village of Okkak, where the Spanish flue killed 216 members of a population of 310 in the autumn of 1918. Photo HARALD LINDOW, 1918.

codfish feed, in August you may pull the great fish out of the water as fast as you can let down your line. Outside the coast, at the *sîna* there is in winter time plenty of seal and sometimes walrus.

Since the tragic ravages of the Spanish influenza in 1918 (p. 459) the place has been deserted and now lies ghostly desolate. Some who survived have formed a little community called Nutakk a little to the south on the same island (cf. 472—478).

The station N a i n, Figs. 325, 326, 199, 236, 238, 239, 195, may be considered the capital of Eskimo Labrador. It is a pleasant place, more civilized than the other stations of Labrador, and more frequently visited by ships in the summer than the other northern stations. It was for a long time the seat of the Moravian Bishop of Labrador and the German Consul. It lies at the head of a little bay, nestling at the foot of steep hills that make a half circle around it. The climate is much pleasanter than that of the stations nearer the open ocean with its arctic maritime influence. The grey slopes are splashed with green and it has been possible to plant small vegetable and flower gardens near the houses. The trim white Mission house and church, a long line of grey

Fig. 325. The Reverend P a u l H e t t a s c h and his son and Assistant-Missionary. In the background Nain (536 a).

Eskimo huts and the jetty make a very pleasant picture of the little settlement. The main street winds along the water-line, and one gets the idea that some of the huts have tried to elbow their way to the front between their neighbours: it means much to an Eskimo to live in the front line; at high tide his boat can come to his door and in winter the great broad road — the frozen sea — lies flat before him. On the hillside there is more room and the huts stand in solitary ones and twos. The huts are handed down from one generation to the next, though tumbledown and half covered in snow (472, 474, 761, 475, 478).

Fig. 326. Nain, looking from the south. In the foreground on the wharf, Hudson's Bay Company's store houses, and behind them the two houses of the Moravian Mission. In the house farthest to the left live the Hudson's Bay Company's staff. Photo ALEXANDER FORBES, Aug. 1935.

Fig. 327. Hopedale. The large buildings near the wharf are those of the Moravian Mission and the Hudson's Bay Company. Beyond are the homes of the Eskimoes. At the left edge, near the shore was the old Eskimo igloo-village (Fig. 235). In the foreground lies the famous park. The small island is crossed by two diabase dikes. Aerophoto the FORBES-GRENFELL EXPEDITION, July 1931.

Fig. 328. Hopedale, looking from the east (536 a).

The station Z o a r was abandoned some decades ago.

The H o p e d a l e Station -- Figs 237, 168, 327 and 328 -- was originally set up to make contact with the Indians who used to come in groups to the coast there (cf. p. 482). The Eskimoes then feared them so much that they fled at the first glimpse of one. In 1790 CRANZ says the fear began to decline and several Naskaupee families came to Kippekak to trade with the European factories there, 20 miles from Hopedale. In 1799 the missionaries had some talk with a few Indians who had been converted by a French priest; these had greatly feared the Eskimoes. The Indians came no more to the mission.

According to an old diary this station was founded where the first tent was pitched in 1752 to establish a mission. Numerous foundations of Eskimo igloos, earth huts, and the remains of three great-family peat houses half a kilometre north of the station prove that Hopedale was early an Eskimo centre; cf. Fig. 235. Hopedale is now a mission and trading station of considerable size and visited in summer by the mail steamer or a schooner carrying the mail. Nature there is less friendly than in Nain, it is visited by the same snow storms and blizzards as the northern stations and hemmed in by the same frozen sea in winter time. But there is a little wooded park in Hopedale (478).

Makkovik is the most southerly of all the Moravian Mission stations on the Labrador coast and was established to look after the halfbreeds and some white settlers who, in varying numbers, began to inhabit the surrounding islands. Makkovik lies in a well-wooded bay, Fig. 329, and is seemingly a good place for the hunter, though better for the trappers of fur animals than for the hunter of seals. The church here was planned and made in sections and then shipped out and put together in a pleasant spot. At Christmas 1897 the first service was held in the new church with a temperature of 47° C of frost outside. The boarding-school at Makkovik has become one of the most valuable parts of the Mission's work among the halfbreed settlers and the Eskimo. The pupils come from far and near, not a few are Eskimo, but most are children from settlers' homes and there-

Fig. 329. The Moravian Mission station at Makkovik, looking from north-west (536 a).

fore English is their native tongue.   Some young Englishwomen are teaching
the children what they need for their future life as sons and daughters
of Labrador: they shall be good and loyal citizens and good and loyal
Christians (478).

SPIRITUAL AND TEMPORAL WORK.

The missionary's difficult and complex task required wisdom and patience.
It is scarcely possible for him to have permanent control of such children of
nature only with the ten commandments and philanthrophy, especially when
they live in such a wild state and in such isolated, peculiar social conditions
as the Eskimo on the Labrador coast. He was obliged to show interest also
for the temporal welfare and ways of living of his charges in order to win their
lasting devotion, and even in that sphere show himself superior to them. Yet
he had to keep the spiritual apart from the temporal.

The first task of the mission was to try to tame the minds of these wild
people and modify the barbarian, almost asocial habits which seem to have
been specially characteristic of the pagan forefathers of the Labrador Eskimo. For this purpose it was necessary to win their confidence, a none too easy
task. Even the old Nordic sagas speak of the white men's savage behaviour
to the S k r a e l i n g s — Eskimoes — and in the brief history of the coast
the evil white man has played a conspicuous role by his iniquities among the
aborigines. But the loving work of the missionaries wrought a miracle. They
looked after the poor, gave food to the helpless and disabled, cared for the
sick and helped to bring up the children. In spite of great difficulties at the
beginning the Moravian Brethren have really been apostles to the Eskimo
in Newfoundland-Labrador. By means of great sacrifices of life and goods they
have won a brilliant victory over the barbarous paganism of olden days. Cf.
Dr HUTTON's penetrating descriptions.

Moreover they have taught the Eskimoes to continue to provide for themselves, to remain Eskimoes, keeping to their ways of living as adapted to the
surrounding natural conditions. This was very important for the mere existence of the race.

At the beginning of the eighteenth century the southern Eskimo were
trading with French fishermen and settlers in the Straits of Belle Isle region,
and since that time their desire for European goods has been intense. The
practices of these traders were not humanitarian and no idea of mutual benefit
entered into their calculations when trading with the Eskimoes, GOSLING
notes: Rum and tobacco soon became the chief articles given in exchange. To

preserve their flock from this contaminating influence the Moravians started trading. If they not had begun the trade for the benefit of the Eskimoes themselves and for the support of the Mission, the people would have flocked south and would have shared the fate of the tribes that once inhabited the south coast (p. 482 sq.). The traders began to push up the coast in increasing numbers in the nineteenth century, 'spreading spirituous liquors, quarrels and brutal lust among the natives'. How far-sighted HAVEN had been when he reserved areas for the Mission and the authority of the Justice of the Peace! In another respect too the trading was an advantage. Many times both Eskimo and missionaries had to live on the verge of starvation. It was therefore a great blessing to the community that the policy of the Brethren was to keep in hand a year's supply of all necessaries, fearing some such contingency as that the mission ship would be wrecked.

DAVEY (205) has given a comprehensive and fascinating picture of the Brethrens' spiritual pioneer work (cf. also the bibliography).

At Nain Brother DRACHARDT preached his first sermon in 1770 to an assemblage of about 800 people. Long periods of probation were, however, necessary before they could consider the applicants genuinely converted. In 1776 for instance they adjudged a man to be worthy of baptism, but in 1789 he relapsed after a visit to the south: he took to himself two or three wives that he required to man his boat. He finally left the settlement. The temptation to travel south in the summer remained for years. The goods the white traders there offered in exchange were most attractive.

At the close of the eighteenth century, after thirty years of administration, there were in the Moravian settlements 85 persons professing Christianity. At this period twenty-six missionaries had been employed. This may show the difficulty of their task.

At the beginning of the nineteenth century the efforts to convert the Eskimoes had not been very successful. However, after the conversion of a noted Eskimo sorcerer 1804 the Brethren's work began to bear fruit. In 1824 KOHL-MEISTER reported that Nain and Hopedale were practically Christian settlements, but Okkak was still a mission among the heathen. The missionary REICHEL, stated in 1861 that of the estimated number of 1500 Eskimoes living on the Labrador 1163 were under the influence of the Moravians, if not all actually converted.

Through long years of patient labour, unpraised and unrewarded, the Brethren gradually won the Eskimo population to the Christian belief, civilized and instructed them and preserved the race from extinction. At the close of the nineteenth century there were only some thirty pagan Eskimoes in the far north. In our day there are no pagans at all.

The psyche of mankind is, however, in many respects an enigma. How difficult it is to root out the superstitions or to change the customs of long ages! The missionaries could give many instances of how still to-day superstition temporarily gains the upper hand. Sorcery lingered long among these people of nature.

In the fishermen and the traders the Brethren had, however, to face a more insidious enemy than the Eskimo sorcerer, the *Angekok*.

It has been suggested that the Mission could scarcely have reached such thorough and brilliant results merely by spreading Christian teaching only among the primitive pagans. The very simple old customs of the Moravian Church were continued and soon took firm root among the Eskimo, it is true. But there was nothing in the Moravian church's ceremonies and external customs to fascinate the Eskimoes in the same way as the Catholic Mass caught the fancy of the Montagnais (p. 606); the Moravian missionary does not even wear a black coat and a clergyman's collar or anything like the 'robe noir' of the Indian missionary, and in winter he wears Eskimo clothes of the Eskimo cut which are warm and convenient for the bitter weather. Therefore it is probable that it was indeed the power of the irreproachable example and purely practical knowledge which the Missionaries gave to their charges that in this respect contributed so much to the miracle of conversion.

It is of course the personality of the teacher which wins the ear and mind of the audience as much as the doctrines he puts forward. The Moravian missionaries in Labrador were men of kindly, understanding hearts and of deep piety, but at the same time men of wide culture. They were also well informed on many practical subjects; this latter fact greatly attracted their charges. Many of the missionaries were skilful masters of some useful handicraft, one might say Jacks-of-all-trades: doctor, surgeon, dentist, architect, gardener, poultryman, electrician, engineer, painter, plumber, tinsmith, blacksmith, baker, carpenter, upholsterer, etc. etc. *ad lib.* Nor must one forget tailoring and knitting, translating and teaching. They can mend a motorboat, fashion a boat, indeed turn their hands to most things with success.

Their wives, the Moravian Sisters, teach the native women to clean, cook and do various kinds of work. They tend the sick and nurse those suffering from disease. Earlier they had also to help in the schools. When the work of the day was over they joined their husbands in teaching the settlers' children. The more things they could turn their hands to, the better were they able to have a comfortable house and a little pleasure in life, and the more respect they earned from the natives (478, 761).

This life is thus not less trying for the missionary's wife than for her husband. These noble ladies bear their children and tend them through illness, usually without medical aid. At seven years of age their children are sent home to the mission schools; this often means the practical separation of parents and children for a lifetime; how can father or mother hope to find their old place in their child's love and confidence after ten years of absence? PRICHARD asks with good reason. Even communication with home is restricted at most of the stations to one or two mails a year. Their boys and girls will be on the threshold of manhood and womanhood before they next see their parents. Self-abnegation can scarcely go further; nothing but the strongest sense of duty and a deep piety can enable them to carry on their noble work on the Labrador, HUTTON wrote.

The Brethren generally spend the whole active part of their life on the coast; that is some thirty to forty years. This life of devotion and exile has a few breaks, as for instance when once in ten years they return to their homes in Europe on a visit. But under these conditions the missionaries learn to know thoroughly the life of their charges and its demands on them, also the psyche of each individual which, combined with their good knowledge of the language, makes it possible to get very close to them. They lose themselves in the absorbing interest of their work and are ever ready with sympathy for the shortcomings and sorrows of the people. In this lies the secret of the missionaries' unparalleled success. The missionary loves the Eskimoes and they love the missionary; all day long a stream of folk throngs the steps of the Mission house during the great festivals.

In general the Eskimoes of our days are very religious.

During the season of open water the attendance at church services is comparatively small, for the Eskimoes are away at their hunting and fishing places, perhaps twenty miles or more from the village. A good number of those who are nearer come by boat for the Sunday service or the religious festivals: then hardly anyone stays at home, the church is packed. There is a dignity and sincerity over the whole service. The missionary has the people well in hand and they listen eagerly to what he has to say and read. The organist plays and the community join lustily in the hymns in perfect tune and stately time. Every eye is fixed on the speaker (476).

The winter time from Christmas to Easter is a very busy season for the missionary. Daily services, school, confirmation classes, festivals and missionary journeys; besides being 'at home' to the Eskimoes at any hour of the day to hear their complaints, their troubles, temptations, and struggles. The Eskimo is said to turn to his missionary with the same feelings as a child to his father, and he gives him a few sympathetic words. Often in the end it all amounts to very little, but the natives feel the need of opening their hearts and of having a good talk; some of the apparently trivial things are possibly mountains in the Eskimoes' lives, and they wish to sit and cogitate over them. The missionary must indeed look after everything in the public life of the Eskimo: he must baptize him, teach him, be his representative when proposing marriage, he must marry him and he must bury him. Added to that he must comfort him when he is terrified by the powers of darkness. Further he must look after his worldly business, if in silence yet still to his advantage. If necessary he must give him medical care. It is about as much as a man can manage (478).

The missionary has, however, also to visit the scattered families and it is no enjoyment to travel in winter time, plodding along on snow-shoes in front of the flagging dogs, or helping the heavy sledge over frozen brooks, sleeping on the hard floor of some lonely hut and subsisting on the frozen food (476).

Success of the Moravians' civilizing work.

The missionaries' school-teaching has been most successful. It was reported already in 1801 that many Eskimoes could read tolerably well (cf. p. 558 sqq.). The first book printed in their language — a history of Passion Week — was said to be eagerly studied and read aloud in their homes. In our days there are at least some 22 books printed in Labrador Eskimo. As far as I know all Eskimoes under the Moravians' control can read, write and reckon to-day. They were taught sloyd. The culture of the home was developed in a more hygienic form (p. 521). The Eskimoes' love for music and singing was observed and developed by the missionary very early. In 1828 Okkak was gladdened by the present of an organ. Later on they began to play instruments of various kinds, and were even able to form both brass and string bands (p. 568).

A better way of living was introduced by means of new ideas. The Brethren taught the Eskimo to catch cod; they bought their catch in summer and sold it to them again when they were most in want and almost starving; in this way the people gradually learned the value of cod-fishing. As early as 1806 the Mission introduced seal nets as they were in use on the south coast, and in this way the supply became more regular. The trapping of fur-bearing animals was encouraged long before the Hudson's Bay Company settled on the coast.

The most important means by which the Eskimo's standard of living has been raised to meet the demands of the time has been trade. As mentioned above it was necessary to preserve the natives both from exaction by the bootleggers and from falling into a kind of economic slavery to the traders who knew only too well the different forms of the truck system. This proved possible by the missionaries themselves controlling the trade. The trade done by the mission was thus of great value to the Eskimoes. To the Mission itself on the other hand it brought many a troublesome hour.

The Eskimoes, and in later times the settlers, brought their fish, furs, sealskin boots and other articles of barter to the mission's store and could there provide themselves with salt and different kinds of necessary gear for fishing and hunting, provisions, garments, in fact all the supplies that they needed. The prices paid were fair and even liberal. In the case of valuable pelts such as silver and black fox the storekeeper paid a deposit at once and when the skin had been disposed of on the London market, he passed over the whole amount obtained to the hunter less a small percentage. The possession of a round sum of money was not of the same value to the Eskimo formerly as it would be to-day. They soon spent it in making useless purchases from the schooner-men who traded along the coast in summer. It therefore happened that for instance a man who had drawn 300 to 600 dollars applied to the Mission funds two or three months later as a starving pauper (761).

The mission has often been criticized because of the trade it carried on. These animadversions are however unjust. The whole proceeds have been devoted to the upkeep of the Mission and have rarely been more than sufficient to defray the heavy expenses incurred by the trade (cf. p. 531 sqqq.). If the Mission had not had shops the people would have been obliged to tramp perhaps double or treble the distance to some private trader, with the added temptation of finding it possible to buy undesirable things. One of the foremost experts in this matter regarding Labrador, Sir WILFRED GRENFELL, states that the Moravian Mission stations just as trading posts have been of enormous importance for the welfare of the people.

Now, there is one unfortunate circumstance: jolly people generally live above their income, and the sly and speculative Eskimo are no exception to this rule. It was not long before the majority of them were in debt to the mission shops. Theoretically there should be no bad debts, but gradually the debts accumulated. The Mission's trade was undertaken principally in the interests of the Eskimo. It has also kept them together. But there was nothing else to be done from time to time than to cancel their debts and then start again with a clean sheet. The credit system was finally refused to those

who made no effort to pay their debts; after a while even the Eskimo admitted
that the new rules were founded on justice (761). This form of welfare has of
course been disapproved by traders and many others. Meanwhile GRENFELL
has definitely and successfully met all the superficial and egoistic criticisms
which they and other occasional visitors have expressed concerning the high
prices of goods, etc. (cf. 364).

In one other respect too the trade was not always advantageous for the
natives. With the policy of 'the open door' the Eskimoes developed demands
hitherto unknown amongst them. As early as 1861 REICHEL remarked that
few families were then content with the food that had satisfied their ancestors.
Molasses, sugar, biscuits and other white man's foods had become almost a
necessity and were then principally obtained from the fishing schooners or
the traders (cf. p. 533 sq.). The Mission did not succeed in completely preventing the traders' smuggling.

In these days when journalism flourishes criticism is often more plentiful
than founded on fact. Opinions are expressed in many quarters regarding the
results of the Moravian Mission's work. Some who express unbounded admiration for the primitive freshness and ethnological characteristics of the natives
do not hesitate to criticise this 'taming' of the Eskimoes. If, however, it is
remembered that the original Eskimoes in Labrador very much resembled the
gipsies of the Old World and therefore not only lacked social organization,
but also seem to have been rather asocial and practically speaking amoral
any thinking world citizen may be thankful for the educative work which has
been done in this forgotten corner to raise the level of these people so highly
gifted by nature. The statement often made by chance passengers that the
Mission was injuring the Eskimoes' way of living and earning capacity by
persuading them to live in settled dwellings, only shows complete ignorance
of the Mission's purpose and the life lived by the Eskimo at the present
time (p. 561). Similarly wanting in knowledge has been the criticism of
the Eskimo's schooling. The blessed instruction both on the school benches
and in practical life has undoubtedly facilitated the adaptation of these
people to the Euro-American form of culture, and aimed at making their
unavoidable transference to it as painless as possible (cf. KNUD RASMUSSEN
p. 562).

The positive advantages have been incontestable. Through the Moravian
missionaries the Eskimo have been preserved from extinction, have been
civilized and educated and brought to a knowledge of their Creator, GOSLING
wrote. One of the sailor veterans on the coast, NICOLAS SMITH (862), a Labra-

dorman, writes: To those missionaries who have served in the past and to those who are carrying on at the present time I desire to pay tribute for their self-neglecting work for humanity. The effect of the Moravian Mission on Labrador has been very profound, more so than is apparent to the casual visitor, writes FORBES (275). The Moravians have solved a hard and dangerous problem and deserve full credit for the happy relations that exist to-day. SIR WILFRED GRENFELL's opinion is: 'To no other people on earth does the lonely Labrador owe one-half the debt it does to these devoted servants of the Moravian Mission.' And the charitable work carried on by SIR WILFRED himself in Labrador has made his word decisive in this matter.

Let us now glance at SIR WILFRED's own work.

The International Grenfell Association's work

for the well-being of the people of the Labrador Coast.

The Moravian Brethren did what they could even for the large floating white population which visited the Atlantic coast and the liveyeres. But they could not do very much for them; the Eskimoes were their particular clients. Fortunately the liveyeres and the visiting fishermen found a champion in Dr. WILFRED GRENFELL, who in 1892 started a new era for the white population of the Labrador coast. For the missionaries in the north too, this grand philanthropist was a powerful ally.

Before his arrival a floating population had gradually invaded the coast: fishermen, furriers and traders. The western part of the Straits of Belle Isle had been to some extent settled from the beginning of the nineteenth century and the white people had advanced as far north as near Hamilton Inlet. Not before 1860 did Newfoundland fishing-schooners more generally reach Hopedale. But in 1863 they were at Hebron and soon they sailed up to Cape Chidley. The business grew rapidly to some 1500 to 1800 schooners with 15,000 to 20,000 people about the year 1900. With them came many troubles, diseases and abuses.

ORIGIN AND PURPOSE OF THE ASSOCIATION.

In England rumours were soon heard of need and disease in Labrador. It was expected that the fishing population there was in want of the same aid and support as that given to the North Sea fishermen by the »Mission to Deep

50

Sea Fishermen». For this purpose Dr. GRENFELL was sent out to investigate this matter.

Dr WILFRED T. GRENFELL was born in Scotland 1865; Fig. 330. He was appointed British Medical Officer to the 'Royal National Mission for Deep Sea Fishermen'; the task of this association is to improve morally and physically the condition of the North Sea fishermen, partly by giving them medical care, partly by protecting them at sea from victimisation by bootleggers and cardsharpers and other kinds of exploitation — the so-called bumboats. After his first training in the North Sea he was sent out in the Mission's ship in 1892 to Labrador to investigate the conditions.

Dr. GRENFELL found the white population, both resident and floating, in the greatest need of help. Long years of isolation, privation, ignorance and neglect had reduced the residents of the country to the depths of misery (325). He was deeply impressed by the terrible poverty, the al-

Fig. 330. The late Sir W i l f r e d T. G r e n f e l l, surgeon, navigator, Justice of the Peace and philanthropist, the founder of the *International Grenfell Association*, at his work-table. The inserted picture taken by the author at Woods Hole September 19th 1939.

most permanent half-starvation which prevailed among many, the wretched habitations, the total absence of the simplest sanitary ideas, the ignorance and apathy of both sexes living from hand to mouth and burdened by the truck system. All this formed a vicious circle from which the people could not get away with their own power. What remedy was there for this evil? To move them into a country with better conditions?

Poor people become extraordinarily attached to their homes, unattractive as they may appear to inhabitants of more favoured countries. The liveyeres of

Labrador are no exception; they will not leave for places where they could earn a livelihood and be in touch with civilization. I know trappers in Labrador in an excellent financial position, who after some years' experience in the States have returned to their former homes in Labrador's freedom and rugged wastes. We have found some of these families living in a rough kind of plenty (at Sandwich Bay and North West River), whereas the greater number living near the coast northwards up to Makkovik and Hopedale are only just above the poverty line. But the margin easily disappears: an accident, illness, bad fishery or unsuccessful furring can plunge an independent family into the direst poverty from which they cannot extricate themselves unaided. Who does not know the coast cannot figure what poverty is. GRENFELL tells us that in 1909 he found a family on an island in Hamilton Inlet living in an absolutely destitute condition: the mother of Scottish descent, the father a half-breed Eskimo and five or six children. They were half-clad and had no provisions, they had neither gun, nor axe, nor fishing gear; yet the children seemed to be in fairly good condition; they lived on berries. GRENFELL took several of the children to his headquarters in St. Anthony and he helped the family to make another start in life. This is only one of the stories of fantastic poverty; in GRENFELL's books we find others. I have myself had the opportunity to travel with the hospital boat for two weeks along the coast and I was surprised still to find so much and such varieties of disease and accidents among the 'floaters' after nearly forty years of devoted charitable, surgical work. I was astonished to find how many weak and sickly folk there were on the coast, and I could now realize for the first time how terribly bad the people's health must have been (tuberculosis, beriberi, anaemia, avitaminosis) before the start of GRENFELL's work.

Moreover the liveyeres along the outer coast are still few in number and live so far apart that they can afford each other but little support. The medical and social need of this population is still crying; what can it have been when GRENFELL began his work! The only settled practising surgeon on the Atlantic Labrador coast is said to have been an Eskimo living in Hopedale. In summer time a doctor, who travelled up and down the coast on the mail steamer, supplied them with some medical help. This was naturally ineffectual, and if people got seriously ill they just died (325). There had been no one to look after the interests of these poor people: floaters and planters; no magistrates nor police lived on the coast. They were an inarticulate group when GRENFELL came to their rescue.

Dr. GRENFELL had already gained an extensive knowledge of the sea, and this specially enabled him to take up the work in Labrador. He soon procured also a master's certificate and has since pursued the higher study of nautical science, even to the making of charts and accurate surveying. For this reason he could also act as agent for Lloyd's for the coast of Labrador, and has been able to bring to justice several notable offenders.

For lack of business the Circuit Court for Labrador (p. 764) had gone out of existence a generation before GRENFELL came to the coast, and it had not been found necessary to have even a policeman stationed there (before 1935).

GRENFELL therefore soon became a magistrate for the coast. His duties were happily not onerous, the inhabitants are not a vicious people, as we have already seen. Quarrels about trap berths formed the bulk of the cases brought before him. They were settled by him with prompt justice. One of his achievements was that the liquor traffic was practically suppressed; in his position as magistrate he eagerly hunted out every offender.

Dr. GRENFELL decided to found an international association for the welfare of the Labrador people as a result of recognizing the handicap under which the people lived. This plan has been realized with rare success. Large voluntary contributions have streamed in. In 1939 the staff comprised a hundred persons, including voluntary workers.

Fig. 331. The salient features of the International Grenfell Association's activity presented about 1935 by a 'whop', i.e. one of the volunteer summer workers of the Association.

The Association's vessels now sail under their own, internationally recognized flag.

On the schematic picture, Fig. 331, drawn by one of the »wops», i.e. a voluntary summer worker in Labrador with the International Grenfell Association, one will find the salient features of the I. G. A's philanthropic work on the coast. We cannot here enter into details, but I will give some indications as far as the Newfoundland-Labrador coast is concerned. For details the reader is referred to the literature (235, 338—382, 390, 428, 504, 560, 566, 615, 717, 719, 721—730, 732, 733, 734, 739, 747, 897, 938, 987, 988, 991, 1039).

MEDICAL WORK.

Two or three hospital ships patrol the coast all the summer. They go from their base in St. Anthony up to Hopedale, sometimes, if time permits, even to Hebron, and visits are regularly made both to the harbours and the schooners at their anchorages at out-of-the-way places as well as to the liveyeres and the planters, giving them medical relief if needed, carrying their mail, etc. The principle is to help the fishermen not to lose time on superfluous voyages. In difficult cases of illness which need special care or operation the patients are taken to one of the hospitals for treatment, where they can be better attended to than at sea. The appearance of the hospital ship »*Maraval*», the »*Strathcona*» (Fig. 104) and the other Grenfell vessels has been a source of much thankfulness and joy in many a lonely harbour.

In winter time the doctors visit the liveyeres, travelling with dog teams all along the coast up to Makkovik, where the Moravians' sphere of interest begins.

At Okkak alone a Moravianh ospital has been at work some years (363, 472—478); it has had to suffice for a population dispersed over some 300 miles of coast. But there was no health patrol along the coast. Considering the bad communications there were of course many cases in which patients were compelled to wait for weeks, or perhaps months, before help or alleviation could arrive. However this hospital, a gift of charity, could not be sufficiently highly esteemed, because they had had no real medical assistance before (761).

We find now on the coast the following h o s p i t a l s with modern equipment, of which five are still in use: North West River (Fig. 292), Indian Harbour (abandoned in 1939), Cartwright (Fig. 332), St. Mary's River (Fig. 333), Battle Harbour (burned in 1931 and abandoned), Harrington and St. Anthony.

Besides these there are small cottage hospitals called n u r s i n g s t a- t i o n s in use during the fishing season at Indian Harbour, Spotted Island, Boulters' Rock, George's Cove, Red Bay, Forteau, Mutton Bay, Flowers' Cove, Black Duck Cove and Conche. They are operated by a staff of trained nurses working under the direction of a doctor, and these give the sick and injured excellent first aid.

The doctor however spends most of his time visiting the different stations, harbours and other settlements, travelling in one of the hospital ships belonging to the Association. During these journeys health propaganda and the care of children is promoted both by practical treatment and the distribution of literature. Some support in the form of wholesome food, especially such as contains vitamins, is given in large amounts and gratis to the poor and sick.

3　　　　　6　2　4　　5　　1
↓　　　　　↓　↓　↓　　↓　　↓

Fig. 332. 'The Grenfells' at Cartwright (upper picture); to the left the hospital (3) to the extreme right the school (1) the industrial shop (4) Mrs Keddie's house (5) and Dr Forsyth's house (6). The Cartwright village (lower picture); on the headland to the left the Hudson's Bay Company's houses (3), in the foreground to the right the Church of England church (2), to the extreme right the wireless station (4). Aerophotos the author, July 1939.

The great importance of these establishments is easily understood when I mention that the absolute state of health in the country is alarming, especially the incidence of tuberculosis. In 1937 7 % of the population of the colony Newfoundland were found to have active, progressive pulmonary lesions. The infected population, however, shows the most surprising resistance; they are only very slowly succumbing to the scourge; somehow or other they must develop their own immunization. The annual death rate for the period 1932—36 was about 12.7 per thousand for Newfoundland including Labrador (585).

Fig. 333. The Grenfell hospital at Mary's River (536 a).

The distances are long, transportation difficult and the visits made by the itinerant doctor or nurse are still perforce insufficient. From Belle Isle Straits to Hebron, is a stretch about five hundred miles in length as the crow flies. JUNEK, after having studied the community of Blanc Sablon very closely, states that the educational and clinical work accomplished by the Grenfell Mission is meagre, sporadic, and most ephemeral in its influence upon the lives of the people. It must be enlarged. In the absence of such aid the people resort of necessity to their own local remedies. On the whole the most important of these is resignation and fatalism. The Association will undoubtedly find a remedy for all these evils if only it can continue to obtain the necessary funds and trained assistants.

EDUCATIONAL WORK.

Besides the purely medical work the Association also carries on charitable and educational work.

Children of poor parents and orphans are taken into homes where they are carefully tended by qualified nurses and taught weaving, carpentry, etc.

The public education (except on the coast north of Hamilton Inlet where it is under the Moravians) is entrusted in the main to the Association. There are now modern schools with boarding-houses attached in North West River, Cartwright, St. Mary's River, Red Bay, Forteau, Mutton Bay, Flowers' Cove, and St. Anthony. Access to all schools is free, both for study and practical training, and in these schools the adherents of the Anglican and Nonconformist churches find no difficulty in receiving a common education. Small libraries are distributed all along the coast.

These few notes may give some impression of the great understanding the medical and teaching staff have had of their task.

The Association took upon itself, says Dr GRENFELL, to make Labrador's population happy. This 'the Mission' should attain more easily through the power of example, he considered, than by issuing regulations: 'We do not work to form any sect, or to teach any intellectual dogma; we endeavour to do for the workers of the sea the practical service they may need in the circumstances of Labrador; we wish to improve their social milieu and to save them from an economic system [the truck system] which is responsible for their wretchedness and poverty.' In 1922 Dr GRENFELL said 'the method aimed at is to illustrate in practice the attitude Christ would assume to-day in the varying phases of the fishermen's life.' In agreement with this members of the staff have always been chosen from the laity.

MODEL CONSTRUCTIVE CHARITY.

The humanitarian scheme is more widespread.

Sir WILFRED's attention had been early directed to the assistance of 'the submerged'; without his help they must have gone under. In his work inspired by compassion he wished on the one hand to avoid the *'destructive charity'* that encourages dependence, decreases self-reliance, and pauperizes, and on the other to encourage the *'constructive charity'* that instructs, helps the unfortunate round a tight corner, puts him on his feet and inspires him with renewed courage, writes ALEXANDER FORBES (275). To avoid pauperizing those whom he assists he has required that some work or service be given in return. He aptly likened responsibility to vitamins: the paralysis that comes from a dole without responsibility to that which results from food lacking in certain vitamins. For example the stocks of clothes and other necessary articles are available for those who really need them, but they can only be obtained in return for work; in this way, therefore, they are not dependent upon the Mission. This method expresses one of the principles of the practical aims of the Association's work.

In Labrador there have been among the traders in bygone days many an octopus who used tricks to entrammel his client and then continually extort his money or goods. Dr. GRENFELL soon discovered the current system of extortion in Labrador. In his opinion the evil of such a system of supply could be overcome to some extent by cooperative associations. The first cooperative store was established in Red Bay in 1896. This was a risky enterprise. If such stores had some temporary success they were insufficiently provided with capital to carry them through periods of adversity. Dr. GRENFELL soon found that one of them had to be closed because of insolvency. However, the value of these stores in promoting the independence, self-reliance and honesty the population was found so great, that the system was continued.

In 1901 a small cooperative lumber mill was established.

In many ways the Association has shown the people how to use what they have to better advantage: fish, fur, soil. These are sufficient to enable the population of the coast to improve their living conditions, and when the necessary technique and enterprise for utilizing these resources have been developed the liveyeres will have more to fall back on. Much has been done, but much more remains to be done (275).

In 1907 Dr. GRENFELL introduced from Norway 280 domesticated reindeer with some Laplander pastors. These animals it was thought would add much to the comfort, health and wealth of the Labrador people if suitably combined with fishing and trapping, just as is done so successfully in Northern Fennoscandia. In this matter Dr. GRENFELL unfortunately was not successful: a systematic plan for the pasturing of the herd had not been worked out in advance by competent people and the population was not instructed in the

Fig. 334. The late Dr H a r r y L. P a d d o n, surgeon, Justice of the Peace, Great Gentleman of Labrador, during some thirty years a local leader of the I. G. A:s work at northern Labrador. Photo the author, July 1937.

importance of this experiment before the herd was brought to Newfoundland. The people, perhaps taking the deer for caribou, soon exterminated the herd. The present author has not a little experience in the reindeer industry, and my personal opinion is, that Dr. GRENFELL's idea was excellent in principle, and I am sure that sooner or later new experiments in this matter must be made and will have practical success (835, 364, 380, 1098).

The funds required for Dr. GRENFELL's magnificent charitable organization are great. They have been provided partly by the Newfoundland governement (in 1940 $ 5000) but chiefly by his generous friends in the United States, Canada and England. In the great American cities interested people have formed themselves into supporting and branch associations which they call Grenfell Associations, with a central bureau in New York (cf. the series »Among the Deep Sea Fishers»).

Some days after the above had been written SIR WILFRED GRENFELL expired. The International Grenfell Association however remains for all times a brilliant memorial to what a strong and intensive personality can accomplish in the service of the good. His seed is already bearing fruit in Labrador. The self-imposed duties of the great philanthropist must now to a great extent fall upon his faithful collaborators under the leadership of Dr. CURTIS. The necessary funds must largely become the responsibility of the Newfoundland Government, for it is to be feared that the outside material support will in consequence of the new world war gradually fall away when the genial master himself and his faithful and experienced co-worker, Dr. HENRY L. PADDON, Fig. 334 are no longer among the active members of the Association. If the structure so carefully erected by SIR WILFRED were allowed to collapse it would be a terrible calamity for the Labradorians.

The final aim of the Association's efforts was given by SIR WILFRED in the following words: 'We are working for the time when no 'mission' need work among these men of Labrador, for they will be self-sustaining and powerful in their simple, wholesome life by the sea.' May the men who continue his work tend carefully the flame of self-sacrificing love for mankind, which he lit, see that the ideas inspiring him are not lost to sight, and make sure that the International Grenfell Association with its rapidly increasing numbers of institutions does not appear to the generous donors as a bureaucratic body and an end in itself!

Like myself the Labrador people will remember SIR WILFRED as a cheerful soul, full of life, loving a joke and a good story, and a great sportsman. Looking back upon his grand life's work in Labrador one can say with GOSLING: It has been truly apostolic: to heal the sick, to cloth the naked, to feed the hungry, to teach the ignorant, to protect the fatherless. One of the salient features of his life was his religious sense of responsibility. His life was devoted to practical Christianity, someone has said. It was one of the most simple, direct, and vital applications of the Gospel to human needs that modern times have seen, says Dr. HENRY VAN DYKE. This explains his fatalistic conviction that faith can remove mountains which characterised the execution of his enterprises. 'I shall get the money for a new hospital', he would say, 'but I don't know how'; and he got it.

* * *

The generally high moral status of the Labradorians in our days is in all essentials the result of the self-sacrificing work carried on by the unselfish and frugal-minded personalities of the Moravian Mission and the International

Grenfell Association. But it must not be forgotten that ever since Bishop FEILD consecrated the first church of southern Labrador at Francis Harbour, July 8, 1853, a number of devoted young clergymen and scholars of different confessions have tended and nurtured the good seed. We find among them representatives of the Church of England (768), Methodists (1062, 1063) and the Roman Catholic Church (130, 191, 198, 226, 250, 251, 454, 496, 498, 569, 793, 809, 904, 944, 1067, 1068, 1080, 1127). There are few amongst the white people who are illiterate or so nearly illiterate as only to be able to sign their names, since schools have also been established outside the Moravian Mission and the International Grenfell Association. The percentage of illiteracy in the whole colony was in 1921 21.3 %, in 1935 18.3 % (222). In a general way elementary education is provided by the churches, aided financially by the state which almost wholly pays the salaries of the teachers (585).

Industry and Trade.

In passing some space must be devoted to the industries at present running or planned for the future in Labrador. Naturally the resources of the sea so easily available were the first of which use was made.

Fisheries.

Fishing: 'tis an honest trade; 'twas the Apostle's own calling, said James I once (325). In our days the attitude is somewhat more complex (cf. 13, 58, 102, 148, 913; 221, 341—343, 346, 392a, 368, 453, 862, 1033).

Cod-fishery

has been since olden times the staple of Labrador and its abundance unequalled in any portion of the globe. The codfish of Labrador have been known since the days of the first explorers by sea. This harvest of the deep has also been quantitatively the greatest source of income on the coast. Even in our days the salted and dried fish from that country is in favour, though of comparatively low quality and produced at high cost in a very unprofitable way (p. 760). This latter product is popular only in West India, the Mediterranean countries and Portugal. However, to-day a crisis has arisen (cf. p. 762) (1033), and if the cod-fishing cannot become more profitable by using new methods to preserve the fish, it is probable that the rather primitive Labrador fishery will no longer be able to compete with the modern fisheries of Scotland, Norway and Iceland. Nowadays it is quite clear, LODGE says, that the cod-fishery has ceased to be the economic mainstay of the colony and especially of the Labrador. Many people believe that not even the erection of artificial drying and curing stations in the area of production would help very much. I do not know whether any estimates have been made with regard to an industry for transforming the inexhaustible schools of cod into fish meal. I think LODGE has convincingly shown that an expansion of the Labrador cod-fishery is not compatible with sound economy if present methods are con-tinued, seeing that the competition on the world market both in price and quality is so very keen.

Salmon-fishery.

The salmon-fishing on the east coast of Labrador is confined to that por-tion lying between the Straits of Belle Isle and Hopedale. Further north salmon appear to be exceptional if not altogether absent until one comes to Ungava Bay (761).

The salmon-fishery has been systematically developed by the Hudson's Bay Company. There is no better salmon in the world than that of Labrador. The catch of the individual fishermen is collected by the company's boats, frozen on a special vessel, Blue Peter, and is now being sent to the British market in cold storage and can be recognized by the distinctive trade mark: 'Hubay' (288). It might be imagined that salmon and trout canneries would be a suitable local industry. However, it seems doubtful whether the fish could be caught in sufficient profusion. It is also a question whether floating, freezing and preserving plant would be suitable for the northern Labrador coast. The production of the salmon-fishery can scarcely be increased, it is considered (cf. p. 722).

Herring-fishery.

The herring-fishery south of Domino Run might be thought to have vast possibilities. But I have no exact information concerning the experience there of an American company in recent years.

Whaling.

Whaling was carried on in olden times on a very extensive scale from the Straits of Belle Isle up to Hudson Strait (325, 889, 625). This was done by Basques, Dutch and people from New England, later also by English and to-day by Norwegians. Huge piles of whale bones have been found on some islands, which is rather surprising, because they were once an important article of trade. At the end of last century the whale-fishery became very unsuccessful and for several years past the plants at Schooner's Cove and at Gready have been closed. After the first World War two factories were started at Cape Charles and Hawke's Harbour (364) but in 1921 both were temporarily closed owing to the low prices of the oil.

At present only one factory is running on the coast: Hawke's Bay, Figs 335 and 336. It is supplied by fast small steamers furnished with the most modern equipment, with harpoon guns handled by Norwegians trained to the

Fig. 335. Hawkes Harbour with the whaling station in the background. Photo the author, Aug. 1939.

work. Regarding the whale oil manufacture the reader is referred to GREN-FELL's description (364).

It is evident that the whales have progressively decreased in number, and from a value of $ 73.440 in 1904 the return has continuously diminished, and will probably soon lose its place as a potent factor in the economy of the coast.

Sealing.

Sealing has since time immemorial been an important industry on New-foundland where many hundred thousands of them are slaughtered every spring (45 a). Also on the Gulf Coast masses of seals are killed, panned and sculped (skinned) every spring. Already in 1774 CURTIS (196) strongly recommended that the seal fishery should be more widely followed on the coast. However, it has not developed much. For instance, about 1860 at the four Moravian Mission stations the average annual catch was only from 3000 to 4000 seals

(787 a). The ice-belt off the coast, which often breaks up, and the large pans of moving drift-ice which the many herds of seal follow toward the south in November, are unfavorable for the industrialization of seal-hunting. Another discouraging factor is the low price of seal oil. This is due of course to the industrialization of the hunt of the big sea animals. Lately I have seen a notice that the Hudson's Bay Company will actively take part in the sealing

Fig. 336. Whale oil plant at Hawkes Harbour. Copyright WILMONT T. DE BELL, Aug. 1937.

industry. Whether this will have some reaction on the sealing in Labrador I am not able to say.

Fur-farming.

One industry which has been much developed elsewhere under the same natural conditions as prevail in Labrador is f u r - f a r m i n g. This industry was once tried out in modern ways, for instance at Voicey's Bay. The result however was not very encouraging and the attempt was given up. There is a chronic lack of land animal's meat in Labrador, and experience has shown that if the foxes are given more than 15 % of fish in their food they do not thrive and the skins are not first class. This industry has therefore not gone beyond a trial and, according to what one of North America's foremost experts in this sphere told me in conversation, has no future in Labrador (71).

Forestry.

The largest and perhaps most valuable of the natural resources of Labrador which can be continuously utilized lies in the spruce woods (cf. p. 408 sqqq.). With reasonable investment it can be gradually exploited with the labour power of the country itself and developed into a large, permanent industry with sawmills and cellulose plants. Timber for lumbering is available

Fig. 337. The »Labrador Development Company» at Hope Simpson greeting the expedition with the flag of Finland (536 a).

in large quantities near the shores of many bays and rivers (p. 410 sq.), which offer swift conveyance of the logs and the lumber from the interior to mills on the seashore.

Yet to-day the lumbering industry is surprisingly little developed. Only one undertaking on any important scale, 'The Labrador Development Co', has since 1935 secured a medium term loan from the government and embarked on the cutting of pit props for the Welsh market. But nobody has as yet exhibited any anxiety to follow the example set by this firm (585). The forest industry's vanguard is this Development Company's plant at Hope Simpson in Alexis Bay, Fig. 337. Another small private establishment is to be found in Kipekak Bay. Here there is planning for export, which if it succeeds, may be of great importance for the future of the country. Many sawmills, however, were started earlier but closed after a few years, and it is not difficult to conceive the cause: a high custom tariff has resulted in an abnormal high cost of living, and this reduces the competitive power. A firm which began work about 1910, of which the buildings still stand in the Mud Lake district, was compelled to give up; likewise another at Epinette Point. But hitherto efforts in this direction have been of the very simplest type: the export of boards and some pit props for English collieries. Demand is still great. At the present time concentrations of population, such as those at Hope Simpson and at North West River, may become a factor in the rationalization of this branch of industry. In the future there is little need to fear that, as happened a few years ago, the work must be abandoned because the imported workers, dissatisfied with conditions and pestered by the flying insects, simply departed to countries with happier conditions. There are at present only a few — about ten — small sawmills for the domestic needs of the coast. Much less than the annual growth of timber is taken. The primeval forests therefore degenerate and decay because no attempt is made to look after them (p. 407 sq.). For all this, sooner or later, a forest industry will come into existence. The basis for the utilization of the forests must then be a systematic development of the necessary transport ways and floatable water- ways on the one hand and on the other the employment of local workers. Both these depend on a reasonable land settlement (p. 823 sq.). No great capital investment is necessary for a gradual development of the industry, the capital expenditure should be on the direct employment of local labour (585).

Minerals.

Indications of the deposits of minerals containing copper and iron in Newfoundland-Labrador have been previously mentioned (p. 111). At the present state of our knowledge, however, they have no economic value.

Iron ore fields.

On the other hand, in 1895 Low (588) brought back the first authentic report of treasures in Labrador's hinterland, especially of the enormous iron ore deposits. To-day it may be judged that these latter in the heart of the peninsula, on the Lake Plateau, on both sides of the boundary between Newfoundland's and Canada's parts of Labrador, will be of great importance. There are large deposits of magnetite with a content of mangan and layers of hematite and some siderite both separate and mixed. When Low found them during his exploration work in 1892—1895 the Indian canoe in summer and the snowshoes in winter were the only means of communication in that part of the country, and not until the airplane became a practical method of travelling was it possible to carry out thorough geological prospecting for ore. Since 1936 this has been carried on by the *Labrador Mining and Exploring Company*, and its fieldwork is entrusted to the Canadian geologist Dr. J. A. RETTY; cf. Figs 338 and 339. The investigations have been most successful, and the leader told me he was preparing a comprehensive survey of the results. Unfortunately owing to difficulties due to the new World War I have been unable to get access to this report so I can only give some indications of what it may

Fig. 338. Meeting at Ashuanipi Lake with Dr. J. A. R e t t y, the leader of the »Labrador Mining and Exploring Company's'» geological expedition to Labrador in 1937 and the prospector T o p i S e p p ä n e n from Finland. Photo the author, Aug. 1937.

contain deduced from what I heard during my visit to the prospectors and read in an authorized newspaper article 1938 (261; cf. also 919).

At that time three iron ore fields of high quality had been located, viz:

1. Sawyer Lake Field, bearing hard hematite ore, lies about 284 miles (c. 457 km) northwards from Seven Islands on the north shore of the Gulf of St. Lawrence and about 245 miles west-north-west from North West River at the west end of Lake Melville.

2. Attikamagen Field lies about 35 miles north-west of the former, and

3. Ruth Lake Field north-west of the two previous ones near the Canadian border.

The iron content varies somewhat, but the general tests give the following round figures:

Sawyer Lake .... 60—70 %
Attikamagen .... 40—45 %
Ruth Lake .... 57—64 %

The phosphor and sulphur content is reported to be specially low, and that agrees with what I myself saw in the samples.

At Sawyer Lake the deposits are considered exceptionally rich and fine, and are said to surpass, from a metallurgical point of view, the world's best

Fig. 339. Visit to the prospectors' encampment in the heart of Labrador, north-west from Petitsikapau Lake. Photo the author, Aug. 1937.

hematite ore. The Attikamagen Field is poorer as regards iron content, whereas the tonnage is said to be extremely large; the actual tonnage was estimated in 1938 at more than one billion tons. According to the results obtained by 1938 the supply of ore worth mining in these three fields alone may be estimated at several thousand million tons. The extensive prospecting and development work will give a far more definite basis for estimates of the potential reserves than was available before the beginning of the World War.

If we now sum up we may conclude that the supplies of iron ore give certain promise of securing there a great production of pig iron and steel for a very long time ahead. One may even dare to say that for many future years the continent can be self-sufficient in regard to iron from the Labrador fields.

No other iron deposits have since been discovered in the northern part of North America so plentiful and so rich as those in Labrador (288). It is thus to be expected that just based upon this supply of ore a great local iron production will arise, probably one of the largest in the world.

Sources of power.

To these enormous resources of ore is added another extraordinary favourable circumstance for the setting up of an industry here: the possibility of cheap production of power on the spot. Only 85 miles south-east of the Sawyer Lake field there is an inexhaustible source of energy in one of the world's greatest waterfalls: the Grand Falls in Hamilton River; Figs 74 and 75. It was discovered a hundred years ago, but because of its inaccessibility has been visited by only a few whites, and its enormous power has never yet been utilized. Since the end of the ice-age the roar of the Grand Falls, the lonely rival of Niagara, the second greatest cataract of North America, has thundered across the woodland, VARICK FRISSELL writes. It is more than twice as high as Niagara, and, if harnessed, it could not only turn the wheels of industry in all eastern Canada, but would electrify the other sleeping goods of Labrador's wealth (288). No wonder that it has stirred the imagination of many industrial minds. Many explorers, commercial and industrial leaders, believe the Grand Falls region to be the key to Labrador's development.

In the uppermost part of the Grand Falls the masses of water from the Lake Plateau drop almost vertically 101 (588) [114] m, resp. 312 ft (288) down the north side of a canyon, at the bottom of which it continues twelve miles as a steep rapid between precipitous cliffs 100—150 m high. The total fall between the lowest lake of the Lake Plateau and the mouth of the canyon valley where the connected rapids come to an end is said to be 245 m. The whole of this stretch of falls and rapids which together form the so-called Grand Falls can be exploited without technical difficulties.

The Hamilton River leads a very considerable mass of water into Grand Falls; I was told that it amounted to 50,000 cub. ft. per sec. or 1414 sec. cub. m at average low water (!). When VARICK FRISSELL in July 1925 visited the falls, the discharge was about 80,000 cub. ft. per sec. (288). Calculated in hydraulic energy the former quantity of water should give a total of

a. 1,600,000, HP. from the chief fall of 101 m.
b. 3,850,000 HP. from the total fall of 245 m.

In considering the size of these figures I have checked them somewhat. The precipitation area of the Lake Plateau is about 92,400 km²; these figures of course stand or fall by the reliability of the map material. The annual precipitation on Labrador's Lake Plateau is some 550 mm. There appear to be no systematic measurements of the discharge of this area. Yet, if one dares to proceed from the assumption that in two large districts which correspond completely morphologically and which have the same vegetation habitus ought to indicate an agreement in moisture conditions, and I think one is justified in so doing, the conclusion may be drawn that hydrologically the Lake Plateau can best be compared with the central northern part of Lapland with an average discharge coefficient of 0.65 or perhaps at the most 0.75. The lower figure gives an average discharge of about 1000 (1040) cub. m per sec. which would give respectively

$$a_1. \quad — \quad 1,150,000 \text{ HP}$$
$$b_1. \quad — \quad 2,750,000 \text{ HP}$$

The difference between the total amount of possible energy calculated thus and that above for the average amount of low water is very great. Yet both agree fairly well in regard to order of size.

One favourable circumstance for the utilization of the water power is that the Lake Plateau of Labrador is a completely wild country, so that this store of water can be regulated exactly as is found best. VARICK FRISSELL thinks that by damming the lakes and a certain 'pirate' river farther upstream, the flow could probably be brought to well over 100,000 cub. ft. per sec. This means that from 3,000,000 HP upward is available (288). When, for comparison, we realize that the total harnessed water power of Sweden is only something more than 2.5 millions HP and in Finland some 0.75 million HP, a slight conception may be had of the unrivalled scale of Grand Falls and the enormous amount of energy going to waste here. The exploitation of this source of power will further be fairly cheap and easy because nature herself has done the chief work by wearing out the deep canyon in which the water runs and which is easily dammed up; cf. Figs 74 and 76.

The solution of the problem of transportation is a vital one for the harnessing of Grand Falls. After the conclusion of the prospecting in the three ore fields mentioned it is assumed that the first task will be to connect Sawyer Lake Field by a railway about 300 miles (480 km) long with Seven Islands on the Gulf of St. Lawrence; this is necessary also for the transporting of building material and mining machinery. A side track for the same purpose, 70—85 miles long, will be built from this railway to Grand Falls.

But there are also other projects. FRISSELL reports that the Provincial Government of Quebec has already granted a charter for a railway via Grand

Falls and connecting Quebec City with Hamilton Inlet. An ocean port for Canada is an old idea. The Labrador transpeninsular railroad would form the shortest route between Europe and America; the route from Hamilton Inlet is 750 miles shorter than that from New York and 450 miles shorter than that from Montreal. Paper, pulp and iron would be the visible and most important freight on the railway (cf. 288). The experience of railway building through primeval country of similar geological formation as that, e.g. of Canada's ore-bearing country is an encouraging precedent.

The actual value of Grand Falls depends, of course, on whether or not there is a ready market for the energy it could develop. Fortunately in the same region there is besides the iron ore also spruce in inexhaustible quantities which only await transformation into pulp and paper etc. The outlook for the manufacture of steel having been revolutionized by the development of electric smelting, the Labrador iron deposits have become infinitely more valuable through the tremendous energy of Grand Falls. Though iron will give the Grand Falls its initial load there is no doubt that the weight of pulp and paper mills will be felt soon after (288). The situation of the country would favour the establishment of industries, e.g. nitrogenous fertilizers, carbide etc., which involve high energy consumption.

The industrial enterprises in central Labrador would be protected against enemy attacks, considered from England's point of view, as well by two thousand miles of ocean and by the Monroe doctrine (585).

For the present it is not possible to reckon upon any other m e t a l-l u r g i c a l or m i n i n g i n d u s t r y in Newfoundland-Labrador than that upon the Lake Plateau.

On the other hand s o u r c e s of h y d r a u l i c e n e r g y are available in many places and these could be regulated in any suitable way, because in most of the valleys there are lakes which can easily be dammed and their storing capacity thus enlarged. But for the present no demand for energy exists.

Communications.

Much has been done to improve the means of communication in Labrador for the purpose of giving support to the occupations of the fishermen from Newfoundland.

Nearly four decades ago a t e l e g r a p h line was laid along the north shore of the Gulf of St. Lawrence to Forteau Bay, and during the first World

War this was continued to Henley's Harbour. In 1939 there were wireless Marconi stations at seven places, and these are in connection with commercial and amateur stations in the interior and in the north. At present it is generally possible to get an answer to telegraphic questions within 24 hours from the most distant European places.

The m a i l is mostly under the care of the Hudson's Bay Company's and the Moravian Mission's stations. In the summer the post comes every second or third week. Every little boat can however be said to be a mail-boat, and they take the post from the stations to e.g. the fishing schooners, etc. in a short time. When the steamer traffic ceases in late autumn the post is sent from the Gulf coast two or three times in the winter to stations on the Atlantic coast. Occasional aeroplanes also take the post with them; I myself once took with me to North West River a Montreal newspaper only 20 hours old.

S t e a m e r s. In 1877 the Moravian Missionaries were given a steam launch to ply between their stations. In 1880 the Newfoundland Government first sent a mail steamer along the coast as far as Hopedale. At the present time passenger and freight traffic is very well arranged in summer by the mail steamer *'Kyle'*, Fig. 311, which operates the service from St. John's to Hopedale. There a schooner hired by the state takes (1939) the mail and passengers travelling further to Nain and Hebron. The *'Kyle'* can be most highly recommended; the passengers are well looked after and enjoy all conveniences. It makes some fifty ports of call from Battle Harbour to Hopedale, and the season runs from June through to October. This cruise occupies approximately two weeks and the service is the only one regularly offering access to the area north of Battle Harbour. It is, however, a disadvantage that the *'Kyle'* cannot follow a time-table, for not even Captain CONNOR can control the capricious ocean fogs. On the Labrador coast, north of Hamilton Inlet, experience is the only safe guide. There are few marks and no buoys in the channels; the most able captain could never navigate this by chart (614; cf. however 273). People and cargo are also conveyed by the Hudson's Bay Company's motor-schooners, e.g. *'Fort Garry'* and others.

The residents one meets here are very obliging in every way even towards strange travellers, and ordinarily there seems also to be room for an occasional visitor in their homes. On the other hand only very seldom can one reckon upon hiring a boat in the summer; this should be arranged during the early spring by correspondence in St. John's; the boats on Labrador are few and much in demand during the salmon- and cod-fishing season.

Trade.

In our days the trade north of Indian Harbour is practically speaking exclusively in the hands of the H u d s o n's B a y C o m p a n y. South of it there is a slight penetration of traders and co-operative stores during the fishing season, e.g. in Battle Harbour, Fig. 340, and this is said to have produced some competition which helps the local population to buy and sell or

Fig. 340. Battle Harbour. Looking from south-west (536 a).

barter under somewhat more favourable conditions than would otherwise be possible. Yet the Company dominates the trade.

In Labrador the Company has been a very important instrument in the cultural transition, and some account of it here is therefore justifiable.

HUDSON'S BAY COMPANY,

incorporated 2nd May, 1670, is a privileged British trading firm originally founded for the purchasing of pelts in the countries round Hudson Bay; its motto: *Pro Pelle Cutem*, speaks for itself. It is the only private firm which still has the right to bear its own emblem: H. B. C., in the British national flag, and its officials wear a uniform cap of military style. In 1870 trade in articles of consumption was added to the fur traffic and now the company supplies these, and almost exclusively, to the population living in arctic and sub-arctic North America (cf. 51, 92, 66, 110, 181, 264 a, 404, 431, 462—465, 566, 608, 610, 611, 622, 668 b, 811, 1001, 1027, 1037, 1108).

In Labrador, as in other places, the Hudson's Bay Company headed by indefatigable men has been a harbinger and pioneer of civilization in the wilds. It has jokingly been said that H. B. C. means that they were 'Here Before Christ'. And it is true that it was they who first penetrated and explored the endless Canadian wilderness, after which the missionaries followed in their tracks (903).

The H. B. C. was followed by other fur trading companies (66, 194, 609, 634, 751 a, 984, 1027) but particularly since its amalgamation in 1821 with the North-West Company it has controlled the boreo-arctic part of the continent from coast to coast. To-day it still continues its commercial operations with the liveyeres and the aborigines all over this territory, thus extending the fur trade into the less accessible parts where wild life flourishes (1036, 510). In the life of Newfoundland-Labrador on the other hand the Company is a relatively recent factor.

As far as concerns Labrador the land was purchased from the successors to the rights originally obtained by CARTWRIGHT (325). McLEAN (610) certainly pointed out about the year 1831 that the Company having learned from a pamphlet published by the Moravian Missionaries that this country produced excellent pelts was induced to make use of the knowledge. It seems however more probable that CARTWRIGHT's proposals to establish hunting and trading posts on the Labrador were carried out by the H. B. C. after his death.

Many trading-posts were gradually founded by the Company in eastern Labrador: F o r t C h i m o (1827), F o r t G e o r g e R i v e r, E r l a n d-s o n P o s t (probably at the northern end of Indian House Lake), F o r t R i g o l e t (1837), N o r t h W e s t R i v e r P o s t, F o r t N a s k a u p e e (on Petitsikapau Lake, 1840—1873; cf. 610), F o r t W i n n i k a p a u (on Hamilton River, 1866—1873; cf. 588), F o r t M i s h i k a m a u, S a g l e k B a y P o s t and N a c h v a k B a y P o s t (after 1866 or somewhat later; abandoned some time ago; cf. 275), F o r t A i l l i k (about 1866; cf. 147), C a r t w r i g h t, D a v i s I n l e t, V o i c e y's B a y, F r e n c h m a n's I s l a n d and P o r t B u r w e l l, the bleakest of all the Hudson's Bay Company's posts in Labrador (Figs 79, 80 and 342). Lately the Company expanded northward. By the agreement of 1925 with the Moravian Mission the trading rights and stores, wharfs and office spaces at the stations M a k k o v i k, H o p e d a l e, N a i n and H e b r o n were transferred to the Company for a period of 21 years expiring 1947.

When one comes to a village the Company's neat, white buildings with their red roofs stand out at once from among the other houses and huts; cf. Figs 341, 169 and 204. The buildings, workshops, vessels etc. which are

Fig. 341. Hudson's Bay Company's store and its manager's bungalow at Cartwright. Note the boxes with salmon to the right. Photo the author, June 1937.

so well managed and looked after have served as models for the population. In the stores, too, perfect order reigns and only first class commodities were stocked wherever I went. Also the friendliness and care which the Company's local managers, clerks and other assistants showed to their clients really greatly impressed a stranger from the Old World. It is clear that in all cases the staff is most carefully selected. Many of the men who have first contributed to the building up of the Dominion of Canada started in the service of the Hudson's Bay Company; Lord Strathcona, the founder of the Canadian Pacific Railway began his apprenticeship in Rigolet and managed the North West River station for several years.

In the shops it is possible to obtain all the goods necessary for a household in Labrador. They are necessarily high in price, they must pay a customs

Fig. 342. Harbour of the Hudson's Bay Company's trading-post of Port Burwell, north-easternmost Labrador. Photo ALEXANDER FORBES, Aug. 7th 1935.

duty for everything imported (up to 64 % *ad valorem*) and furthermore
the risks of storage and unavoidable credit are great under the prevailing
circumstances. But an inestimable advantage is that everything sold is of
good quality. On the other hand the Company buys the products of the
trappers and fishers and pays high prices, especially for the pelts. The internal
price level and the cost of living are therefore rather high in Newfoundland-
Labrador for a foreigner.

In Labrador you will hear much adverse criticism of the Company's trading.
Yet everywhere in the world where fishing is practised on a great scale there
are evil critics of merchants and supply. Meanwhile in most cases the views of
the fishermen are wrong. I remember from my many cruisings in northernmost
Norway at the beginning of this century — and there is not much difference between Norway and Labrador in this respect — how the fishermen bitterly pictured the rapacity of the merchant who bought their catch endeavouring to get
the unhappy fisherman into his toils. People would not realize that such an
uncertain business as cod- or salmon-fishery, which presupposes a large amount
of credit from the merchant to the fishermen during the bad seasons, makes
it necessary for him to have a reasonable profit in advance during the periods
when high prices prevail in the fish market to secure the continuance of his
enterprise when prices fall. On the other hand my own impression was that
the average fisherman was often on the look-out for the merchant who would
give him most of his outfit and food on credit; if they had saved cash few of
them would use it to purchase supplies; the merchant had to run all the risks.
This matter has undoubtedly been very similar in Labrador in bygone days;
(cf. Governor KING's remarks in 1792 in325; HIND; GRENFELL). Irregularities
may have occurred in the business, but they must be exceptions. In many
cases the Company's task has been confused with that of the philanthrophic
associations operating with voluntary contributions. The task of the Hudson's
Bay Company's stations is to act coolly and carefully like a sober and practical
merchant should do if he looks after his firm's advantage well and secures
continuity for it. It is only natural that the Company has no reason to pursue
philanthropic aims and to encourage the penetration of the wilder parts of the
country; the fur traders ought, of course, to endeavour to reserve the country for
the trappers and protect the hunting of fur animals against intruders. And it
seems that the Hudson's Bay Company, having been entrusted with the regulation of Labrador's economic life, has done it well as far as I understand it.
But there seem, of course, to be some circumstances in connection with the
trade which might, with certain modifications, be made more advantageous
for the people without any loss to the Company.

All things considered from what I have seen and heard and read of the
Hudson's Bay Company (153, 610, 264 a, 810) it is clear to me that as an element for the furtherance of order and education among the Labrador population H. B. C. has had and still has an importance which cannot be over-
estimated.   Its managers and staff have pacified the wild people, and
organized and introduced system into the economic life of the whole population. Their only means of compulsion has been trade.

Future of Newfoundland=Labrador.

Labrador is often called the land of promise. This is however a truth which needs modification. Nobody should believe that the permanent population of abode in Labrador can transform its natural resources into clinking coins without great pain and fatigue. It is well known that the great fisheries are run at a loss. The whaling gives a very small margin of profit. The seal-hunting has almost been given up. The capture of fur-bearing animals and salmon-fishing combined with seal-hunting and cod-fishing for household needs are the only regular sources of income worth mentioning at present. The people live in the main from hand to mouth on these predatory means of earning a livelihood. The country has developed its fishing and fur industries to the complete neglect of its agriculture, someone has said; as soon as things go wrong with the present natural industries the people are in a bad way and require food from abroad. It is naturally dangerous to neglect any opportunity for food production. — (Cf. also 352, 370, 421, 585, 720, 918 a).

'The viewpoint has been expressed that the men of Labrador should be left to sink or swim, but', says Dr PADDON, 'it is deplorable if any potentially prosperous section of the community is left to go under for lack of readjustments which are not hard to discover or to effect'. A stranger soon gets the impression that the people, for want of an economic plan, are driving heedlessly toward an uncertain future. Even if the enormous supplies of ore on the Lake Plateau are exploited it will scarcely make the circumstances of the coastal population easier (cf. p. 804 sqq.). The same can be said of the exploitation of the water power and the forest supplies under present conditions.

The future development possibilities of Newfoundland-Labrador can be judged from two points of view.

It may remain an exploitation colony where the utilization of the great natural resources, especially the forests, the ores and the hydraulic power sources, as well as the arrangement of the tasks necessarily connected with them, will be conceded to private, capitalistic enterprises, while the state restricts its activity chiefly to the maintenance of order, jurisdiction, some means of communication and the collection of taxes.

It is however also thinkable that the state without suffering any harm may give up its passive financial policy in Labrador and in the future take part directly in the organizing of an intensive exploitation of the permanent natural resources. These latter have precisely the same possibilities there as for instance in Fennoscandia where this has been done to the advantage of both the local people and the state (cf. my study: L'utilisation économique du territoire de Petsamo. Moyens et buts. Fennia, Vol. 49, No. 3, 1926). There are both advantages and disadvantages to be considered.

Before entering into the details of this question one circumstance must be recalled which is often forgotten and which it is necessary to know for the understanding of the backward condition of Labrador.

The boundary between the possessions of Canada and Newfoundland in Labrador remained for a long time uncertain, and therefore the judicial conditions were vague. No one could say precisely how far the sphere of Canada's jurisdiction extended to the east and where Newfoundland's laws began to be valid in the west. This of course gave rise to a state of insecurity which prevented capital and initiative entering this 'No Man's Land' where no one could use pick or shovel and feel sure that he would have a title to his claim on the morrow, FRISSELL wrote (288). The enormous ore deposits lay unused, millions and millions of horse-power ran to waste in the numerous waterfalls, the forests died away from the roots and there was no cultivation of the soil worth mentioning. The state's passive financial policy was clearly no blessing but is comprehensible when one remembers the uncertain conditions.

Finally in 1926 the decision of the Privy Council definitely fixed the frontier between the rival Dominions and put an end to the insecurity (303, 325, 334, 407, 408, 438, 455 b, 502, 602, 605, 765, 766, 937, 863, 1025, 1026, 1109, 1126). The area governed by Newfoundland's Timber and Mining Laws, etc. is now quite clear (cf. TAIT's Synopsis in 913).

After the settlement of the old frontier disputes greater interest in Labrador began to assert itself because of its vast natural resources. The enormous deposits of ore on the Lake Plateau were subjected to a modern investigation by 'The Labrador Mining and Exploring Company'. An evaluation of the raw material for a pulp and paper industry in the forests of the water system of Lake Melville was drawn up by means of aerophotographs, presumably on the initiative of Bowater-Lloyd. Concessions of enormous forest areas were held by various interests; in 1939 no less than ten firms. These continued to pay small taxes to the government in the hope that some day the world demand for wood would reach such a point as would make their holdings remunerative (585).

This policy of granting concessions has meanwhile up till now been quite unable to show any success worth mentioning in Newfoundland-Labrador's industrial life.

I do not believe that the Labrador population has any reason to bewail this fact. It would rather have been a misfortune for them if large scale industrial activity had been started there before planned colonization of the area had begun. Big industry would certainly have hindered a sound development of colonization before it was well started; with its auxiliary industries claiming the labourers the latter would have been attracted from important seasonal work while the colonization was in progress.

According to the second alternative stated above, which might be called the Fennoscandian working principle, the future of Labrador could be gradually built up in the first place upon these sources of livelihood which guarantee a continued run: the soil, with hunting, trapping, and fishing. When the colonization established upon these resources has been consolidated and indigenous workers are available who are accustomed from childhood to control the severity of nature, the time will have come for a progressive introduction of a rational forestry. On this other industrial activities could be founded. Confirmation of this opinion is: In the boreo-arctic regions the permanent income from well-managed forests and grass-land and the auxiliary occupations possible there are better than gold mines.

POSSIBILITIES FOR MIXED FARMING.

Is there then any justifiable hope for rationalized agriculture in Labrador (cf. 376, 585, 840, 842, 844, 845)? Let us go into the matter a little more in detail.

JUNEK (510) has divided the north shore of the Gulf of St. Lawrence into three ecologic zones. In the first zone from Tadoussac to Betsiamites the dominant industry is agriculture (cereals, hay, various vegetables, tobacco for home consumption). In the second zone from Betsiamites to Seven Islands lumbering with pulp-production predominates. The third zone involves our territory from Seven Islands to the eastern limit of the Labrador coast. But in this zone he distinguishes two physiographic sub-zones divided at Natashquan: the western has a more luxuriant vegetation consisting of occasional heath plants, stunted pines and the like, the eastern sub-zone is bleakly devoid of any leafy or woody growth other than undersized conifers which appear only in depressions and freshwater accumulations. In the third zone only fishing and hunting are found, the former predominates in the spring and summer, the latter is pursued in autumn and the long rigorous winter. Farming is impossible mainly owing to short vernal seasons and mid-summer frost. Two months of growing is not enough to repay the labour and expense.

With JUNEK's conclusion regarding the easternmost sub-zone the author cannot agree. The productiveness of Newfoundland-Labrador is not to be disregarded. The district is not precisely a Garden of Eden, but the experience gained for instance from g a r d e n i n g is impressive.

As regards the coastal tracts near the 57° parallel HUTTON wrote (476, 478):

For many years the missionaries have made gardens. There is not much soil, probably the old missionaries had to carry soil in barrows and build it into gardens before they could get their vegetables to take root and thrive. Six or seven feet down the ground is permanently frozen. But the blanket of snow preserves in winter some of the roots and so the rhubarb comes up year after year.

At Killinek even the keen missionary gardener could get little or nothing for his labour. The garden was a tiny railed-in patch of earth, scraped together from among the stones; a few pallid leaves and shoots showed where the lettuce and the turnips were trying to grow, having a sad struggle against the bleak air and the chilling fogs and rains.

At Okkak one gets the snow cleared away in May and then leaves the ground to thaw in the sunshine. But the actual planting out does not take place until July, and in the meantime the vegetables are growing in the house or under frames. It was here that the missionaries were compelled to give up their greenhouse after futile efforts to grow early vegetables. Not until the nights begin to get a trifle milder can one dare to put the cress and lettuce, cabbage and potatoes in the open air. The potatoes, turnips and cabbages would not be of eatable size until well on in September. In August one gets some lettuce carefully grown in window boxes or under glass in frames, covered with sacking against the night frost. No night is safe from frost in northern Labrador; snow may fall in August. Evening by evening the missionary looks at the thermometer, and if night frost is likely, covers each tiny struggling plant.

The dogs, the mice and the frost are always hostile to the gardens. By the help of wooden palings one manages to keep the dogs out. In years when mice are plentiful one covers each shoot with an empty meat-tin overnight. The frost one fights every night by covering each row with a wooden framework over which sacks are spread. In this way the hardier sorts of vegetables grow to a moderate and eatable size even in Okkak before the ground freezes again in October.

PRICHARD (671) in 1910 found the garden beds in Nain bright with hardy flowers, mostly pansies and poppies, and from the greenhouse, where roses were blooming, they cut lettuce. Later in the season lettuce matures in the open, while rhubarb grows not only sturdily but luxuriantly.

All these northern mission stations are near the ocean and this strip of the country suffers from some real physical disadvantages, the distance from the Gulf Stream makes the winter severe, and the nearness of the cooling Labrador Current and the attendant clouds and fogs, which shut out the life-giving sunshine, make the summer on the coast rather cold.

Conditions further south are better, especially after leaving the outer coastal strip the climate becomes much better than its reputation. Growth may begin relatively late, but when it does begin it soon makes up for lost time. The head of the Hamilton Inlet may be termed the garden of the Atlantic coast of Labrador, already HIND wrote (434). In the middle of July I found the gardens with vegetables (lettuce, carrots, beetroots, turnips and cabbage) growing in profusion in North West River. There is the great insolation of the relatively southerly latitude and moisture in the soil at just the moment when they are most required. — It is just this which makes the forest re-production so striking inland (585). — Is it not surprising that agriculture at the head of Hamilton Inlet is commonly three weeks ahead of agriculture in St. John's (731)!

At Salt Pond, close by Sebashkau (Lake Melville) the Michelin family is flourishing to the tune of ten barrels of potatoes, and enough cabbage, turnips and other common vegetables for winter eating (Labrador souv. suppl. to the 'Daily News', St. John's, October 19, 1938, p. 15).

Following a 'Memoir concerning Labrador, 1715—1716', by an unknown writer, de Courtemanche had at Bradore a large garden where all sorts of vegetables grew: peas, beans, roots, herbs, and salads. He had sown barley and oats which grew well. He kept horses, cows, sheep and pigs (325).

We can thèrefore state that in the southern interior of Newfoundland-Labrador, where the soil is suitable, the climate not only allows potatoes, turnips and cabbages and other common culinary vegetables to grow, but also to give a yield which cannot be much less than the average in the English Midlands (585).

The growing of potatoes and other vegetables is of course important for the people; they give a quick return from the land and the necessary variation of food and vitamins.

But as a basis for a rational colonization of Labrador farming on a large scale is demanded. It has been stated that the natural conditions, specially as regards vegetation, are much like those in northern Fennoscandia. Now, well-managed f a r m s w i t h m i x e d f a r m i n g a n d g a r d e n s in Finland's Lapland clearly show that agriculture can be successful in such an environment. As far as I can judge it will in appropriate forms give also the Labrador people in some sections a chance to grow an important part of their food and to earn a little money by helping to supply the demands of other sections.

Good guidance in this matter is to be had from the experience gained years ago in northern Fennoscandia of which the soil, climate and vegetation are similar to those of Labrador; both are rather continental in type. Consequently the manner in which the land is utilized in the former may be some help for the latter. On the other hand I heard it argued in Labrador that a considerable amount of agreement in this respect was to be found between the Atlantic coasts of Labrador and Norway, and thus the best ways of living should be almost the same in both countries. This is wrong. I think the routine in the coastal districts of Norway cannot hold good for Newfoundland-Labrador. Norway has an extremely maritime coastal climate and the vegetation period develops of course in a different way there from that in the continental Labrador. Under these circumstances agriculture in northern Fennoscandia is not only suitable as an object of reference in considering the colonization of Labrador, but it is the only relevant one because the external conditions are very similar (p. 401). It seems therefore appropriate to follow as far as possible the Fennoscandian principle. The model is to be found among e.g. the rural population in northernmost Finland until agriculture has as far as possible been stabilized and has worked out its own routine in Labrador.

There is no land asking for c r o p s in Labrador. Yet, barley (seed of the hardy Lapland type) would undoubtedly grow rather well in the continental climate of the Hamilton River valley, but scarcely oats. In 1843 experiments were made by the Moravian missionaries with seeds of barley and other cereals, as well as of trees from the Himalayas. The barley came up but was cut down by frost before it had attained much growth, while the tree seeds did not even germinate. This shows, as was to be expected, that grain will not grow in the outer coastal strip. However barley will probably develop satisfactorily in the interior. Yet I think it would not be very profitable to plant cereals; flour would certainly be more cheaply imported from the American agricultural districts with their warmer climate and better soil.

The chief desiderata for Labrador is l i v e-s t o c k, and they should be fed on the produce of the land.

Cows were formerly kept in Labrador for their milk and meat, but were given up later on.

In 1857 it is reported that Mr Donald Smith, later LORD STRATHCONA, when residing in North West River had four head of cattle, besides sheep, goats, and fowls; there was milk in plenty, and the Moravian missionary ELSNER from Hopedale, who reported this, for the first time on Labrador tasted fresh roast beef, mutton and pork (325). At the Hudson's Bay Company's post Rigolet there were about seven acres 'under crop', and the farm boasts of twelve cows, a bull, some sheep, pigs and hens, HIND wrote in the 1860's (434). However, it seems that HIND has confused Rigolet with North West River.

— HIND writes: It will not excite surprise that until 1860 the only cow on the Gulf coast east of Eskimo Point was at Natagamiou; the happy proprietor obtained but little profit from his charge, for the impression gained ground among the simple people that cow's milk was a cure for all imaginable maladies. From far and near they sent for a 'drop of milk' when sickness was upon them, and the owner had no milk for himself. —

For what reason cattle-rearing has not developed since, the author does not know. If I was correctly informed, in Newfoundland-Labrador north from Red Bay were in 1937 only the low total of 12 cattle, 2 calves and (in summer) 2 horses. According to this information most of the cattle belonged to the International Grenfell Association, and their feeding stuffs (hay, grain) were imported by steamer from Canada. The cattle in North West River eighty years ago certainly did not live on imported food, just as little as do the two cows which the trappers in Kenamish were said to have procured some years ago. Cf. Fig. 188.

For live-stock purposes p a s t u r e is a fundamental demand; its area will determine the number of them. Pasture can be laid down in the first place on the clayey ground at the shores of the lakes and rivers if the thickets of willow and alder now occupying them are removed. In this way large areas, e.g. around the Hamilton River's mouth, round Goose Bay, Mud Lake Tikkorlak, etc. could be transformed into pasture. If there is only a slight indication of grass in the scanty damp woods, especially on the river sediments, the cattle themselves will transform them in a short time into grass-land, and it is only necessary to cut down the trees and perhaps scatter some seed on the soil to transform it into permanent pasture land. The manuring resulting from the grazing will create permanent pasture which can be changed into arable land. Already in 1860 HIND stated that near Mingan wild peas and vetches and hay are found in abundance on the coast a little distance from the shore and in the rear of the first belt of timber, affording natural pasturage for the cattle. Although the soil is poor, there is yet so much available manure in the form of fish offal that farming on a small scale might be very easily associated with fishing operations, and a settled population gradually establish themselves on the coast (434).

Grass-land and live-stock seem therefore to be the main essentials if one takes a long view. Cattle, goats and sheep can be kept all over the southern half of the country. It would be most advantageous to use the cattle for d a i r y i n g. Then in the summer they may also be turned loose, they will manage to feed themselves and in the fall they can be rounded up and either kept over the winter or eaten. — The hardy and unpretentious, very productive race of 'North-Finnish cattle' would perhaps be the most appropriate and easily managed under the conditions prevailing in Newfoundland-Labrador.

The summer in Labrador is relatively short (p. 299). The period during which the cattle can graze out in the open can scarcely be longer than from the 25th June to 30th September. For the rest of the year they must be stallfed and well fed, if they are expected to produce plenty of milk. Thus there is need for a considerable amount of stall-feed, especially hay.

The necessary w i n t e r f o o d is to be had partly on the natural grass-land and fen-land rich in *Carices*, and partly by raising hay and green-fodder on cleared ground.

The districts near the outer coast are forbidding, barren in the extreme. Even on the outskirts of the forested areas the acreage of productive land is mostly very restricted. Such regions of course must be avoided by modern society, and people who do live there have special reasons: seasonal abundance of cod, salmon and seal (510). Experience has shown that permanent dwellings are not possible there; in the winter the coastal districts north of Red Bay are also practically destitute of people.

But around the inner parts of the fiords nature is milder. A large portion of that area consists of lakes, rivers and semi-barrens with meagre soil covered with forests but between are peat deposits and fens on damper soil. These latter are natural h a y f i e l d s rich in grass, principally *Carices*. By clearing the shores of the brooks and lakes in the forested area one can also get superb hay fields for harvesting. Already in 1913 BRYANT (107) assumed that the clay and the sand beds in the Hamilton River valley could be successfully cultivated.

The coarse mineral soil, the morainic gravel, is not very profitable for agriculture in Labrador. It contains many big stones and it would be very expensive to remove them. Added to that it deteriorates and forms layers of limonitic hardpan (p. 407), which are very thick and near the surface in Labrador, and which hinders the circulation in the soil. These coarse soils would moreover require lime and live-stock manure to give them reasonable fertility. Probably in this way even the scrub-covered mineral soil could be transformed into grass-land, e.g. by means of the nutritious wild white clover which makes its appearance after the cattle have been over it. Yet work on this soil would be extremely expensive. The considerable capital required for this process would be more profitably used for market-gardening or potato crops.

It is, however, not necessary to clear and cultivate the good forest lands where peat-land and especially rich fens (p. 398) can be successfully drained and transformed into arable land and used for raising g r e e n f o d d e r. This method of cultivation is fully stabilized and has proved the only one economically defensible when taking up new land in northern Finland and Sweden. By cultivating the fens here, too, the expensive land clearing and the arrangement of intermediary pasture can be avoided. Hay from natural grass-land will later be gradually replaced by the cultivation of h a y a n d g r e e n f o d d e r o n a r a b l e l a n d when financially possible. The yield of greenfodder can be stored either as air-dried or made into silage in the young leafy stage.

In the interior valleys of Newfoundland-Labrador there are extensive fen-land areas; I estimate the cultivable grassy fens at many tens of thousands of acres. But as extensive parts of the country have been photographed from the air there should be no difficulty in making a closer calculation (cf. p. 401).

The natural grass-land and the cultivable fens are not limited to the lower and middle parts of the great river valleys. On the Lake Plateau very extensive areas of grassy fen-land can be seen from the airplane, and along the shores of the brooks the grass grows very high, for instance on the 'otter meadows'. I see no reason why even there fen culture could not be used and green-fodder grown on it. On the Lake Plateau to-day there is not a single settlement. But perhaps there will be.

The results of cultivating easily drained fens have been excellent in Fenno-scandia and it may be at least worth a trial on an inexpensive scale in Labrador — if the problem is not tackled on a bigger scale — to discover its profitability.

It seems thus perfectly possible to keep cattle, goats, sheep and poultry on a paying basis in the inner districts of Labrador south of 55°-lat. It is also possible to be self-sufficient as regards all common vegetables — except perhaps potatoes which are relatively susceptible to frost — every year, and as regards potatoes eight years out of ten (731). The people should therefore be allowed, encouraged and urged to live on the bounties of nature subsidized by their own production and preserving as far as possible. Every inducement should be given to agriculture and small scale live-stock farming, also the late Dr PADDON concluded (731) after having closely studied the people's life and wants during twenty-six years' residence.

On the other hand nobody will argue that Newfoundland-Labrador is an ideal country in which to start farming. But the trapping-grounds are limited and probably after some years the Canadian policy as to the Indians must be applied also here. Then auxiliary occupations are needed for the white people, trappers and liveyeres. These people want to live in their home country, and propositions for mass emigration would meet with a very cool reception. If no other permanently productive, subsidiary occupation is to be found for the fishermen and trappers, there is the cultivation of land for livestock (cf. 585). A fisherman who has also fields on which he can grow enough hay for his livestock in winter and enough potatoes, turnips and cabbages for his family all the year round will not go hungry or cold whatever happens to the main fishery. There is fresh fish to be had more than half the year, there is game and a supply of wood for building and fuel, LODGE concludes.

It is said to be difficult to impose a farming life on people who have no inherited or traditional flair for farm work. My personal impression from discussions with white Labradoreans is that they understand fairly well the future prospects and that sooner or later a subsidiary occupation is needed for the hunter and the fisherman: 'when the fisherman cannot obtain a living from fishing he must find it in some other activity.' I imagine therefore that to-day only slight encouragement is needed for the introduction of a suitable form of agriculture. Perhaps it suffices to provide the residents with some instruction by competent persons not only in agricultural technique but also in other matters relating to rural industry as a whole. Some trappers themselves pointed out to me that they have never before had contact with agriculture, and experiments in it can be expensive if carried out at their own risk (cf. 918 a). The most competent experts are to be found amongst the residents of northern Finland and northern Sweden, I think.

FARMERS AND FORESTRY.

There are therefore possibilities for mixed farming in Labrador. The total area thus rendered available for colonization would be considerable; but for the development of communications and cooperative efforts it should be centralized in small model groups. This would mean that life would be very far removed from privation.

The establishment of such small farms presupposes however initiative from outside, and the using of favourable circumstances, also some capital and equipment. In the plan for northern Fennoscandia the government secures employment to the settlers so that they can obtain a fair money income in the winter or the summer as day labourers, principally as workers in the f o r e s t i n d u s t r y (preparing booms, skidways, embankments and roads, logging, etc.), if they do not get sufficient work with the private forest firms. Much of such income is invested by the settlers in the holdings which the state has granted them (clearing arable land, construction). Cash contributions are also made by the state in the form of rather large premiums for cleared areas. When the holder of such a small allotment has, for a certain period (20 years), cultivated a certain area and put up all the buildings necessary upon it according to his contract with the state, everything passes automatically into his unrestricted ownership.

The same arrangement could perhaps be made in Labrador. My impression is that in the present state of things the income of the people in North West River and Dove Brook leaves a fairly good surplus which could be used for the gradual preparation of small farm holdings in the way mentioned above. It is evident that the more numerous the occupations on which the rural life of a country is based the less dangerous is the effect of failure in any one of them for rural industry as a whole.

Under these circumstances plans for the colonization of Labrador on a large scale must aim at intimate, reciprocal action between the white inhabitants and the forest industry. In a wider sense the income from the forests must pay the expenses of the state for the administration and provide the cash needed by the settlers. But this cannot be done until the state following a constructive plan in collaboration with the settlers has developed the technical means for a rational forest industry (the construction of permanent roads, booms, embankments, log-houses etc.). Then in return the expense of felling and conveying the timber to the river mouths will fall progressively and a profit-yielding timber industry arise with free competition between the greater and the smaller enterprises. The preparatory work (cf. also p. 801) is to be done by the white inhabitants here as in northern Fennoscandia.

The complex forest work will thus undoubtedly ensure considerable earnings to the liveyeres and trappers as a part-time occupation when fishing and trapping is not in season, and thus solve their employment problem. An annual cut of 200,000 cords has been calculated to employ probably 1500 men, and probably a smaller number in the subsequent operations of hauling and log-floating (585). If fur or salmon supply fail, and that time will unfortunately come, the trappers and fishers will always have organized labour in the forest industry.

Aerial surveys made by private interests over large areas in the Lake Melville — Hamilton River Valley and their surrounding districts in 1937 showed that Labrador possesses enormous supplies of soft timber, spruce and other conifers. In 1938 it was said that the Bowater-Lloyd group was interested in these forests, but since then nothing has been heard of plans there. As the important consumers of wood show no great enthusiasm for securing concessions or extending their holdings it is fair to assume that either there are difficulties in their rational exploitation or that the resources and the quality are not of such preponderant importance. However, the total stand is unlikely to amount to less than tens of millions of cords (one cord = 128 cub. ft.); cf. p. 408.

Conclusion.

It may seem audacious to put forward the above suggestions for the possible development of this neglected outpost of humanity after a knowledge only acquired during journeys in the country. Yet I have found them supported by people who have lived there for years, and whose opinion is in the main the same as to the necessity of improving the life of many of the inhabitants.

There is nothing in the climate of Labrador to preclude comfort and enjoyment for anyone, said the late Dr PADDON, after his long residence there and his intimate contact with the inhabitants and their problems as well as the resources and forces of nature (731, 824). Apart from the few weeks of diabolical activity on the part of winged pests the period of open navigation can be perfectly delightful, especially west of the fog-zone, which is usually limited to a narrow coastal belt.

One cold fact is that the known resources for a settled population are limited in both number and extent (142 a, 228, 277 a, 288, 290, 291, 302, 370, 376, 585, 682, 678, 670, 814, 864). Newfoundland-Labrador can certainly never be an agricultural exporting country. But it may be expected to produce food from live-stock in far larger amounts than the farmers can consume. It can never be converted into a self-sustaining agricultural country supplying all its own wants. But it can supply a good deal and become as self-supporting as possible in the matter of food-stuffs. And fresh food-stuffs are necessary in order to avoid the common nutritional diseases. Under the present tariff conditions this surplus can maintain a forest-working population in a much healthier and cheaper way than if food supplies were imported from Canada and the States.

In a country in which potatoes, turnips, cabbages and carrots can be successfully grown, in which cattle and sheep, poultry and goats can be raised, in which there are ample supplies of wood, of which houses can be built and which can serve as fuel in the winter, no human being need go hungry, unclothed or unsheltered except as a result of a failure of human organizing ability. If that community produces fish and small game and furs of a value sufficient to exchange for the clothing, wheat, sugar and tobacco required to supplement the native products, every necessity for a tolerable life is available. Cattle will give milk for the children and in the end meat and boots for the family; sheep will give them wool, skin and meat; poultry will give them eggs and meat (cf. 585). The avitaminoses and consumption will gradually disappear.

When a rational forestry has been started the country will be colonized progressively by a small-holding working population. Then the conditions will be secured for a continuous industrial life in quite another way than at present, when 'the ghosts of the wilds' will scare away the alien people employed in industry.

Then, in time, with the wood-working industries will come the mining and iron industries and perhaps also others founded on cellulose as raw material.

*

Labrador is certainly a region of restricted opportunities. But it is potentially relatively rich in natural resources: soil, timber, iron ore and hydraulic energy, and these will enable the introduction of modern civilization even into the darkest wilderness (cf. 287, 352, 379, 422, 452, 585, 651, 668 a, 669, 720, 731, 835, 840, 842, 844, 845, 919, 1098, 1101). Sooner or later Labrador will come into her own.

*   *   *

A BIBLIOGRAPHY

OF

LABRADOR

(SPECIALLY NEWFOUNDLAND-LABRADOR)

BIBLIOGRAPHY.

Reasons have already been given in the introduction to this treatise for the pains taken by the author to search out and note down the printed material which, in part or as a whole, more or less directly concerns Labrador, and which is important especially for a knowledge of Newfoundland's share of the great peninsula. That is why the author has undertaken research in many libraries:

The University Library in Helsingfors (where ADOLF ERIK NORDENSKIÖLD's great geographical library is preserved), and

The Scientific Societies' Library in Helsingfors; also, during the author's travels, in

The Royal Library in Stockholm;

The Danish National Museum's Library,

The Royal Library and

The University Library in Copenhagen;

The British Museum's Library,

The Royal Geographical Society's Library,

The Moravian Mission's Library (Fetter Lane) and

The Hudson's Bay Company's Library in London;

The Gosling Memorial Library in St. John's, Newfoundland;

The Dalhousian College Library in Halifax, Nova Scotia;

The Yale University's Library in New Haven, Conn.;

The Harvard University's Library in Cambridge, Mass.;

The Library of Congress,

The Smithsonian Institution's Library and

The National Geographical Society's Library in Washington, D.C. and finally

The American Geographical Society's Library,

The Columbia University's Library,
The Public Library and
The American Museum's for Natural History Library in New York.

Unfortunately, in most cases, there was not time for such a thorough examination as was desirable of the enormous material contained in periodicals, and the author has been obliged to leave practically speaking the whole of the manuscript collections for future research. It can therefore be assumed that there are considerable gaps in the bibliography. The literature which has seen the light since the summer of 1939 has been almost completely inaccessible for the author by reason of the new dis-order produced in Europe by the war.

Yet the very nature of the work implies that a bibliography of the kind and extent here required can hardly ever be complete unless made by a body of specialists, for an all-round man who has really expert knowledge of the literature of all the sections is scarcely to be found in our day. Nor would it have suited my purpose to endeavour here to include all that more or less indirectly concerns Labrador, for then the bibliography alone would have become a book in which probably it would have been impossible to see the trees for the wood. As an illustration of this it may be said that for the details referring to the Vinland voyages alone, BRÖGGER gives a list of some 250 and JOSEF FISCHER of 322 works in their bibliographies. In my list, on the other hand, are included some works which I found quoted in the literature dealing with Labrador, even although I have been unable to trace the originals. For this reason, some such works have only incomplete titles, e.g. Nos 268 a, 385 a, 453 a, 457 a, 770 a. Yet in this way they are rescued from oblivion, for a specialist can easily find them in the great American libraries, and the present bibliography corresponds to the aim to constitute a framework into which supplementary and new literature may be intelligently fitted. In any case, the author ventures to believe that the writings here listed, numbering far more than thousandthreehundred, together with the previous explanatory material and illustrations and the maps listed at the end of the book, will give the reader, even one who has only slight scientific training, fairly satisfactory information regarding the conditions prevailing in Labrador, both at present and in former times. This is necessary, for instance, in planning an expedition to this country in which travelling is so difficult.

Articles dealing with Labrador, and especially Newfoundland—Labrador, are scattered through monographs, journals and magazines, for the most part published in America and extremely difficult, if not impossible, to discover in the libraries of Europe, and especially in the Northern countries. The whole matter, therefore, cannot be studied exhaustively in Fennoscandia. The author has repeatedly come to the conclusion that one shall not sit in a library armchair in Helsingfors and write about Labrador, but as far as possible do such work in America. If the reader wishes to extend still further his knowledge of the literature concerning Labrador, this can only be done successfully in the great libraries of the United States, with their brilliant organisation and their exceedingly helpful staff, members of which are always ready to rescue even a stranger from being drowned in the catalogues.

The author takes this opportunity of expressing his sincere gratitude to the officials of the libraries listed above, where he has been the object of the most friendly helpfulness while working out the present bibliography. To Dr JAMES B. CHILDS, Chief of Division in the Congress Library, Washington, D.C., and Dr CARL BJÖRKBOM, Librarian at the Royal Library, Stockholm, who have greatly assisted me with literature and films reproducing chapters in books I extend my warmest thanks.

I want to tender my thanks also to the many explorers and authors who have visited Labrador for many good advice; the sources have been acknowledged in each case.

ABBE, ERNEST C. Botanical Results of the Grenfell-Forbes Northern Labrador Expedition. Rhodora, Vol. 38. Boston 1936. — **1.**

»　 Phytogeographical observations in northernmost Labrador. In Forbes: Northernmost Labrador. No 275. — **1 a.**

ACHARD, EUGÈNE. Les Northmans en Amérique I. Les Vikings des grandes étapes. Coll. du Zodiaque, No. 35. Montréal 1935. — **2.**

AGAR, WILLIAM M The East Coast of Labrador. Canad. Min. Journal, Vol. 43, p. 709. Ottawa 1922. — **3.**

ALEXANDER, STEPHEN, and OTHERS. Report to the superintendent of the U. S. coast survey on the expedition to Labrador to observe the total eclipse of July 18, 1860, organized under Act of Congress approved June 15, 1860. With maps and sketches. Report U. S. coast survey for 1860. Washington 1861. —**4.**

ALKMAN, MISS. The Moravians in Labrador. London 1833. — **5.**

ALLEN, GLOVER M. See TOWNSEND, CHARLES W.

AMONG THE DEEP SEA FISHERS. The official organ of the International Grenfell Association.

AMY, LACEY. Labrador, home of the iceberg. Travel. Vol. 27, pp. 24, 52. New York 1916. — **6.**

AMY, W. L. The Liveyeres: Labrador's permanent population. Canad. Mag., Vol. 38, p. 455 — Toronto 1912. — **7.**

ANDERSSON, R. B. America not discovered by Columbus. A historical Sketch of the Discovery of America by the Norsemen in the tenth Century. Chicago & London 1874. — **8.**

ANDERSON, R. M. Mammals of Eastern Arctic and Hudson Bay. In Canada's Eastern Arctic, p. 67. Ottawa 1934. — **9.**

ANDRAE, K. R. Ueber die Heilkunde der Indianern. Ciba. Zeitschr. [d. Gesellsch. f. Chem. Industrie, Basel.] 1934, pp. 329, 335, 343. Basel 1934. — **10.**

ANDREWS, CHARLES M. Guide to the Materials for American History, to 1783, in the Public Record Office of Great Britain. Vol. I & II. Washington 1912 & 1914. — **11.**

ANDREWS, CHARLES M. and DAVENPORT, FRANCES G. Guide to the Manuscript Materials for the History of the United States to 1783, in the British Museum, in Minor London Archives and in the Libraries of Oxford and Cambridge. Washington 1908. — **12.**

ANSPACH, LEWIS AMADEUS. A History of the Island of Newfoundland. Containing a description of the Island, the Banks, the Fisheries and Trade of Newfoundland and the coast of Labrador. London 1819. — **13.**

—»— Geschichte und Beschreibung von Neufundland und der Küste Labrador. Neue Bibl. d. wichtigsten Reisebrschr. u. s. w. Herausg. v. Bertuch, Bd. XXX. Weimar 1822. — **14.**

—»— History of the island of Newfoundland . . . and the coast of Labrador. London 1827. — **15.**

ANTEVS, ERNST. Maps of the Pleistocene Glaciations. Bull. Geol. Soc. America, Vol. 40, p. 631. New York 1929. — **16.**

—»— Physiography of Northern Labrador. Geogr. Review. Vol. 24, p. 488. New York 1934. — **16 a.**

—»— Climatic Variations During the Last Glaciation in North America. Bull. Amer. Meteor. Soc., Vol. 19, p. 172. Milton, Mass. 1938. — **16 b.**

ARI THORGILSSON. The Book of Icelanders, edit. and transl. by H. HERMANNSSON. Ithaca, N.Y. 1930. — **16 c.**

ASHBURN, P. M. How disease came with the white man. Hygeia, T. XIV, pp. 205, 310, 438, 514, 636. Chicago 1936. — **17.**

ASHE, E. D. Journal of a voyage from Quebec to Labrador. Nautical Magazine, 1861, p. 1. London 1861. — **18.**

—»— Journal of a voyage from New York to Labrador. Transact. Lit. and Hist. Soc. of Quebec, Vol. IV, append., p. 1. Quebec 1861. — **19.**

ASHER, G. M. Henry Hudson the Navigator. The orginal documents in which his career is recorded, collected, partly translated, and annotated, with an introduction. Hakluyt Society. Publ., Ser. 1, Vol. 27. London MDCCCLX. — **20.**

AUDUBON, JOHN JAMES. Birds of America. Vol. I—VII. New York 1840—44. — **20 a.**

—»— Delineations of American scenery and character. New York 1926. — **21.**

AUDUBON, MARIA R. Audubon and his journals. I & II. London 1898. — **22.**

AUER, VÄINÖ. Botany of the Interglacial Peat Beds of Moose River Basin. Canadian Geol. Survey., Summary Report, 1926. Part C. Ottawa 1927. — **22 a.**

—»— Peat bogs in south-eastern Canada. Can. Geol. Survey, Mem. 162. Ottawa 1930. — **22 b.**

AUSTIN, Jr, OLIVER LUTHER. The Birds of Newfoundland-Labrador. Memoirs of the Nuttall Ornithological Club. No. VII. Cambridge, Mass. 1932 — **23.**

—»— Migration Routes of the Arctic Tern. Bull. Northeastern Bird-Banding Assoc., Vol. IV, N:o 4. Oct. 1928. — **23 a.**

BABCOCK, W. H. Early Norse Visits to North America. Smithsonian Miscell. Collect.,
     Vol. 59, Nr. 19. Washington 1913. — **24.**
—»— Markland, otherwise Newfoundland. Geogr. Review, Vol. 4, p. 309. New York
     1917. — **25.**
—»— Recent History and Present Status of the Vinland Problem. Geogr. Review, Vol. 11,
     p. 265. New York 1921. — **26.**
BABEL, R. P. O. M. I. Lettre à Monseigneur l'Archevêque de Québec. Escoumains 1852.
     See Streit, T. 3. — **27.**
BACH, RUDOLF. Labrador. Reiseskizze. Deutsche Rundschau f. Geogr. u. Statist., Jahrg.
     XIX, p. 164. Wien u. Leipzig 1897. — **28.**
BADDELEY, F. H. Geology of a portion of the coast of Labrador. Transact. Lit. and Hist.
     Soc. of Quebec, Vol. I, p. 72. Québec 1829. — **29.**
BAFFIN, WILLIAM. The voyages of William Baffin, 1612—1622. Ed., with notes and
     introduction by Clements R. Markham. Works issued by the Hakluyt Soc., 1 ser.,
     Nr 63. London 1881. — **30.**
BAILEY, A. M. Along Audubon's Labrador trail. Bird Study Today and a Century ago.
     Natural Hist., Journ. Amer. Mus. of Nat. Hist., Vol. XXXIII, p. 638. New York
     1933. — **31.**
BAKER, E. H. B. Winter Surveyings in Labrador. The Hydrographic Review, Vol. XIII,
     p. 11. Monaco 1936. — **32.**
BALCH, F. N. On a new Labradorean species of *Onchidiopsis*, a genus of molluscs new
     to eastern North America; with remarks on its relationships. Proceed. U. S. Nat
     Mus., Vol. 38, p. 469. Washington 1911. — **33.**
BALDWIN, HANSON W. The Cartwright Venture. Among the Deep Sea Fishers, Vol. 27,
     No. 3, p. 103. New York 1929. — **34.**
BALLANTYNE, ROBERT M. Hudson's Bay; or Everyday Life in the Wilds of North
     America. Edinburgh & London 1848. — **35.**
—»— Ungava: a tale of Esquimaux Land. London 1860. — **36.**
BANGS, OUTRAM. A List of Mammals of Labrador. Amer. Nat., Vol. XXXII,
     No. 379, p. 489. Boston 1898. — **37.**
—»— Notes on some Mammals from Black Bay, Labrador. Proceed. New England Zoo
     logical Club. Vol. 1, p. 9. Cambridge Mass. 1899. — **38.**
—»— List of the Mammals of Labrador. In Grenfell's Labrador. The Country and the
     People, p. 484. New York 1922. — **39.**
BANNERMAN, H. M. See GILL, J. E.
BARBEAU, MARIUS. The Native Races of Canada. Transact. Roy. Soc. Canada. 3rd Ser.,
     Vol. XXI, sect. 2, p. 41. Ottawa 1927. — **40.**
—»— The kingdom of Saguenay. The Macmillan Company. Toronto & London 1936.
     — **41.**
BARNES, H. T. Some Physical Properties of Icebergs and a Method for their Destruction.
     Proceed. Roy. Soc. of London, Ser. A., Math. and Phys. Sci., Vol. 114, p. 161.
     London 1927. — **42.**
BARROW, John. A chronological History of Voyages into the Arctic Regions. London
     1818. — **43.**
—»— Voyages of Discovery and Research within the Arctic Regions from the year 1818.
     London 1846. — **44.**

BARTLETT, DAVENPORT and GIBSON. Newfoundland and Labrador [On the Geology, Zoology and Botany of] Supplement. U. S. Hydrograph. Office, No. 78. Washington 1886. — **45.**

BARTLETT, ROBERT A. The sealing saga of Newfoundland. Nat. Geogr. Mag., Vol. LVI, p. 91. Washington 1929. — **45 a.**

—»— Sails over ice. New York 1934. — **45 b.**

BARTON, G. H. Evidence of the former extension of glacial action on the west coast of Greenland in Labrador and Baffin Land. Amer. Geologist. Minneapolis 1896. — **46.**

BASSETTI, GIANNI. Viaggio al confini del mondo. Nel Labrador tra gli »scorridori di boschi». Le vie del mondo, Anno VII, No. 5, p. 519. Milano 1939. — **47.**

BAULIG, HENRI. Amérique Septentrionale. P. I, p. 190. Géographie Universelle, T. XIII. Paris 1935. — **48.**

BAXTER, J. P. A Memoir of Jacques Cartier. New York 1906. — **48 a.**

BAYFIELD, ADMIRAL. Sailing Directions for the Gulf and River St. Laurence. Quoted by 434. — **48 b.**

BAYFIELD, CAPT. Notes on the Geology of the North Coast of the St. Laurence. Transact. Geol. Soc., Vol. V. London 1833. — **48 c.**

BEAUVOIS, E. Les colonies européennes du Markland et de l'Escociland (Domination canadienne) au XIVᵉ siècle. Compte-rendu d. trav. du Congr. Int. d. Américanistes, T. I, p. 25. Luxembourg 1877. — **49.**

—»— Les skrælings, ancêtres des esquimaux, dans les temps pré-colombiens. Revue orient. et américaine, nouv. sér., T. II, p. 1. Paris 1879. — **50.**

BECKLES, WILLSON. The Great Company. London 1900. — **51.**

BELL, ROBERT. Notes on some geological features of the northeastern coast of Labrador. Canadian Naturalist. Ottawa 1878. — **52.**

—»— Observations on Geology, Mineralogy, Zoology and Botany of the Labrador Coast, Hudson's Bay and Strait. Ann. Rep. Geol. Survey. Ottawa 1882—84. — **52 a.**

—»— Report on the Geology of the Northern Shore of Labrador. Rep. of Hudson Bay Exped. under Lieut. A. R. Gordon. Dep. of Interior. Ottawa, 1884. — **53.**

—»— Beaches in Labrador. Geol. Surv. Canada, Ann. Rep. New ser., Vol. I, p. 7. Ottawa 1885. — **54.**

—»— Observations on the Coast of Labrador and on Hudson Strait and Bay. Geol. Surv. of Canada Rep. 1885. Ottawa 1888. — **55.**

—»— The Labrador Peninsula. The Scottish Geogr. Mag., Vol. XI, p. 335. Edinburgh 1895. — **56.**

—»— The Geographical Distribution of Forest Trees in Canada. Ibid., Vol. XIII, 1897. —**57.**

—»— An exploration on the northern side of Hudson Strait. Geol. Surv. Canada, Ann. Report 11, N.S., Pt M, p. 9. Ottawa 1898. — **57 a.**

BELLET, ADOLPHE. Les Français à Terre-Neuve et sur les côtes de l'Amérique du Nord. La grande pêche de la mourue à Terre-Neuve depuis la découverte du Nouveau Monde par les Basques au quatorzième siècle. Paris 1902 — **58.**

BELLIN, J. N. Remarques sur le détroit de Belle-Isle, et les côtes septentrionales de la Nouvelle France, depuis la rivière S. Jean, jusqu'au Cap Charles: tirées des journaux de navigation qui sont au Dépôt des cartes & plans de la marine, pour le service des vaisseaux du Roi, par le sieur Bellin . . . le premier mars 1758. [Paris 1758]. — **59.**

BENT, A. C. Notes from Labrador. Bird Lore, Vol. XV, p. 11. Harrisburg, Pa 1913. — **60.**

BENT, A. H. The Unexplored Mountains of North America. Geogr. Review, Vol. VII, p. 403. New York 1919. — **61.**

BERNARD, SIR FRANCIS. Account of the coast of Labrador [16. Febr. 1761]. Mass. hist. soc. coll., 1 ser, Vol. 1, p. 233. Boston 1792. — **62.**

BETTS, RACHEL M. Bibliography of the Geology of Newfoundland 1818—1936. Depart. of Natural Resources, Geol. Sect., Bull. No 5. St. John's 1936. — **63.**

BIART, L. A travers l'Amérique: nouvelles et récits. 1. Paris [ca 1870]. — **63 a.**

BIDDLE, R. Memoirs of Sebastian Cabot, with a review of the history of maritime discovery. Illustrated by documents from the rolls, now first published. Philadelphia 1831. — **64.**

BIEGELOW, HENRY B. Birds of the North Eastern Coast of Labrador. Auk, Vol. XIX, p. 24. New York 1902. — **65.**

BIGGAR, H. P. Early Trading Companies of New-France. Toronto 1901. — **66.**

—»— Precursors of Jacques Cartier. Ottawa 1911. — **66 a.**

—»— The Voyages of Jacques Cartier. Publications of the Public Archives of Canada, No. 11. Ottawa 1924. — **67.**

—»— A Collection of Documents relating to Cartier and Robeval. Ottawa 1930. — **68.**

BILBY, J. W. Among unknown Eskimo. An account of twelve years' intimate relations with the primitive Eskimo of ice-bound Baffin land. London 1923. — **68 a.**

BIGGAR, H. P. See CARTIER, J.

BINNEY, GEO. and OTHERS. The Eskimo Book of knowledge (Aglait Ilisimatiksat). London 1931. — **69.**

BIRD, C. D. Trapping. Ungava. Magazine of Labrador,Vol. 4, No 1, p. 19. Cartwright 1939. — **70.**

BIRD, JUNIUS. (Mention of work in Labrador under heading 'Science in the Field and in the Laboratory'. Natural History, American Mus. of Nat. Hist., Vol. 34, p. 763. New York 1934. — **70 a.**

BIRDSEYE, CLARENCE. The truth about fox farming. Among the Deep Sea Fishers, Vol. 12, No. 2, p. 34. New York 1914. — **71.**

—»— Camping in a Labrador Snow Hole. Among the Deep Sea Fishers, Vol. 11, No. 4, p. 30. New York 1914. — **72.**

BIRKET-SMITH, KAJ. Skrælingerne i Vinland og Eskimoernes Sydost-Grænse. Geogr. Tidskr, Bd 24, p. 157. København 1918. — **73.**

—»— A geographic study of the early history of the Algonquian Indians. Internat. Arch. f. Ethnographie, T. XXIV. Leiden 1918. — **74.**

—»— The Caribou Eskimos. Rept. of the Fifth Thule Exped. 1921—24, Vol. 5. Copenhagen 1929. — **74 a.**

—»— Folk wanderings and culture drifts in northern North America. Journ. d.l. Soc. d. Américanistes de Paris, N. S., T. XXII. Paris 1930. — **75.**

—»— Eskimoerne. Med Forord af Knud Rasmussen. København 1927. — **76.**

—»— Über die Herkunft der Eskimos und ihre Stellung in der zirkumpolaren Kulturentwicklung. Anthropos, Vol. XXV. Mödling 1930, — **77.**

—»— The question of the origin of Eskimo culture: a rejoinder. The Amer. Anthropologist, N. S., Vol. XXXII, p. 608. Menasha 1930. — **78.**

—»— The Eskimos. London 1936. — **79.**

—»— Eskimo cultures and their bearing upon the prehistoric cultures of North-America and Eurasia. In: Early man. Edited by George Grant Mac-Cardy, p. 293. J. B. Lippincott Company. London 1937. — **80.**

BIRKET-SMITH, KAJ. Anthropological Observations on the Central Eskimos. Report of the Fifth Thule Exped. 1921—24, Vol. III, No. 2. Copenhagen 1940. — **80 a.**

BISHOP, H. The Austin Collection from the Labrador Coast. Rhodora, Vol. XXXII, p. 59. Boston 1930. — **81.**

BJØRNBO, AXEL ANTHON. En nordisk Columbus Aar 1476? Berlingske Tidende, Aftenavis 17 Juli 1909. København 1909. — **82.**

—»— Cartographia Groenlandica. Meddel. om Grønland, Bd XLVIII. København 1912. — **83.**

BJØRNBO, AXEL ANTHON and PETERSEN, C. S. Die echte Carte-Real-Karte. Peterm. Mitt. 1919. Gotha 1919. — **84.**

BOAS, FRANZ. Über die ehemalige Verbreitung der Eskimos im arktisch amerikanischen Archipel. Zeitschr. d. Gesellsch. f. Erdk., Bd XVIII, p. 118. Berlin 1883. — **85.**

—»— Notes on the Geography of Labrador. Science, Vol. XI, p. 77. New York 1888. — **86.**

—»— Zur Anthropologie der nordamerikanischen Indianer. Zeitschr. f. Ethnologie, Bd 27, p. 366. Berlin 1895. — **86 a.**

BOGORAS, W. Early Migrations of the Eskimo between Asia and America. XXI Congr. Int. d. Américanistes. Göteborg 1925. — **87.**

BOILIEU, LAMBERT DE. Recollections of Labrador Life. London 1861. — **88.**

BOLTON, CHARLES KNOWLES. Terra Nova: The Northeast Coast of America before 1602. Annals of Vinland, Markland, Estotiland, Drogeo, Baccalaos and Norumbaga. Boston 1935. — **89.**

BONNICK, S. O. Flying over Grand Falls, Labrador's Niagara. Travel (U.S.A.), Sept. New York 1933. — **90.**

BONNYCASTLE, RICHARD HENRY. Newfoundland in 1842. A Sequel to 'The Canada in 1841'. London 1842. — **91.**

BONNYCASTLE, R. H. G. Hudson's Bay Company, Canada's Fur Trade. Transact. of third North Amer. Wildlife Confer. 1938. Amer. Wildlife Instit., Investment Building, Washington 1939. — **92.**

BORDEN, L. E. List of Plants collected in 1904 during the cruise of the Neptune. By L. E. Borden, M.D, and named by Mr. J. M. Macoun. Appendix III, p. 320 to A. P. Low's Cruise of the Neptune. Ottawa 1906. — **93.**

BOUCHARD, GEORGES. Vieilles choses, vieilles gens. Montreal 1929. — **94.**

BOUFFARD, J. La frontière entre la Province de Québec et la colonie de l'Ile Terre-Neuve sur la côte du Labrador. Bull. Soc. Géogr. de Québec, Vol. 13, p. 7. Québec 1919. — **95.**

BOURQUIN, THEODOR. Grammatik der Eskimo-Sprache ... an der Labradorküste. — 1891. Quoted in Wheeler 2nd (1018). — **95 a.**

BOVEY, WILFRID. The Vinland Voyages. Transact. Roy. Soc. Canada, Ser. 3rd, Vol. XXX, sect. II, p. 27. Ottawa 1936. — **96.**

BOWEN, B. F. America discovered by the Welsh in 1170. Philadelphia 1876. — **96 a.**

BOWEN, NOEL H. The social condition of the coast of Labrador. Transact. Lit. and Hist. Soc. of Quebec, Vol. IV, p. 329. Quebec 1856. — **97.**

BOWMAN, PAUL W. Study of a peat bog near the Matamek River, Quebec, Canada, by the method of pollen analysis. Ecology Vol. 12, pp. 694—708. Brooklyn 1931. — **97 a.**

BOYLE, DAVID. Notes on the Discovery of the Grand Falls of Labrador. Transactions of the Canad. Institute, Vol. II, p. 336. Toronto 1892. — **97 b.**

BREWSTER, WILLIAM. Notes on the Birds observed during a summer cruise in the Gulf of St. Lawrence. Proceed. Boston Soc. Nat. Hist., Vol. XXII, p. 364. Boston 1883. — **98.**

BRITISH ADMIRALTY. Arctic Pilot, Vol. II & III. Sailing directions for Davis Strait, Baffin Bay, Smith Sound and channels to Polar sea, Hudson Strait and Bay; also for passages connecting Baffin Bay with Beaufort Sea, through Lancaster Sound. London 1915. — **98 a.**

—»— List of lights and visual time signals. Part VIII. Western side of North Atlantic Ocean (Canada, Newfoundland and Labrador). London 1928. — **98 b.**

—»— St. Lawrence Pilot: Comprising the Gulf and the River St. Lawrence, including the Banks of Newfoundland, and approaches to the Gulf by Cabot Strait, the Strait of Belle Isle, and the Gut of Canso. London 1929. — **98 c.**

—»— See **680.**

BRITTON, N. L. and BROWN, A. An Illustrated Flora of the Northern United States Canada, and the British Possessions. Vols I—III. New York 1913. — **98 d.**

BROOKS, C. E. P. The Meteorology of Hebron, Labrador, 1883—1912. Quart. Journ. Roy. Meteorol. Soc., Vol. 45, p. 163. London 1919. — **99.**

BROOKS, CHARLES F. and WARD, ROBERT D. C. The Climates of North America. First Part. Mexico, United States, Alaska. Handb. d. Klimatologie, Bd. II, Teil J, Lief. 1. Berlin 1936. — **100.**

BROWN, A. See BRITTON, N. L.

BROWN, J. Forest Fires and Fur. Composium of the Views of Seven Practical Men who Have Spent Many Years in the North Woods. The Beaver, Vol. II., No. 5, p. 2. Winnipeg 1922. — **101.**

BROWNE, P W. Where the Fishers go. The Story of Labrador. New York 1909. — **102.**

BRUCE, E. L. Mineral Deposits of the Canadian Shield.Toronto 1933. — **103.**

BRUNET, L. O. Notes sur les plantes, recueillis en 1858 par M. l'Abbé Ferland sur les côtes de Labrador, baignées par les eaux du Saint-Laurent. Québec 18—. — **104.**

BRYANT, HENRY. Remarks on some of the birds that breed in the Gulf of St. Lawrence. Proceed. Bost-Soc. of Natur. Hist., Vol. VIII, p. 65. Boston 1861. — **104 a.**

BRYANT, HENRY G. A Journey to the Grand Falls of Labrador. Bull. Geogr. Club of Philadelphia, Vol. 1, p. 37. Philadelphia 1894. — **105.**

—»— Notes on Early American Arctic Expeditions. Geographical Journal, 1909, p. 72. London 1909. — **106.**

—»— An Exploration in Southeastern Labrador. Bull. of the Geogr. Soc. of Philadelphia, Vol. XI, No. 1, p. 1. Philadelphia 1913. Lancaster, Pa 1913. — **107.**

—»— An Expedition in South-Eastern Labrador. Geogr. Journ., Vol. XLI, p. 341. London 1913. — **108.**

—»— A Visit to the Natasquan Indians of Labrador. Ibid., Vol. XI, No. 3, 1913. — **109.**

BRYCE, GEORGE. The Remarkable History of the Hudson's Bay Company including that of the French traders of north-western Canada and of the Northwest, XV., and Astor fur companies. London 1900. — **110.**

BRÜCKNER, EDW. Meteorologische Beobachtungen der deutschen Polarstationen. Meteorolog. Zeitschr. Braunschweig 1888. — **111.**

BRØGGER, A. W. Vinlandsferdene. Oslo 1937. — **112.**

BROGGER, A. W. Vinlandsferdene. Foredrag, Norsk Geogr. Tidskr., Bd V., p. 65. Oslo 1937. — **113.**

BUGGE, THOMAS. In Gerhard Schøning. Norges Riges Historie. Deel III, p. 419. Kjøbenhavn 1781. — **114.**

BUGGE, ALEXANDER. Spørsmaalet om Vinland. Mål og Minne. Kristiania 1911. — **114 a.**

BUSCH, KATHARINE J. Catalogue of Mollusca and Echinodermata dredged on the coast of Labrador by the expedition under the direction of Mr. W. A. Stearns, in 1882. Proc. U. S. National Museum, Vol. VI, p. 236. Washington 1883. — **114 b.**

BUTEUX, J. P. Epistola P. J. Buteux ad P. Generalem M. Vitelleschi. Tria Flumina 1640. See Streit T. 2. — **115.**

BUTLER, ETHAN. Certain Medical Problems of the Labrador demanding a non-medical solution. Among the Deep Sea Fishers, Vol. 8, No. 2, p. 26. Toronto 1911. — **116.**

BUTLER, SAMUEL R. Labrador Plants. Canadian Naturalist, Sept. Ottawa 1870. — **117.**

CABOT, JOHN. An extract taken out of the mappe of Sebastian Cabot, cut by Clement Adams, Concerning his discouerie of the West Indias, which is to be seene in her Maiesties privie gallerie at Westminster, and in many other ancient merchant houses. Hakluyt: The Principal navigations, etc. p. 511. London 1589. — **118.**

CABOT, WILLIAM B., In Northern Labrador. London 1912. — **119.**

—»— Labrador and why one goes there. Appalachia. The Journ. of the Appal. Mount. Club., Vol. XII, p. 224. Boston 1911. — **120.**

—»— Labrador. Boston 1920. — **121.**

—»— Labrador. London 1922. — **122.**

—»— The Indians. In Grenfell's Labrador. The Country and The People, p. 184. New York 1922. — **123.**

CALLENDER, GEOFFREY. The Naval Side of British History. London 1924. — **123 a.**

CAMPBELL, J. F. Frost and Fire. Edinburgh 1865. — **124.**

CANTO, ERNESTO DO. Quem deu o nome do Labrador. Breve estudo. Extrahido do Archivo dos Açores, Vol. XII, pp. 533, 529. Ponta Delgada 1892. — **125.**

CAREGA, GIORGIO. Alcuni dati demographici su gli Esguimesi. Metron, Vol. 7, p. 52. Ferrara 1928. — **126.**

CARLETON, FRED P. Notes on the Labrador Dialect. Among the Deep Sea Fishers, Vol. 21, No. 4, p. 138, New York 1924. — **127.**

CARNE, JOHN. A history of the missions in Greenland and Labrador. New York 1846. — **128.**

CARON, P. JOSEPH LE. Recit de l'hivernement chez les Montagnais 1618. See Streit T. 2. — **129.**

CARPENTER, C. C. Report on the Labrador Mission at Caribou Island, Straits of Belle Isle. Annual report Canada Foreign Missionary Soc., Nos. 1—6 (1858—1863). — **130.**

CARSON, JOHN. Ships of the snows. Among the Deep Sea Fishers, Vol. 34, No. 1, p. 14. New York 1936. — **131.**

CARTIER, JACQUES. Discours du voyage fait par le capitaine Jacques Cartier aux Terres Neufves de Canada, Norembergue, Hochelage, Labrador et pays adjacents, dite Nouvelle-France, avec particulières moeurs, language et cérémonie des habitants d'icelle. Rouen 1598. — **132.**

[CARTIER, J.] BIGGAR, H. P. The Voyage of Jacques Cartier. Publ. of the Publ. Arch. of Canada. Ottawa 1926. — **133.**

CARTWRIGHT, GEORGE. A Journal of Transactions and Events, during a Residence of

Nearly Sixteen Years on the Coast of Labrador; Containing Many Interesting Particulars Both of the Country and Its Inhabitants not Hitherto Known. Vol. I—III. Newark 1792. — **134.**

CARTWRIGHT GEORGE. Sixteen years on the coast of Labrador. Newark 1792. — **135.**

—»— Labrador: A poetical epistle. St. John's [1783] 1882. — **136.**

—»— Captain Cartwright and his Labrador Journal, ed. by. C.W. Townsend . . . D. Estes & Co. Boston 1911. — **137.**

CARY, AUSTIN. Geological Facts Noted on Grand River, Labrador. Amer. Journ. of Sc., Vol. 42, pp. 419, 516. New Haven 1891. — **138.**

—»— Exploration on Grand River, Labrador. Amer. Geogr. Soc. Journ., Vol. XXIV, p. 1. New York 1892. — **189.**

CATHER, WILLA. Shadows on the Rock. New York 1931. — **139 a.**

CAYLEY, EDWARD. Up the River Moisie. Transact. Lit. and Hist. Soc. of Quebec, N. S., Vol. I, p. 73. Quebec. 1863. — **140.**

CELSIUS D.Ä, OLOF. Iter in Americam. Diss. Uppsala 1725. — **140 a.**

CENSUS. Newfoundland and Labrador. St. John's 1901. — **141.**

CHALMERS, Robert. The preglacial Decay of Rocks in Eastern Canada. Amer. Journ. of Sc., Vol. V, p. 273. New Haven 1898. — **142.**

CHAMBERS, E. J. Canada's fertile northland, a glimpse of the enormous resources of part of the unexplored regions of the Dominion: evidence before a Select C:ttee of the Senate . . . In the Rept. based thereon. Ottawa 1908. — **142 a.**

CHAMBERS, R. W. Thomas More. London 1935. — **142 b.**

CHAPMAN, FRANK M. Birds of Eastern North America. 1934. — **143.**

CHAPPELL, EDWARD. Narrative of a voyage to Hudson's Bay in His Majesty's Ship *Rosamond*, containing some account of the northeastern coast of America and of the tribes inhabiting that remote region. London 1817. — **144.**

—»— Reise nach Neufundland und der südlichen Küste von Labrador. Jena 1819. — **145.**

CHARLEVOIX, LE PÈRE DE. Journal d'un voyage, fait par ordre du roi dans l'Amérique septentrionale. Paris 1744. — **146.**

—»— Histoire et description de la Nouvelle France. Paris 1744. — **146 a.**

CHIMMO, W. A Visit to the Northeast Coast of Labrador during the Autumn of 1867 by H. M. S. Gannet. Journ. Roy. Geogr. Soc., Vol. 38, p. 258. London 1868. — **147.**

—»— A visit to the fishing grounds of Labrador by H. M. S. Gannet in the autumn 1867. Nautical Magazine, pp. 113, 187. London 1869. — **148.**

CHITTY, DENNIS and ELTON, CHARLES. Canadian Arctic Wild Life Enquiry. 1935—36. Journ. of Animal Ecology, Vol. 6, No. 2, p. 368. Cambridge 1937. — **149.**

CHOUINARD, F. X. Le nord-est du Labrador. Soc. Géogr. de Québec, Vol. 16, p. 48. Québec 1922. — **150.**

CHRISTY, M. The Voyages of Captain Luke Faxe of Hull and Captain Thomas James of Bristol in search of a Northwest Passage. London 1894. — **151.**

—»— See also KNIGHT, JOHN.

—»— See also WAYMOUTH, GEORGE.

CILLEY, JONATHAN PRINCE, JR. Bodwoin Boys in Labrador. An account of the Bodwoin College Scientific Expedition to Labrador. Rockland 1891. — **152.**

CLARK, SAMUEL. New Description of the World. London 1688. — **152 a.**

CLARKE, J. M. Observations on the Magdalen Islands. New York State Mus. Bull. No 149, p. 134. New York 1910. — **152 b.**

COATS, W. Notes on the Geography of Hudson's Bay. Being the remarks of Captain W. Coats in many voyages to that locality between the years 1727 and 1751. Hakluyt Society. London 1852. — **153.**

COHEN, E. Das Labradorit-führende Gestein der Küste von Labrador. Neues Jahrb. f. Mineralogie, Bd I, p. 183. Berlin 1883. — **153 a.**

COLEMAN, A. P. Mt. Tetragona. A First Ascent in Labrador. Canadian Alpine Club Journal, 1916, p. 5. Banff, Alta 1916. — **154.**

—»— The Building of the Torngats, Ibid., Vol. 7, 1916, p. 67. — **155.**

—»— Five Climbs in the Torngats. Ibid., Vol. 8, 1917, p. 34. — **156.**

—»— La Péninsule du Labrador. Bull. Soc. Géogr. de Québec, Vol. 12, p. 143. Québec 1918. — **156 a.**

—»— Extent and thickness of the Labrador ice sheet. Bull. Geol. Soc. Amer., Vol. 31, No. 3, p. 319. New York 1920. — **157.**

—»— Northeastern Part of Labrador and New Quebec. Geolog. Survey of Canada, Memoir 124. Geol. ser., No. 106. Ottawa 1921. — **158.**

—»— Le Nord-Est du Labrador. Bull. d. l. Soc. de Géographie de Québec, Vol. 16, No. 1, pp. 48, 204. Québec 1922. — **159.**

—»— Physiography and glacial geology of Gaspé peninsula, Quebec. Geol. Surv. Canada, Mus. Bull. 34. Ottawa 1922. — **159 a.**

—»— The Pleistocene of Newfoundland. Journ. of Geology, Vol. XXXIV, p. 193. Chicago 1926. — **159 b.**

—»— Ice Ages, recent and ancient. New York 1926. — **160.**

—»— The Labrador Coast. Geogr. Review, Vol. 18, p. 525. New York 1928. — **161.**

—»— Extent of Wisconsin Glaciation. Amer. Journ. of Sc.; 5th ser., Vol. XX, p. 181. New Haven 1930. — **161 a.**

—»— The Torngats of Labrador. Canadian Geogr. Journ., Vol. XIV, p. 283. Montreal 1937. — **162.**

COLONIAL SECRETARY'S OFFICE. Census and return of the population &c. of Newfoundland & Labrador, 1874. St. John's 1876. — **163.**

—»— Census of Newfoundland and Labrador 1901. St. John's 1903. — **164.**

—»— Interim report on the census of 1911. St. John's 1911. — **165.**

—»— Census of Newfoundland and Labrador 1911. St. John's 1914. — **166.**

—»— Census of Newfoundland and Labrador 1921. St. John's 1923. — **167.** — See further Public Health and Welfare Department.

COMEAU, N. A. Life and Sport on the North Shore of the Lower St. Lawrence and Gulf, containing chapter on salmon fishing, trapping, the folk-lore of the Montagnais Indians and tales of adventure on the fringe of the Labrador Peninsula [with list of birds]. Quebec 1909. — **168.**

CONINE, W. H. See TWENHOFEL, W. H.

CONNOR, A. J. The Climates of North America. Canada. Handb. d. Klimatologie, Bd. II, Teil J, Lief. 2. Berlin 1938. — **169.**

CONVERSE, FRANK H. A Sunday afternoon in Labrador. The Christian Union, p. 391. New York 1884. — **170.**

COOK, J. and others. The North American pilot for Newfoundland, Labrador, the Gulf and River St. Lawrence. London 1775. — **171.**

COOK, JAMES; LANE, MICHAEL; GILBERT, JOSEPH. The North American Pilot for Newfoundland, Labradore, the Gulf and river Saint Lawrence: being a collection of sixty accurate charts and plans drawn from original surveys. London 1784. — **172.**

COOKE, H. C. Studies of the physiography of the Canadian Shield. I. Nature valleys of the Labrador Peninsula. Roy. Soc. Canada Transact, ser. 3, Vol. 23, sect. 4, p. 91. Ottawa 1929. — **173.**

—»— Ibid. II. Glacial depression and postglacial uplift. Idem., Vol. 24, sect. 4, p. 51. 1930. — **174.**

—»— Ibid. III. The pre-Pliocene physiographies, as inferred from the geologic record. Ibid. Vol. XXV, sect. 4, p. 127. 1931. — **174 a.**

—»— Land and Sea on the Canadian Shield in Precambrian Time. Part I & II. Amer. Journ. of Sc., Vol. XXVI, p. 428. New Haven 1933. — **175.**

COOKE, WELLS W. Labrador Birds Notes. Auk. Vol. XXXIII, p. 162. New York 1916. — **176.**

COOPER, JOHN M. The Northern Algonquian Supreme Being. Anthropolog. Series, Catholic University of America, No. 2. Washington 1934. — **177.**

—»— Is the Algonquian family hunting-ground system pre-Columbian? Amer. Anthropologist, Vol. 41, p. 66. Menasha 1939. — **177 a.**

[CORNWALLIS KING, W.] see WEEKES, MARY.

CORTESÃO, ARMANDO. Cartografia e cartôgrafos portugeses dos séculos XV e XVI. Lisboa 1935. — **178.**

COSTA, B. DE. Pre-Columbian Discovery of America. Albany 1889. — **179.**

COTÉ, LOUIS PHILIPPE. Vision du Labrador. Montréal 1934. — **180.**

COTTER, H. M. S. The Birchbark Canoe. An Important Factor in H.B.C. Transport from Earliest Times. The Beaver, Vol. II, No. 9, p. 5; No. 10, p. 10. Winnipeg 1922.—**181.**

—»— The great Labrador gale. The Beaver, No. 2, Sept. Winnipeg 1932. — **182.**

COTTER, JAMES L. The Eskimos of East Main. The Beaver, No. 4, p. 362. Winnipeg 1930. — **183.**

COUES, ELLIOT. Notes on the Ornithology of Labrador. Proceed. Acd. Nat. Sc. Philadelphia, Vol. XIII, p. 215. Philadelphia 1861. — **184.**

COUPER, WILLIAM. Investigations of a naturalist between Mingan and Watchiconti. Labrador. Quebec 1868. — **185.**

COX, LEO. The Golden North. Labrador and North Shore. Canad. Geogr. Journ., Vol. 16, No. 4, p. 203. Ottawa 1938. — **186.**

CRANZ, DAVID. Alter und neuer Brüder-Historie oder kurz gefasste Geschichte der Evangelischen Brüder-Unität. Barby 1772. — **187.**

—»— The Ancient and Modern History of the Brethren. Translated by Benjamin La Trobe. London MDCCLXXX. — **188.**

—»— History of Greenland. Sketch of the Mission of the Brethren in Labrador, Appendix, Vol. II, p. 287. London 1820. — **189.**

CRARY, A. P. See EWING, MAURICE.

CRAWSHAY, L. R. On the Distribution of the Microplankton. Report on the work carried out by the S.S. *Scotia* 1913. Ice observations, meteorology, and oceanography in the North Atlantic Ocean, p. 68. London 1924. — **190.**

CRÉPIEUL, FRANCOIS DE. Remarques touchant la mission de Tadoussac. 1671—1686. [Québec?] 1686. See Streit, T. 2. — **191.**

CRONE, G. R. The alleged pre-Columbian discovery of America. Geogr. Journ., Vol. XXXIX, p. 455. London 1937. — **192.**

CROUSE, N. M. Contributions of the Canadian Jesuits to the Geographical Knowledge of New France, 1632—1675. Cornell Univ. Ithaca 1924. — **193.**

CRUIKSHANK, ERNEST. Early Trades and Trade Routes, 1760—1782. Transact. Canad. Institute, Dec. Vol. III, p. 253; Vol. IV, p. 299. Toronto 1893 & 1895. — **194.**

CURLING, J. J. Contribution to the discussion following Randle F. Holme's Paper on »A Journey into the Interior of Labrador, July—Oct. 1887». Proc. Roy. Geogr. Soc., Vol. X, p. 203. London 1888. — **195.**

CURTIS, ROGER. Particulars of the Country of Labrador, *extracted from the Papers of Lieutenant* Roger Curtis, *of His Majesty's Sloop the* Otter, *with a Plane-Chart of the Coast. Communicated by the Honourable* Daines Barrington. Philos. Transact. of the Roy. Soc. London. Vol. 64, Pt. 2, p. 372. London 1774. — **196.**

—»— Also in FORSTER, J. R. & SPRENGEL, M. C. Beiträge zur Völker- und Länderkunde . . ., Vol. 1, p. 79. Leipzig 1781. — **197.**

DABLON, CLAUDE. Rélation De Ce Qui S'Est Passé De Plus Remarquable Aux Missions Des Pères de la Compagnie de Jésus. En La Nouvelle France, les années 1671 & 1672. Paris MDCLXXIII. — **198.**

DAHLGREN, E. Nya forskningar angående de gamla nordmännens Vinlandsresor. Ymer, bd 8, p. XVI. Stockholm 1889. — **198 a.**

DALE, BONNYCASTLE. Sailing Labrador Seas. Canad. Geogr. Journ., Vol. II, p. 391. Montreal 1913. — **199.**

DALY, REGINALD A. The Geology of the Northeast Coast of Labrador. Bull. Mus. of. Compar. Zoölogy at Harvard College, Geolog. Ser., Vol. V, No. 5. Cambridge 1902. — **200.**

—»— Report on Geology. Bull. of the Geogr. Club of Philadelphia, Vol. III, p. 206. Philadelphia 1901—1903. — **201.**

—»— Post-glacial changes of level in Newfoundland and Labrador. Bull. Geol. Soc. Amer., Vol. 32, p. 53. New York 1921. — **202.**

—»— Post-glacial warping of Newfoundland and Nova Scotia. Amer. Journ. of Sc., 5th Ser., Vol. 1, p. 384. New Haven 1921. — **202 a.**

—»— The Geology and Scenery of the Northeast Coast. In Grenfell's Labrador. The Country and the People, p. 81. New York 1922. — **203.**

—»— Our Mobile Earth. New York and London 1926. — **204.**

—»— The changing World of the Ice Age. New Haven 1934. — **204 a.**

DANISH METEOROLOGICAL INSTITUTE, THE. The state of the ice in the arctic seas. The Nautical-Meteorological Annual. København (appears yearly). — **204 b.**

DAVENPORT. See BARTLETT. —.

DAVENPORT, FRANCES G. See ANDREWS, CHARLES M. Snowshoes. Mem. of the Amer. Philosoph. Soc., Vol. VI, p. 207. Philadelphia 1937. — **204 c.**

DAVEY, J. W. The Fall of Torngak or the Moravian Mission on the Coast of Labrador. London 1905. — **205.**

DAVIDSON, D. S. Notes on Tête de Boule Ethnology. Amer. Antropologist. Vol. 30. p. 18. Menasha 1928. — **206.**

—»— Family Hunting Territories of the Waswanipi Indians of Quebec., Mus. Amer. Ind. Heye Found. — Indian Notes, Vol. 5, No 1. New York 1928. — **207.**

DAVIES, W. H. A. Notes on Esquimaux Bay (Hamilton Inlet) and the Surrounding Country. Transact. Lit. and Hist. Soc. of Quebec, Vol. IV, p. 70. Quebec 1843. — **208.**

DAVIES, W. H. A. Notes on Ungava Bay and its Vicinity. Ibid. 1842. — **208 a.**

DAVIS, JOHN. The second voyage attempted by Master Iohn Dauis with others, for the discouerie of the Northwest passage, in Anno 1586. Hakluyt, p. 781. London 1589. — Same in Hakluyt, Vol. III, p. 163. London 1600. — **209.**

—»— The World's Hydrographical Description. [Not available]. — **209 a.**

—»— The Voyages and Works of John Davis, the Navigator. Ed. with an introduction and notes by Albert Hastings Markham. Works issued by the Hakluyt Society, 1 Ser., No. 59, p. XXXVI. London 1880. — **210.**

DAVISON, C. The Atlantic earthquake of November 18, 1929. Nature, Vol.. 124, p. 859. London 1929. — **211.**

DAWSON, J. W. Notice of Tertiary Fossils from Labrador . . . Canadian Naturalist, Vol. V, Ottawa 1860. — **212.**

DAWSON, RHODA. The Folk Art of the Labrador. Among the Deep Dea Fishers, Vol. 36. No. 2, p. 39, New York 1938. — **213.**

DAWSON, SAMUEL EDWARD. The Voyages of the Cabots in 1497 and 1498; with an attempt to determine their landfall and to identify their island of Saint John. Mém. et comptes rendus de la. Soc. Roy. du Canada. Montréal 1896. — **214.**

—»— Canada and Newfoundland. London 1897. — **214 a.**

—»— The Saint Lawrence basin and its borderlands. London 1905. — **215.**

—»— Brest on the Quebec Labrador. Can. Roy. Soc., Vol. 11, sect. 2, p. 3. Ottawa 1905. — **215 a.**

DAWSON, W. BELL. The currents in Belle Isle Strait. Dept. of Marine and Fisheries of Canada. Ottawa 1927. — **216.**

DEANE, RUTHVEN. Great Auk (*Alca impennis*). American Naturalist, Vol. VI, p. 368. New York 1872. — **217.**

DECKERT, E. und MACHATSCHEK, F. Nordamerika. Leipzig 1924. — **217 a.**

DEFANT, A. Bericht über die ozeanographischen Untersuchungen des Vermessungsschiffes *Meteor* in der Dänemarksstrasse und der Irminger See. Sitzungsber. d. Preuss. Akd. d Wissensch., Phys.-Math. Kl., Bd XIX. Berlin 1931. — **217 b.**

DELABARRE, EDMOND BURKE. Report of the Brown and Harvard Expedition to Nachvak, Labrador, in the year 1900. Bull. Geogr. Soc. of Philadelphia, Vol. II, p. 65. Philadelphia 1901—03. — **218.**

—»— The Flora. In Grenfell's Labrador. The Country and the People, p. 391. New York 1922. — **219.**

DE LAGUNA, F. A Comparison of Eskimo and Palaeolithic Art. Amer. Journ. of Archaeol., Vol. XXXVI, p. 477 & Vol. XXXVII, p. 77. Concord. 1932—33. — **220.**

DEPARTMENT OF NATURAL RESOURCES [NEWFOUNDLAND]. Fisheries. Report of the Commission of enquiry investigating the seafisheries of Newfoundland and Labrador other than the sealfishery. Economic Bulletin, No. 13. St. John's 1937. — **221.**

DEPARTMENT OF PUBLIC HEALTH AND WELFARE [NEWFOUNDLAND]. Tenth census of Newfoundland and Labrador, 1935. Vol. I & II. St. John's, NFLD. 1937. — **222.**

DEUTSCHE SEEWARTE. Deutsche Überseeische Meteorologische Beobachtungen, herausgegeben von der Deutschen Seewarte. Hamburg 1883—1902. — **222 a.**

—»— Deutsche Überseeische Meteorologische Beobachtungen, gesammelt und herausgegeben von der Deutschen Seewarte. Hoffenthal, Zoar, Nain, Okak, Hebron, Rama: Bd. **1.** IX. 1883—XII. 1884; Bd. **2.** 1885; Bd. **3.** 1886; Bd. **4.** 1887; Bd. **5.** 1888; Bd. **6.** 1889. Hoffenthal, Zoar, Nain, Hebron: Bd. **7.** 1890. Hoffenthal, Zoar, Nain: Bd. **8.**

I—VII. 1891; Bd. **9.** VIII—XII. 1891. F e r n e r M o n a t s m i t t e l aus den Terminbeobachtungen. Bd. **14.** Hoffenthal: I. 1892—VI. 1894; Nain: I. 1892—III 1899; IX. 1899—XII. 1904. Hebron: I. 1892—VII. 1900; X. 1901—III. 1902; V. 1902 —XII. 1904. Bd. **15.** u. **16.** Nain: I.—VI. 1905; X. 1905—XII. 1906. Hebron: I. 1905—XII. 1906. Monatsmittel für Nain und Hebron: Bd. **17.** 1907; Bd. **18.** 1908; Bd. **19.** 1909; Bd. **20.** 1910; Bd. **21.** 1911; Bd. **22.** 1912; Bd. **23.** 1913. — **222 b.**

DEWEY SOPER, J. Intimate Glimpses at Eskimo Life in Baffin Island. The Beaver, Outfit 266, No. 4, p. 34. Winnipeg 1936. — **223.**

DEWITZ, A. VON. An der Küste Labradors. Oder: Innere Mission im Gebiet der Heidenmission. Bearbeitet nach schriftlichen Mittheilungen der Missionare F. RINDERKNECHT and H. RITTER; von A. VON DEWITZ. Niesky 1881. — **224.**

DINWOODIE, MRS. HEPBURN. All the World and I. A Novel of Labrador. Boston 1940. — **224 a.**

DIONNE, N. E. Inventaire chronologique des cartes, plans et atlas relatifs à la Nouvelle France et à la Province de Québec. 1508—1908. Ottawa 1909. See Proceed. and Transact. Roy. Soc. Canada. Sér. 3, Vol. 2., Part. 2. 80—87. — **225.**

DISSEGAND, H. P. The Labrador Mission church in the Colonies, No. 26. London 1850. — **226.**

DOBBS, ARTHUR. An Account of the Countries Adjoining Hudson Bay in the north-west part of America. London 1744. — **227.**

DOMINION OF NEWFOUNDLAND AND LABRADOR: Some information about the resources of the ancient colony. St. John's 1921. — **228.**

DOXSEE, W. W. See HODGSON, E. A.

DRESSER, J. A. Part of the district of Lake St. John, Quebec. Geol. Surv. Canada, Mem. 92. Ottawa 1916. — **229.**

DUCKWORTH, W. L. H. and PAIN, B. H. An account of some Eskimos of Labrador. Proceed. Cambridge Philos. Soc., Vol. X, p. 286. Cambridge 1900 — Also No. 34 in Studies from the Anthropolog. Laboratory, the Anatomy School, p. 268; see p. 196 for corrections. Cambridge 1904. — **230.**

—»— A contribution to Eskimo craniology. Journ. of the R. Anthrop. Instit., Vol. XXX, London 1900. — **231.**

DUMAIS, P. H. Formation du Saguenay. Le Naturaliste Canadien, Vols 21, 22 & 23. 1894—1896. — **231 a.**

DUNBAR, CARL O. see SCHUCHERT, CHARLES.

DUNCAN, N. Dr Luke of the Labrador. New York & Chicago 1904. — **232.**

DUNCAN, NORMAN. The Labrador »Liveyere». Harper's Mag., 1904, p. 514. New York. 1904. — **233.**

—»— The Fleet on »The Labrador». Ibid., p. 857. 1904. — **234.**

—»— Dr Grenfell's Parish; the deep sea fishermen. New York & Chicago 1905. — **235.**

DUNN, BARBARA CROCKER. The geography of Labrador. Vassar Journ. of undergrad. studies, Vol. VIII, p. 116. New York 1934. — **236.**

DURGIN, G. F. Letters from Labrador (1858—1905). Concord, New Hampshire 1908. — **237.**

DÖLL, LUDWIG. Klima und Wetter an der Küste von Labrador. Aus dem Arch. d. Deutschen Seewarte, Bd 57. No. 2. Hamburg 1937. — **238.**

EATON, D. I. V. Meteorological Observations in the Labrador Peninsula, 1893—1894 and 1895. Appendix VII in Low's report 1896. (**588**), — **238 a.**

EDGELL, J. A. Survey on the Labrador Coast. Geogr. Journ., Vol. LXXXI, p. 59. London 1933. — 239.

EIDMANN, H. Beiträge zur Kenntnis der Fauna von Südlabrador, insbesondere des Flussgebietes des Matamek River. Zoogeographica, Bd 2, p. 204. Jena 1934. — 240.

EIFRIG, C. W. G. Ornithological results of the Canadian »Neptune» expedition to Hudson Bay and northward, 1903—1904. The Auk, Vol. XXII, p. 233. Cambridge, Mass. 1905. — 241.

ELLIS, HENRY. A Voyage to Hudson's Bay, by the DOBBS GALLEY and CALIFORNIA In the Years 1746—1747 For Discovering a North West Passage ... London MDCCXLVIII. — 242.

ELLIS, MINA BENSON HUBBARD. See Mrs LEONIDAS HUBBARD JR.

ELTON, C. Fluctuations in wild Life. In Canada's Eastern Arctic, p. 62. Ottawa 1934. — 243.

ERDMANN, F. Eskimoisches Wörterbuch (Labrador). Budissin 1864. — 243 a.

ELTON, CHARLES. See CHITTY, DENNIS.

ESPEJO, ANTONI DE. New Mexico, otherwise the voyage of Antoni de Espejo, who in the years 1583 with his company go to the land termed the Labrador. London 1587. — 244.

EWING, MAURICE; CRARY, A. P. and RUTHERFORD, H. M. Geophysical Investigations in the Emerged and Submerged Atlantic Coastal Plain. I. Bull. Geol. Soc. of America, Vol. 48, p. 753. Washington 1937. — 244 a.

—»— WOOLLARD, G. P. and VINE, A. C. Ibid. III. Ibidem, Vol. 50, p. 257. 1939. — 244 b.

EYRIÈS, J. B. B. Voyages ... Abrégé des voyages modernes depuis 1780. Vol. 8. Terre de Labrador. Paris 1822—24. — 245.

FAIRCHILD, H. L. Post-glacial uplift of Northeastern America. Bull. Geol. Soc. Amer., Vol. 29, p. 187 [226]. Washington 1918. — 246.

—»— Post-glacial continental uplift. Science, N. S., Vol. 47, p. 615. New Haven 1918. — 247.

FARNHAM, C. H. Labrador. Harper's Monthly Magazine, Sept., Oct., New York 1885. — 248.

FARRELLY, T. S. . . . Slabland reunited; recent political coalescence of Labrador and Newfoundland follows their geographical integrity . . . Journ. of Amer. history, Vol. 27, No. 1. New York 1933. — 249.

FEILD, E. A visit to Labrador in the autumn of 1848 by the Lord Bishop of Newfoundland. [St. Johns?], 1849. — 250.

—»— Journal of the . . . voyage of visitation and discovery on the south and west coasts of Newfoundland and on the Labrador, in the church ship »Hawk» ... 1848. Soc. for the Propagation of the Gospel. Church in the Colonies, No. 21. London 1851. — 251.

FERLAND, JEAN BAPTISTE ANTOINE. Le Labrador, notes et récits de voyage. Montréal 1858 [?] — 252.

—»— Le Labrador. In Litterat. canade. Vol. 1, Québec 1863. — 253.

—»— Opuscules. Québec 1877. — 254.

FERNALD, M. L. Notes on the Plants of Wineland the Good. Rhodora, Vol. 12. Boston 1910. — 255.

—»— A botanical expedition to Newfoundland and Southern Labrador. Rhodora, Vol. XIII. 109. Boston 1911. — 256.

FERNALD, M. L. and SORNBORGER, J. D. Some recent Additions to the Labrador Foral. Ottawa Naturalist, T. XIII, p. 89. Ottawa 1899. — **256 a.**

FERNALD, M. L. The natural history of Ancient Vineland and its geographic significance. Bull. Amer. Geogr. Soc., Vol. 47, p. 687. New York 1915. — **257.**

—»— Persistence of Plants in Unglaciated Areas of Boreal America. Amer. Acad. of Arts and Sc., Vol. 15, No. 3, p. 237; Mem. of the Gray Herbarium of Harvard University, Vol. II. 1925. Boston 1925. — **258.**

—»— Unglaciated western Newfoundland. Harvard Alumni Bull., Jan. 25. Boston 1930. — **259.**

—»— Recent discoveries in the Newfoundland flora. Rhodora, Vol. 35, Nos 409—420, New York 1933 — **260.**

FERRIER, W. F. Notes on the Microscopic Structure of some Rocks from the Labrador Peninsula. Appendix V to Low's Report (**588**). Ottawa 1896. — **260 a.**

FIELD, F. F. Labrador is reported scene of rich iron ore discovery. Montreal Gazette, November 10. Montreal 1938. — **261.**

FISCHER, Josef. Die Entdeckungen der Normannen in Amerika. Unter besonderer Berücksichtigung der kartographischen Darstellungen. Freiburg in Breisgau 1902. — **262.**

FISHER-MÖLLER, K. Skeletal Remains of the Central Eskimos. Report of the Fifth Thule Exped.; Vol. III, No. 1. Copenhagen 1937. — **262 a.**

FISK, W. W. The Glazier party in the wilds of Labrador. Boston[?] 1902. — **263.**

FISKE, John. The discovery of America. Vol. I & II. London 1892. — **264.**

FITZGERALD, J. E. An Examination of the Charter and Proceedings of the Hudson's Bay Company. London 1849. — **264 a.**

FLAHERTY, ROBERT J. Two Traverses across Ungava Peninsula, Labrador. Geogr. Review Vol. VI, p. 116. New York 1918. — **265.**

FLAHERTY, R. J. and F. H. My Eskimo Friends, Nannook of the North. New York 1924. — **266.**

FLETCHER, JAMES. List of diurnal lepidoptera and coleoptera collected by Mr. J. S. Cotter at Moose Factory in 1888, and by Mr. J. M. Macoun on the south coast and islands of James' Bay in 1887. Rept. Geolog. and Nat. Hist. Survey of Canada for 1887—88; Vol. III, pt. 2. Appendix II. Ottawa 1889. — **267.**

—»— List of insects collected in the Interior of the Labrador Peninsula, 1894. Appendix IV to Low's Report (**588**). Ottawa 1896. — **267 a.**

FLINN, T. SMITH. The Humanness of the Liveyere. Among the Deep Sea Fishers. Vol. 12, No. 4, p. 137. New York 1915. — **268.**

FLÜCKINGER, OTTO. Glaziale Felsformen. Petermanns Geogr. Mitteil., Ergänzungsheft, p. 218. Gotha 1934. — **268 a.**

FONTE, ADMIRAL DE. Labrador: the great probability of a N-W passage ... Appendix consuming account of a discovery of part of the coast and inland country of Labrador made in 1753. London 1768. — **269.**

FORBES, ALEXANDER. Surveying in Northern Labrador. Geographical Review, Vol. XXX, p. 30. New York 1932. — **270.**

—»— An aerial Survey in Northern Labrador. Harvard Alumni Bull., p. 917. Boston 1932. — **271.**

—»— A Northern Labrador Cruise. Yachting, Vol. LIII (March—May). New York 1933. — **272.**

FORBES, ALEXANDER. Note on Labrador Sailing Directions. Among the Deep Sea Fishers, Vol. 31, No. 4, p. 163. New York 1934. — **273.**

—»— Completing the Northern Labrador Survey. Ibid., Vol. 38, p. 704. 1936. — **273 a.**

—»— A Flight to Cape Chidley, 1935. Geogr. Review, Vol. XXVI, p. 48. New York 1936. — **274.**

—»— Northernmost Labrador mapped from the Air. Amer. Geogr. Soc. Spec. Publication No. 22. New York 1938. — **275.**

—»— Northernmost Labrador. Plates and Navigational Notes. Ibid. — **276.**

—»— Flight to Labrador. Among the Deep Sea Fishers, Vol. XXXVII, No. 4, p. 158. New York 1940. — **277.**

—»— Rivers of the South Shore of Lake Melville, Labrador. Geogr. Review, Vol. XXX, p. 394. New York 1940. — **277 a.**

FORBUSH, WM. Pomiuk, a Prince of Labrador. London 1903. — **278.**

FORD, W. E. Chemical and optical study of a labradorite. Amer. Journ. of Sc., Vol. 30, p. 151. New Haven 1910. — **279.**

FORGUES, C. E. Survey of the rivers St. John, Mingan, Natashquan, and Esquimau. Ottawa 1890. — **280.**

FORSTER, J. R. Allgemeine Geschichte der Entdeckungen und Schiffahrten im Norden. Frankfurt a. O. 1784. — **281.**

—»— History of Voyages and Discoveries. London 1786. — **282.**

FORSTER, GEORG. Geschichte der Reisen die seit Cook an der Nordwest- und Nordost-Küste von Amerika und in dem nördlichsten Amerika selbst von Meares. Dixon, Portlock, Coxe, Long u.a.m. unternommen worden sind. Berlin 1791. — **283.**

FOSSUM, A. The Norse Discovery of America. Minneapolis 1918. — **284.**

FRAZER, M. A. An Ornithologist's Summer in Labrador. Ornithology and Oölogy, Vol. XII, pp. 1, 17, 33. 1887. — **285.**

FRENCH, C. H. Animal Traps and Trapping. The Beaver, Vol. IV, No. 9, p. 324. Winnipeg 1924. — **286.**

FRISSELL, VARICK. Explorations in the Grand Falls Region of Labrador. Geogr. Journ., Vol. LXIX, p. 332. London 1927. — **287.**

—»— Light in darkest Labrador; a future made bright by hydroelectricity. World to-day, Vol. 50, p. 624. London 1927. — **288.**

—»— Penetrating the Grim Heart of Labrador. Travel (U.S.A.) Apr. 1927. New York 1927. — **289.**

—»— The Grand Falls of the Labrador. Among the Deep Sea Fishers, Vol. 25, No. 1, p. 3, New York 1927. — **290.**

—»— A Niagara in the heart of Labrador. Windsor Mag., p. 58. London 1927. — **291.**

FROBISHER, MARTIN. See STEFANSON, V.

FULLER, M. L. The elevated beaches of Labrador. Science, N. S., Vol. 25, p. 32. Cambridge 1907. — **292.**

GANONG, W. F. Crucial Maps in the early Cartography and Place-nomenclature of the Atlantic Coast of Canada, 1. Transact. Roy. Soc. Canada, 3rd Ser., Vol. XXIII, Sect. II, p. 135. Ottawa 1929. — **293.**

—»— Ibid. II. Vol. XXIV, Sect. II, p. 135. 1930. — **294.**

—»— » III. Vol. XXV, Sect. II, p. 169. 1931. — **295.**

—»— » IV. Vol. XXVI, Sect. II, p. 125. 1932. — **296.**

GANONG, W. F. Ibid. V.   Vol. XXVII, Sect. II, p. 149. 1933. — **297.**
—»—   » VI. Vol. XXVIII, Sect. II, p. 149. 1934. — **298.**
—»—   » VII. 3rd Ser., Vol. XXIX, Sect. II, p. 101. 1935. — **299.**
—»—   » VIII. 3rd Ser., Vol. XXX, Sect. II, p. 109. 1936. — **300.**
—»—   » IX. 3rd Ser. Vol. XXXI, Sect. II, p. 101. 1937. — **301.**

GARDNER, GÉRARD. Les resources minérales du Labrador. Actualités economiques, II (5). Mars 1936, p. 439. Montréal 1936. — **302.**

—»— La frontière Canada-Labrador, Rev. trimestr. Canad. 24ᵉ année, p. 272. Montréal 1938. — **303.**

GARTH, THOMAS A. A Comparison of Mental Abilities of Nomadic and Sedentary Indians on a Basis of Education. Amer. Anthropologist, Vol. 29, p. 206. Menasha 1927. — **303 a.**

GATHORNE-HARDY, G. M. The Norse discoverers of America. The Wineland Sagas translated and discussed. Oxford 1921. — **304.**

—»— A recent journey to Northern Labrador. Geogr. Journ., Vol. LIX, p. 153. London 1922. — **305.**

—»— Alleged Norse Remains in America. Antiquity, Vol. 6, p. 420. Gloucester 1932. — **306.**

—»— See PRICHARD.

GATHORNE-HARDY, ROBERT. Summer in Labrador. Blackwood's Magazine, Vol. CCXXX, p. 365. New York 1931. — **307.**

GATSCHET, ALBERT A. The Beothuk Indians. Proceed. Amer. Philos. Soc., Vol. XXII, p. 408; Vol. XXIII, p. 411. Philadelphia 1885, 1886. — **307 a.**

GAUTIER. Notice sur les observations météorologiques faites sur la côte du Labrador par les Missionaires Moraves. Bibliothéque universelle. Arch. des sciences physiques et naturelles, T. 38, p. 132. 1870; T. 55, p. 39. 1875; T. 60, p. 392. 1877. Genève 1870—1877. — **307 b.**

GEETE, R. Var låg Vinland det goda? Ymer, bd 5, p. 317. Stockholm 1885. — **307 c.**

GELCICH, E. Der Fischfang der Gascogner und die Entdeckung von Neufundland. Nach den »Disquisiciones nauticas» von CESAREO FERNANDEZ DURO bearbeitet von EUGEN GELCICH. Zeitschr. d. Gesellsch. f. Erdkunde, Bd. 18, p. 249 f. Berlin 1883. — **308.**

GENEST, A. T. Ungava. Soc. Géogr. de Québec, Vol. 4, p. 77. Québec 1910. — **309.**

GIBBON, MURRAY J. The Coureur de Bois and his Birthright. Transact. Roy. Soc. Canada, Ser. 3rd, Vol. XXX, Sect. II., p. 61. Ottawa 1936. — **310.**

GIBSON. See BARTLETT.

GIBSON. See GILLPATRICK.

GILBERT, J. Terra Labrador, 1768. Büschings Nachrichten, Vol. III, p. 224. Berlin 1775. — **311.**

GILBERT, JOSEPH. See COOK, JAMES.

GILL, J. E.; BANNERMAN, H. M. and TOLMAN, C. Wapussakatoo Mountains of Labrador. Bull. Geol. Soc. of Amer., Vol. 48, p. 567. Washington 1937. — **312.**

GILLPATRICK & GIBSON. Newfoundland and Labrador. The coast and banks of Newfoundland. U. S. Hydrograph. Office, No. 73 Washington 1884. — **313.**

GLADWIN, GEO E. Coast and Harbors of Labrador, summer 1876. Sketches — Pen & Ink. Boston 1877. — **314.**

GODSELL, PHILIP H. »Relief» in the sub-arctic. The tragic economic story of the northern

Indians: a thirty-year decline from the freedom of their ancient hunting grounds to a place in the »breadline». Natural History, Vol. XXXVIII, p. 289. New York 1936. — **315.**

GOLD THWAITES, REUBEN. See JESUIT RELATIONS, **496.**

GORDON, CHAS. H. M. The Disappearing Birch-Bark. The Beaver, Outfit 257, p. 12. Winnipeg 1926. — **317.**

GORDON, HENRY. A winter in Labrador 1918—1919. Cartwright c. 1919. — **318.**

—»— The Eskimo Dogs. Among the Deep Sea Fishers, Vol. 20, No. 1, p. 1. New York 1922. — **319.**

—»— Journal (Jan. 1 to March 31, 1924). Among the Deep Sea Fishers, Vol. 22, No. 3, p. 125. New York 1924. — **320.**

—»— Life in Labrador. United Empire. The Roy. Colonial Institute Journ., Vol. XIV (N. S.), p. 285. London 1923. — **321.**

GORDON, A. R. Report of the Hudson Bay Expedition under the command of Lieut. A. R. Gordon, R. N. Toronto 1884. — **322.**

—»— Report on the second Hudson Bay Expedition. Ottawa 1886. — **323.**

—»— Preliminary report of the Hudson Bay expedition of 1886. 19th Ann. Rpt. Depart, of Marine, Appendix No. 27. Ottawa 1887. — **324.**

GOSLING, W. G. Labrador: its Discovery, Exploration and Development. London 1910. — **325.**

—»— Ibid. New York 1911. — **326.**

GOVERNMENT OF CANADA. The Migratory Birds Convention Act and Federal Regulations for the Protection of Migratory Birds. Ottawa 1933. — **327.**

GRAHAM, R. D. Rough passage; being the narrative of a single-handed voyage to Newfoundland, Labrador, and Bermuda in the seven-ton yacht Emanuel and the subsequent return to England with a soldier crew. Edinburgh & London 1936. — **328.**

—»— Rough Passage (extract from book of the same title, published by Wm Blackwood & Sons, Ltd. Edinburgh). Among the Deep Sea Fishers, Vol. 34, No. 4, p. 151. New York 1937. — **329.**

GRAVIER, G. Découverte de l'Amérique par les Normands au 10. siècle. Paris 1874. —**329 a.**

GRAY's New Manual of Botany, 7th ed., rearranged and revised by B. L. Robinson and M. L. Fernald. New York 1908. — **330.**

GRAY, E. F. Leif Eriksson. Discoverer of America A.D. 1003. New York. 1930. — **331.**

GREENE, WILLIAM Howe. The Wooden Walls among the Ice Floes. London 1933. — **332.**

GREGORY, JOHN URIAH. En racontant. Récits de voyages en Floride au Labrador et sur le fleuve Saint-Laurent. Québec 1886. — **333.**

GREGORY, J. W. Geographical considerations in the Canadian-Newfoundland boundary in Labrador. Canada. Privy Council. The matter of the boundary between the Dominion of Canada and the Colony of Newfoundland. Vol. V. Joint appendix, p. 2489. London 1926. — **334.**

—»— The Earthquake South of Newfoundland and Submarine Canyons. Nature, Vol. 124, p. 945. London 1929. — **335.**

—»— The earthquake of Newfoundland Banks of November 1929. Geogr. Journ., Vol. 77. p. 123. London 1931. — **336.**

GRENFELL, Mrs. ANNE ELIZABETH CALDWELL. Le Petit Nord; or, Annals of a Labrador Harbour, by Anne Grenfell and Katie Spalding. New York 1920. — **337.**

GRENFELL, WILFRED T. Vikings of To Day. Life and medical work among the fishermen of Labrador. London 1895. — **338.**

—»— Climate and Travel in Labrador. Climate (publ. by Livingstone College) 1, p. 17. London 1899. — **339.**

—»— Life in Labrador. Blackwoods Mag., Vol. 170, p. 688. London 1901. — **340.**

—»— The harvest of the sea; a tale of both sides of the Atlantic. New York & Chicago 1905. — **341.**

—»— Off the rocks; Stories of the deep sea fisherfolk of Labrador. Philadelphia 1906. — **342.**

—»— Fisher folk of Labrador. Canad. Club Tor., Vol. 4, p. 98. Toronto 1906—07. — **343.**

—»— Experiences on the Labrador. The Century, Vol. 78, p. 233. New York 1909. — **344.**

—»— Land of Eternal Warring. Nat. Geogr. Mag., Vol. XXI, p. 665. Washington 1910. — **345.**

GRENFELL, DR. W. T. and OTHERS. Labrador, The Country and the People. New York 1910. — **346.**

GRENFELL, WILFRED. Down to the sea, yarns from the Labrador. New York & Chicago 1910. — **347.**

—»— Hard Times and Side Tracks in Labrador. Among the Deep Sea Fishers, Vol. 8, No. **3**, p. 25. Toronto 1910. — **348.**

—»— Labrador. The Geographical Journal, Vol XXXVII, s. 407. London 1911. — **349.**

—»— Down North on the Labrador. New York & Chicago 1911. — **350.**

—»— Down North on the Labrador. London 1912. — **351.**

—»— The Future of Labrador. The Outlook, Aug. 1912, p. 957. New York 1912. — **352.**

—»— Dr. Grenfell's Log. Ibid., Vol. 9, No. 4, p. 20. Toronto 1912. — **353.**

—»— That Christmas at Peace Haven. Ibid., Vol. 10, No. 4, p. 28. Toronto 1913. — **354.**

—»— Twenty years in Labrador. Wide World Mag., Vol. 32, pp. 243, 330. London 1913. — **355.**

—»— Adrift on an icepan. Cambridge 1913. — **356.**

—»— Labrador. Oxf. Surv. Brit. Emp., Vol. 4, p. 295. Oxford 1914. — **357.**

—»— The year's work in Labrador. Journ. Natl. Inst. of Social Sc., Vol. 1, p. 106. New York 1915. — **358.**

—»— Tales of the Labrador. London 1916. — **359.**

—»— A Sketch of Labrador & its People. Peoples of all Nations, p. 3758. London 1918. — **360.**

—»— Labrador Days; tales of the sea toilers. Boston & New York 1919. — **361.**

—»— What's in Labrador. Canad. Forestry Magz, Vol. XVI, p. 499. Ottawa 1920. — **362.**

—»— Story of Labrador medical mission. Emp. Club. Can., p. 295. Toronto 1921. — **363.**

GRENFELL, WILFRED T. and OTHERS. Labrador. The Country and the People. New York 1922. — **364.**

GRENFELL, WILFRED F. In Icy Labrador. Current History. Vol. 18, p. 823. Detroit 1923. — **365.**

—»— Northern neighbors; stories of the Labrador people. Boston & New York 1923. —**366.**

—»— A Labrador Doctor. Autobiography. Cambridge. Mass. 1925. — **367.**

—»— Report on Labrador fisheries, 1892—1924. Canada. Privy Council. In the matter of the boundary between the Dominion of Canada and the Colony of Newfoundland, Vol. V. of Joint appendix, p. 2564. London 1926. — **368.**

GRENFELL, WILFRED T. More Leaves from Dr. Grenfell's Diary. Among the Deep Sea
      Fishers, Vol. 24, No. 1, p. 30. New York 1926. — **369.**
—»— Labrador: Its Opportunities Large and Small. Ibid., Vol. 24, No. 2, p. 43. New York
      1926. — **870.**
—»— The Log of the Strathcona for 1927. Ibid., Vol. 25, No. 3, p. 114. New York 1927.—**871.**
—»— The Expert Seamanship of the Fisherman, Ibid., Vol. 25, No. 4, p. 179. New York
      1928. — **872.**
—»— Sir Wilfred's Letter. Ibid., Vol. 26, No. 4, p. 166. New York 1929. — **873.**
—»— Progressive March in Labrador. The Beaver, No. 1, p. 16. Winnipeg 1930. — **374.**
—»— Labrador an asset of Empire. An address to Roy. Empire Soc. United Empire
      Jan. 1931. London 1931. — **874 a.**
—»— The Lure of Labrador, National Rev. Nov. 1931. London 1931. — **875.**
—»— Agriculture in Northern Newfoundland and Labrador. Among the Deep Sea
      Fishers, Vol. 29, No. 1, p. 7. New York 1931. — **876.**
—»— Log of the Strathcona. Ibid., Vol. 29, No. 4, p. 150. New York 1932. — **877.**
—»— Labrador tragedy. Travel, Vol. LIX, p. 36. New York 1932. — **378.**
GRENFELL, Sir WILFRED. The Problems of Labrador. Canadian Geogr. Journ., Vol. VII
      p. 201. Montreal 1933. — **879.**
—»— Forty years for Labrador. London 1934. — **880.**
—»— The Romance of Labrador. New York 1934. — **381.**
—»— A Labrador Logbook. Boston 1938. — **382.**
[W. GL.] Labrador. Encyclopedia Britannica, Vol. XIII, p. 555. London 1929. — **382 a.**
GRENFELL ASSOCIATION. The Grenfell Calendar. Publ. yearly. — **882 b.**
GRESWELL, W. Geography of the Dominion of Canada and Newfoundland. Oxford 1890.
      — **383.**
GRIEVE, JOHN. Battle Harbour Items. Among the Deep Sea Fishers, Vol. 9, No. 2, p. 35.
      Toronto 1911. — **384.**
—»— Birds of Grenfell Land. Among the Deep Sea Fishers, Vol. 30, No. 4, p. 156 &
      Vol. 31, No. 1, p. 14. New York 1933. —**385.**
—»— To the strange »Buttons» . . . Natural History, Vol. 36, p. 133. 1935. [Quoted by
      Forbes, 235.] — **385 a.**
GUNN, HUGH. The British Empire. A Survey in 12 Volumes — each self-contained.
      London 1924. — **386.**
HAKLUYT, RICHARD. The principal Navigations, Voyages and Discoveries of the English
      nation, made by sea or over Land, to the most remote and farthest distant Quarters
      of the earth at any time within the compasse of these 1500 yeares. Divided into
      three several parts, according to the position of the Regions wherunto they
      were directed. London 1589. — **387.**
—»— The Third and Last Volume of the Voyages, Navigations, Traffiques, and Dis-
      coveries of the *English Nation*, and in some few places, where they have not been,
      of Strangers, performed within and before the time of these hundred yeares, to
      all parts of the *Newfound* world of *America*, or the *West Indies*, from 73. degrees
      of Northerly to 57. of Southerly latitude: — — —, *Tierra de Labrador*, — — —.
      London 1600. — **887 a.**
—»— Divers voyages touching the discovery of America and the islands adjacent.
      Collected and published in the year 1582. Printed by Hakluyt Society. London
      1850. — **888.**

HAKLUYT, RICHARD. Hakluyt's voyages. Vol. V. Everyman's Library, No. 338. London & New York 1926. — **389.**

HALL, A. GRATTEN. Doctor Wilfred Grenfell. London [1919?] — **390.**

HALL, CHARLES FRANCIS. Life with the Esquimaux. A Narrative of Arctic Experience in search of survivors of Sir John Franklin's Expedition. With Maps. London Vol. I & II 1864, Vol. I 1865. — **391.**

HALL, CHRISTOPHER. The first Voyage of M. Martin Frobisher, to the Northwest, for the search of the straight or passage to China, written by Ch. H. . . . in the yeare 1576. Hakluyt, Vol. III, p. 24. London 1600. — **392.**

HALLOCK, CHARLES. Three months in Labrador. Harper's Monthly Magazine, Vol. XXII, pp. 577, 743. New York 1861. — **392 a.**

HALLOWELL, A. I. Bear ceremonialism in the northern hemisphere. The American Anthropologist, N. S., Vol. XXVIII. Menasha 1926. — **393.**

—»— The Physical Characteristics of the Indians of Labrador. Journ. d. l. Soc. des Américanistes de Paris. N. S., T. XXI, p. 337. Paris 1929. — **394.**

—»— Kinship Terms and Cross-Cousin Marriage of the Montagnais—Naskaupi and the Cree. Amer. Anthropolog., Vol. 34, p. 171. Menasha 1932. — **395.**

HAMILTON, J. TAYLOR. A History of the Missions of the Moravian Church, Bethlehem. Pa 1901. — **396.**

HAMILTON, R. V. On the Portion of the Coast of Labrador between Blanc Sablon Bay in Latitude 51°20' N and Cape Harrison in lat. 55° N. Proceed. Roy. Geogr. Soc., Vol. IX, p. 131. London 1864—65. — **397.**

HANN, J. Resultate der meteorologischen Beobachtungen an der Küste von Labrador, Rigolet, Hoffenthal. Meteorolog. Zeitschr. Jahrg. XIII, p. 117. Wien 1896. — **398.**

—»— Zum Klima von Labrador. Ibid. p. 359. 1896. — **398 a.**

—»— Zum Klima von Labrador. Okak, Hebron und Rama. Ibid., p. 420. 1896. — **398 b.**

—»— Übersicht über die mittlere Temperatur und den jährlichen Wärmegang an der Küste von Labrador. Ibid., p. 422. 1896. — **398 c.**

HANN, J. v. Handbuch der Klimatologie, Bd III. Stuttgart 1911. **398 d.**

HANTSCH, BERNHARD. Beitrag zur Kenntnis der Vogelwelt des nordöstlichsten Labradors. Journal f. Ornithologie, Bd. 56, pp. 177, 307. Leipzig 1908. — **399.**

—»— Beiträge zur Kenntnis des nordöstlichen Labradors. Mitt. d. Ver. f. Erdkunde zu Dresden, H. 8. p. 168. Dresden 1909. — **400.**

—»— Avifauna of Northeastern Labrador. Canad. Field Naturalist, Vol. XLII, pp. 2, 33, 87, 123, 172, 201, 221, Vol. XLIII, pp. 11, 31, 52. Ottawa 1928—29. — **401.**

—»— Eskimo stone graves in north-eastern Labrador and the collection of anthropological material from them. Transl. by M. B. A. Anderson. Canad. field-naturalist, Vol XLIV. Ottawa 1930. — **402.**

—»— Contributions to the knowledge of extreme N.E. Labrador. Canad. Field Naturalist, pp. 143, 169, 194, 222 & 224. Ottawa 1931. — **403.**

HARDING, CHRIS. The Monetary System of the Far Fur Country. Values of Pelts and Merchandise was Measured in »Made-Beaver» or »Skins»; Many Queer Canards About H.B.C. Trade Tactics Not Founded on Facts. The Beaver, Vol. I, No. 9, p. 2. Winnipeg 1921. — **404.**

HARPE, M. DE LA. Abrégé de l'Histoire Générale des Voyages . . . Vol. XIV. Paris MDCCLXXX. — **405.**

HARRIS, SIR ALEXANDER. Newfoundland and Labrador. In Lagden's Native races of the Empire, p. 181. London 1924. — **406.**

—»— Labrador Boundary. Contemp. Rev., Vol. 131, p. 415. London 1927. —- **407.**

HARRIS, ALEX. C. On the Labrador Boundary settlement. London 1927. See MC GREGOR, Spectator, 2. Apr. 1927. — **408.**

HARRIS, C. A. See ROGERS, J. D.

HARRISSE, HENRY. The Discovery of North America. A critical, documentary and historic investigation. An Essay on the Early Cartography of the New World, including Descriptions of Two Hundred and Fifty Maps or Globes existing or lost, constructed before the year 1536. — — — [Part First: voyages. Part Second: Early cartography. Part Third: Cartographia Americana Vetustissima. Part Fourth: Chronology of voyages a. 1431 to 1492 and b. 1492 to 1504. Part Fifth: Biographies of Pilots and Cartographers. 1492—1550. — Geographical index. London MDCCCXCII. — **409.**

—»— Bibliotheca Americana vetustissima. A Description of Works relating to America published between 1492 and 1551. New York 1866. — **410.**

—»— Additions to foregoing. Paris 1872. — **411.**

—»— Les Corte-Real et leur Voyages au Nouveau-Monde. Paris 1883. — **412.**

—»— Découverte et évolution cartographique de Terre-Neuve et des pays circon-voisins, 1497—1501—1769. Essais de géographie historique et documentaire. Paris 1890. — **413.**

—»— John Cabot, the discoverer of North America and Sebastian his son. London 1896. — **414.**

—»— Decouverte et évolution cartographique de Terre-Neuve et des pays circonvoisins. Revue de Géographie. Paris 1900. — **415.**

HARVEY, M. See HATTON, JOSEPH.

HARVEY, MOSES. Labrador and Newfoundland. Encyclop. Britannica. London 1884. — **416.**

—»— Handbook and Tourist's Guide to Newfoundland. St. John's 1894. — **417.**

HASSEL, GEORG. Vollständige und neueste Erdbeschreibung des Britischen und Russischen Amerika's und der Französischen Fischerinseln mit einer Einleitung zur Statistik dieser Länder Labrador, p. 491. Weimar 1822. — **418.**

HATT, GUDMUND. Kyst- og Inlandskultur i det arktiske. Geogr. Tidskr., Vol. XXIII. København 1916. — **419.**

—»— North American and Eurasian culture connexions. Proceed. of the Fifth Pacific Sc. Congr. Toronto 1935. — **420.**

HATTON, JOSEPH and HARVEY, M. Newfoundland: its history, its present condition, and its prospects in the future. Boston 1883. — **421.**

HAUSER, IRMA. Problems of the Labrador People. Among the Deep Sea Fishers, Vol. 21, No. 4, p. 127. New York 1924. — **422.**

HAUTREUX, A. Les glaces et les brumes de l'Atlantique Nord. Revue Philomathique de Bordeaux et du Sud-Ouest. Bordeaux 1908. — **423.**

HAWKES, E. W. The Labrador Eskimo. Geolog. Survey., Mem. 91, Nr 14. Anthropolog. Series. Ottawa 1916. — **424.**

—»— Round Labrador and Hudson Bay. World Wide Mag., Vol. 38, p. 492. London 1917. — **425.**

HAYDON, ARTHUR LINCOLN. Canada at work and at play, with a chapter on Newfoundland and Labrador. New York and London 1904. — **426.**

—»— Canada, Britain's largest colony. With a chapter on Newfoundland and Labrador. With an introduction by Lord Strathcona. »Our Empire»-series. London 1904. — **427.**

HAYES, ERNEST HENRY. Forty years on the Labrador. The life-story of Sir Wilfred Grenfell. New York & Chicago 1930. — **428.**

HEILPRIN, A. On the direction of glacial movement in Labrador. Science, Vol. 6, p. 388. Cambridge Mass. 1885. — **429.**

HEPWORTH, W. C. The Effect of the Labrador Current upon the Surface Temperature of the North Atlantic, and of the latter upon Air Temperature and Pressure over the British Isles. British Meteorol. Office. Geophys. Mem., No. 1. London 1912. —- **430.**

HERKENRATH, AUG. Canada und die Hudson's bay Company. Phil. Dissert. V. Bonn. 1904. — **431.**

HERMANNSSON, HALLDÓR. The Wineland Voyages: A few suggestions. Geogr. Review, Vol. 17, p. 107. New York 1927. — **432.**

—»— The Problem of Wineland. Islandica. An annual relating to Iceland and the Fiske Icelandic collection in Cornell University Library, Vol. XXV, p. VI. Ithaca. N.Y. 1936. — **433.**

—» — The Vinland Voyages. Le Nord 1940, Fas. 2, p. 9. —- **433 a.**

HEYE, GEORGE G. See SPECK, FRANK G.

HIND, HENRY YOULE. Explorations in the Interior of the Labrador Peninsula, the country of the Montagnais and Nasquapee Indians. T. I & II. London 1863. — **434.**

—»— An Exploration up the Moisie River, to the Edge of the Tableland of the Labrador Peninsula. Journ. Roy. Geogr. Soc.. Vol. 34, p. 82. London 1864. — **435.**

—»— Wanderungen in Labrador. Globus, Vol. V, p. 208. Hildburghausen 1864. — **436.**

—»— Observation on the Supposed Glacial Drift in Labrador Peninsula, Western Canada, and on the South Branch of the Saskatchewan. Canad. Naturalist, 2nd ser., Vol. 1, p. 300. 1864 and Quart. Journ. Geol. Soc.; Vol. XX, p. 122. London 1864. — **437.**

—»— Explorations on the Labrador Coast. [Quoted in GOSLING, **325.**] 1876 [?]. — **437 a.**

H[INKS], A[RTHUR] R. The Labrador boundary. Geogr. Journ., Vol. LXX, p. 38. London 1927. — **438.**

HITCHCOCK, C. B. Physiography of the Cape Chidley Sheet. App. II to Forbes, No. 1938. Geogr. Rev., T. 26, p. 56. New York 1936. — **439.**

HODD, DONALD G. Tuberculosis on the North Shore. Among the Deep Sea Fishers, Vol. 26, No. 4, p. 138. New York 1929. — **440.**

HODDER & STOUGHTON. Labrador's Fight for Economic Freedom. [Self and Society Booklets]. Benn 1929. — **441.**

HODGE, FREDERIC Webb. Handbook of American Indians. Bureau of Amer. Ethnology, Bull. 30, Vol. I & II. Washington 1907. — **442.**

HODGSON, E. A. and DOXSEE, W. W. The Grand Banks earthquake, November 18, 1929. Seis. Soc.-Amer., Eastern Sect., Proc. 1930, p. 72. Washington 1930. — **443.**

HOFFMAN, W. J. The Graphic Art of the Eskimo. Ann. Rep. of the United States Nat. Mus. 1895, p. 739. Washington 1897. — **444.**

HOLAND, HJALMAR R. Westward from Vinland. The Story of Norse Discoveries and Exploration in America, Centuries before Columbus. New York 1940. — **444 a.**

HOLDRIDGE, DESMOND D. Down the Labrador. Fore an' aft, Oct. 1927—April 1928. New York 1927, 1928. — **445.**

—»— Northern Lights. New York 1939. — **446.**

HOLM, G. Small additions to the Vinland Problem. Medd. om Grønland, Vol. LIX, København 1924. — **447.**

HOLME, RANDLE F. A journey in the interior of Labrador, July to October, 1887. With discussions by Rev. J. J. Curling and General Dashwood. Proc. Roy. Geogr. Soc., Vol. X, p. 189. London 1888. — **448.**

HOLMES, J. Historical sketches of the missions of the United Brethren for propagating the gospel among the heathen, from their commencement to the year 1817. London 1827. — **449.**

HORSFORD, CORNELIA. Vinland and its Ruins. Some of the Evidences that Northmen were in Massachusetts in Pre-Columbian Days. Repr. fr. Appelton's Popular Science Monthly, Dec. 1899. New York 1899. — **450.**

HOVGAARD, W. The Voyages of the Norsemen to America. New York 1915. — **451.**

HOWARD, WM. WILLARD. Neighbors Who Need Help. Among the Deep Sea Fishers, Vol. 20, No. 2, p. 89. New York 1922. — **452.**

HOWARTH. Report on the Newfoundland and Labrador fisheries. St. John's 1874. — **453.**

HOWLEY, ARCHBISHOP. Ecclesiastical History of Newfoundland. [Quoted by Gosling, 325]. — **454.**

HOWLEY, JAMES P. The Beothuks or Red Indians. The Aboriginal Inhabitants of Newfoundland. Cambridge Univ. Press. Cambridge 1915. — **455.**

HOWLEY, M. F. Cartier's Course — a Last Word. Trans. Roy. Soc. Canada, 1st Ser., Vol. XII, Part. II, p. 151. Ottawa 1895. — **455 a.**

—»— Labrador boundary question. Can. R. Soc., Vol. 1, sect. 2, p. 291. Ottawa 1907. — **455 b.**

HUARD, V. A. Labrador et Anticosti: Journal de voyage, histoire, topographie, pêcheurs canadiens et acadiens, Indians montagnais. Montréal 1897. — **456.**

HUBBARD, CHARLIE. The Sportsman, for March. 1933 [Quoted by Forbes, **235.**] — **457.**

HUBBARD, Mrs LEONIDAS. Labrador, from Lake Melville to Ungava Bay. Bull. Amer. Geogr. Soc., Vol. 38, p. 533. New York 1906. — **458.**

—»-- My Explorations in Unknown Labrador. Harper's Mag., Vol. CXII, p. 813. New York 1906. — **459.**

HUBBARD, JUNR, Mrs LEONIDAS [MINA BENSON]. A woman's way through unknown Labrador. Journ. of the Manchester Geogr. Soc., Vol. XXIII, p. 169. Manchester 1907. — **460.**

—»— A Woman's Way through Unknown Labrador: An Account of the Exploration of the Nascaupee and George Rivers. New York 1908. — **461.**

—»-- A Woman's Way through Unknown Labrador. London 1908. — **461 a.**

HUDSON'S BAY COMPANY. The Governor and Company of Adventurers of England Trading into Hudson's Bay during Two Hundred and Fifty Years 1670—1920. London 1920. — **462.**

—»— Hudson's Bay Company, incorporated 2nd May 1670. A Brief History. London 1934. — **463.**

HUDSON'S BAY COMPANY. Trading into Hudson's Bay, publ. by the H. B. C., Winnipeg
    1934. — **464.**

—»— The Archives of the Hudson's Bay Company. The Beaver, Dec. 1933 and Can. Hist.
    Rev., Vol. XV, p. 92. Toronto 1934. — **465.**

HULL, EDWARD. Monograph on the suboceanic physiography of the North Atlantic ocean.
    With a chapter on the suboceanic physical features off the coast of North America
    and the West Indian Islands by J. W. W. SPENCER. London 1912. — **465 a.**

HURLBUT, G. C. Geographical Notes on Labrador. Amer. Geogr. Soc. Journ., Vol. 23,
    p. 455. New York 1891. — **466.**

—»— The Name of Labrador. Ibid., Vol. 26, p. 77. 1894. — **467.**

HUSTICH, I. Newfoundland-Labrador. Terra, Bd. 49, p. 186. Helsingfors 1937. — **468.**

—»— Notes on the coniferous forest and tree limit on the east coast of Newfoundland-
    Labrador, including a comparison between the coniferous forest limit on Labrador
    and in northern Europe. Acta Geographica, Vol. 7, No. 1. Helsingfors 1939. — **469.**

HUTTON, J. E. Fire and Snow. Stories of early Missionary Enterprise. London 1908.
    — **470.**

—»— A history of Moravian missions. London 1923. — **471.**

HUTTON, SAMUEL KING. Among the Eskimos of Labrador. London 1912. — **472.**

—»— By Eskimo Dog-Sled and Kayak. A Description of a Missionary's Experiences
    & Adventures in Labrador. London 1919. — **474.**

—»— Health Conditions and Disease Incidence among the Eskimos of Labrador. Poole
    1925. — **475.**

—»— An Eskimo Village. [Soc. for Promoting Christian Knowledge.] New York 1929.
    — **476.**

—»— By patience and the Word. The story of the Moravian missions. London 1935
    — **477.**

—»— A Shepherd in the snow. The Life Story of Walter Perrett of Labrador. London
    1936. — **478.**

HUXLEY, H. M. See RUSSELL, FRANK.

HYDROGRAPHIC AND MAP SERVICE, CANADA. Tide Tables for the Atlantic Coast of Canada
    for the year 1939 etc. Ottawa 1939. — **479.**

IMRAY, J. AND SON. Sailing directions for the island of Newfoundland, and adjacent
    coast of Labrador. Compiled from recent surveys. London 1862. — **480.**

INDRENIUS, ANDR. ABRAHAM, resp., PETRUS KALM . . . Praes. Specimen academicum
    de Esquimaux, Gente Americana. Diss. Aboae 1756. — **480 a.**

IRVING, WASHINGTON. Life and voyages of Christopher Columbus. Vol. IV. London 1828.
    — **480 b.**

ISELIN, COLUMBUS A. The log of the schooner Chance from the port of New Rochelle,
    N.Y., to Boston, Mass. via Cape Chidley, Labrador, July third to September
    twenty-sixth 1926. New York 1927. — **481.**

—»— Report on the Coastal Waters of Labrador. Based on Explorations of the
    »Chance» during the summer 1926. Proceed. of the Amer. Acad. of Arts and Sciences,
    Vol. 66, No. 1. Boston 1932. — **482.**

JACOB, A. M. Birds of Newfoundland (reprinted from Bird Notes and News, the Journ a
    for the Royal Society for the Protection of Birds.) Among the Deep Sea Fisher
    Vol. 34, No. 1, p. 24. New York 1936. — **483.**

JAMES, JOSEPH F. The flora of Labrador. Science, Vol. III, p. 359. Cambridge 1884.
— **484.**

JANNASCH, H. Labrador, Land und Leute. Jahresber. des Frankfurt. Vereins f. Geogr.
u. Statistik. Bd. 70. p. 94. Frankfurt a/M. 1907. — **485.**

JANNASCH, P. Ueber die Löslichkeit des Labradors von der Paulinsel in Salzsäure. Neues
Jahrb. f. Miner., Bd II, p. 42. Berlin 1884. — **485 a.**

JEFFERSON, SELBY. Adventure for Christ on Labrador. London 1933. — **486.**

JEFFERYS, THOMAS. The great probability of a northwest passage deduced from observations on the letter of Admiral De Fonte, who sailed from the Callao of Lima
on the discovery of a communication between the South Sea and the Atlantic
Ocean, and to intercept some navigators from Boston in New England, whom
he met with, then in search of a northeast passage, proving the authenticity of the
Admiral's letter. With three explanatory maps. By Thomas Jefferys, geographer
to the king. With an Appendix. Containing the account of a discovery of part
of the coast and inland country of Labrador, made in 1753. London 1768. — **487.**

JENNESS, DIAMOND. Eskimo string figures. Rep. of the Canad. Arct. Exp., T. XIII,
Pt B, p. 192. Ottawa 1924. — **488.**

—»— A new Eskimo culture in Hudson Bay. Geographical Review. Vol. XV, pp. 428
—437. New York 1925. — **489.**

—»— Ethnological Problems of Arctic America. Repr. Problems of Polar Research, Amer.
Geogr. Soc. Spec. Publ. No. 7, p. 167. New York 1928. — **489 a.**

—»— Notes on the Beothuk Indians of Newfoundland. Ann. Rept. Nat. Mus. of Canada
for 1927, Bull. 56, p. 36. Ottawa 1929. — **490.**

—»— A Demographic Inquiry into the Eskimo Population. Geogr. Review, Vol. 19
p. 336. New York 1929. — **491.**

—»— Who are the Eskimos? The Beaver, No. 2. Outfit 262, p. 267. Winnipeg 1931. — **492.**

—»— The Indians of Canada. Bull. National Mus. of Canada, No. 45. 2nd Ed. Ottawa
1932. — **492 a.**

—»— The Problem of the Eskimo. Chapter X in the American Aborigines. Publ. for 5th
Pacif. Congr. Toronto 1933. — **493.**

—»— Canada's Fisheries and Fishery Population. Transact. Roy. Soc. Canada. 3rd Ser,
Vol. XXVII, sect. II, p. 41. Ottawa 1933 — **494.**

—»— The Indian Background of Canadian History. Bull 86, Canada Dep. of Mines and
Resources. Ottawa 1937. — **494 a.**

JENSEN, AD. S. Uddrag af Professor, Dr. phil. H. P. Steensby's Dagbog om Rejsen til
»Vinland». Med Bistand af Kommandør, Dr. phil. Gustav Holm, Professor, Dr. phil.
& litt. Finnur Jónsson og Cand. mag. Einar Storgaard utgivet ved Ad. S. J. Medd.
om Grønland. Bd. LXXVII. København 1930. — **495.**

JESUIT RELATIONS and Allied Documents. The Travels and Explorations of the Jesuit
Missionaries in New France, 1610—1791. Edited by Reuben Gold Thwaites.
(71 Volumes). Cleveland 1896—1901. — **496.**

JEUNE, P. PAUL LE. Relation De Ce Qvi S'Est Passé En La Nouvelle France En L'Année
1634. Enuoyée au R. Père Prouincial de la Compagnie de Iesus en la Prouince
de France. Paris MDCXXXV. — **497.**

—»— Ibid in Thwaites, The Jesuit Relations VI. Cleveland 1897. — **498.**

JOHANSSON, OSC. V. Den årliga temperaturperiodens egenskaper och typer, främst
i Europa. Acta Geographica. Vol. 2, no. 1, p. 98. Helsingfors 1929. — **498 a.**

JOHNS, ROWLAND. Our friend the Labrador ... New York 1934. — **499.**

JOHNSON, CHARLES W. Insects of Labrador. In Grenfell's Labrador. The Country and the People, p. 453. New York 1922. — **500.**

—»— The Molluscs. Ibid., p. 479. New York 1922. — **501.**

JOHNSON, D. W. Physiographic aspects of the Labrador boundary problem. Labrador Appeal, Vol. 8, p. 3789. London 1926. — **502.**

—»— The New England-Acadian Shoreline. New York 1925. — **502 a.**

JOHNSON, GEORG. The Origin of Labrador. Nat. Geogr. Mag., Vol. XVII, p. 587. Washington 1906. — **503.**

JOHNSTON, JAMES. Grenfell of Labrador. London 1908. — **504.**

JOHNSTON, W. A. Late Pleistocene oscillations of sea level in the Ottawa valley. Geol. Surv. Can., Museum Ball. 24. Ottawa 1916. — **504 a.**

JONES, ARTHUR EDOUARD. Documents Rares ou Inédits. I. Mission du Saguenay. Rélation Inédite du R. P. Pierre Laure, S. J. ,1720 à 1730 ... Arch. du Coll. Ste-Marie. Montreal 1889. — **505.**

JÓNSSON, FINNUR, HAUKSBÓK, udgiven efter de Arnamagnæanske haandskrifter no. 371, 544 og 675, 40 af det Kongelige Nordiske Oldskriftselskab ved FINNUR JÓNSSON. Kjøbenhavn 1892—96. — **506.**

—»— Erik den rødes Saga og Vinland. Norsk Hist. Tidskr., 5 Ræ., Bd. 1, p. 116. Kristiania 1912. — **506 a.**

—»— Opdagelsen af og Rejserne til Vinland. Aarbøg. f. nord. Oldskyndigh. København 1915. — **507.**

—»— The Discovery of Vinland (America). In »Greenland, Published by the Commission for the Direction of the Geological and Geographical Investigations in Greenland», Vol. 11, p. 359. Copenhagen 1928. — **508.**

—»— Ari Thorgilsson. Islendingabók, tilegnet Islands alting, 930—1930, af dansk-islandsk forbundsfond. Ed. Finnur Jónsson. København 1930. — **508 a.**

JUNEK, D. W. Blanc Sablon. A study of an isolated Labrador Community. Journ. of the Roy. Soc. of Arts, Nov. & Dec. 1936 Vol. LXXXIV, p. 1245. London 1936.— **509.**

—»— Isolated communities. A Study of a Labrador fishing village. American Sociology Series. New York 1937. — **510.**

KEEFER, T. C. Notes on Anchor Ice. Read before the Canadian Institute, February 1862. Quoted by **434.** — **510 a.**

KELLY, W. On the Temperature of the Surface Water over the Banks and near the Shores of the Gulf of St. Lawrence. Quoted by **434**. —**510 b.**

KELSON, SAUL. Experiences of two Outdoor Workers. Among the Deep Sea Fishers, Vol. 29, No. 4, p. 163. New York 1932. — **511.**

KENDALL, W. C. The Fishes of Labrador. Portland Soc. Nat. Hist., Vol. 2, pt. 8, p. 207. Portland 1909. — **512.**

—»— Report on the Fishes collected by Mr. Owen Bryant on a Trip to Labrador in the Summer of 1908. Proceed. U.S. Nat. Mus., Vol. 38, p. 503. Washington 1910. —**513.**

KENNEDY, W. P. M. a. Oth. The Cambridge History of the British Empire, Vol. VI, Canada and Newfoundland. Passim. Cambridge 1930. — **514.**

KENTON, see JESUIT RELATIONS.

KENTON, EDNA. The Indians of North America from »The Jesuit relations and allied documents». New York 1927. — **515.**

KERR, J. H. Observations on Ice-Marks in Newfoundland. Quart. Journ. of the Geol. Soc., Vol. XXVI, p. 704. London 1870. Also in The Philosophical Magazine, Vol. XLI, p. 77. London 1871. — 516.

KETTLE, W. R. Sailing directions for the islands and banks of Newfoundland with the coast of Labrador. London 1905. — 517.

KIDDER, A. V. Eskimos and Plants. Proc. Nat. Acad. Sc., 13: 74—75. Washington 1927. — 518.

KINDLE, E. M. Notes on the Forests of Southeastern Labrador. Geogr. Review, Vol. XII, p. 57. — New York 1922. — 519.

—»— Notes on post-glacial terraces on the eastern and western shores of the Gulf of St. Lawrence. Canad. Field. Naturalist, Vol. 36, p. 111. Ottawa 1922. — 519 a.

—»— Unusual type of sand bar. Pan-Amer. Geologist, Vol. 39, No. 1, p. 15. De Moines, Iowa 1923. — 520.

—»— Range and distribution of certain types of Canadian Pleistocene concretions. Bull. Amer. Geol. Soc., Vol. 34, No. 3, p. 609. Washington 1923. — 521.

—»— The Labrador Environment. An Analysis of the Effects of the Climate and Physical Features of Labrador upon its Population. Canad. Magz. Vol. LXIII, p. 471. Dec. 1924. Toronto 1924. — 522.

—»— Geography and Geology of Lake Melville District, Labrador Peninsula. Geolog. Survey Canada, Memoir. 141. Ottawa 1924. — 523.

—»— The Terraces of the Lake Melville District, Labrador. Geogr. Review, Vol. XIV, p. 597. New York 1924. — 524.

—»— Extracts from the report on Geography and Geology of Lake Melville district, Labrador Peninsula, Canada. Privy Council. In the matter of the boundary between the Dominion of Canada and the Colony of Newfoundland. Vol. V. of Joint Append:x, p. 2362. London 1926. — 525.

KLEINSCHMIDT, SAMUEL. Grammatik der groenländischen Sprache mit teilweisem Ein schluss des Labradordialects. Berlin 1851. — 525 a.

—»— Den grønlandske Ordbog. Udg. af H. F. JÖRGENSEN. Kjøbenhavn 1871. — 525 b.

KMOCH, GEORGE See KOHLMEISTER. BENJAMIN.

KNIGHT, JOHN. *The Voyage of Master* John Knight, imployed into Groynland, *us Captaine, the years before by the King of* Denmark, *but now out of* England *to search the* N.W. *passage*, 1606. In Miller Christy: The Voyages of Captain Luke Fox of Hull and Captain Thomas James of Bristol in search of a North-West Passage in 1631 —1632 . . . Hakluyt Society, Ser. I, Vol. LXXXVIII, p. 109. London MDCCCXCIV. — 526.

KOBBÉ, GUSTAV. »Down on the Labrador». The Century. Sept. 1906, p. 672. New York 1906. — 527.

KOCH, K. R. Die Küste Labradors und ihre Bewohner. Deutsch. Geogr. Blätter, Bd. 7, p. 151. Bremen 1884. — 528.

—»— Geschichte der supplementären Expedition unter Dr. K. R. Koch nach Labrador. Die Internationale Polarforschung 1882—1883. Bd. I. Berlin 1891. — 528 a.

KOEPPE, C. E. The Canadian Climate. Bloomington, Ill., 1931. — 529.

KOHLMEISTER, BENJAMIN and KMOCH, GEORGE. Journal of a Voyage from Okak on the coast of Labrador to Ungava Bay, westward of Cape Chidley. Undertaken to explore the coast, and visit the Esquimaux in that unknown Region by Benjamin

Kohlmeister and George Kmoch, Missionaries of the church of the Unitas Fratum or United Brethren. London 1814. — **530.**

KOHT, HALVDAN. The finding of America by the Norsemen. Norweg. Trade Rev., Vol. IX. Kristiania 1926. — **531.**

KOLISCHER, KARL ARTHUR. Zur Entdeckungsgeschichte Amerikas. Die Normannen in Amerika vor Columbus. Eine kritische Studie. Mitt. d. K. u. K. Geogr. Gesellschafti. Wien 1914. — **532.**

KÖLBING. Grönland und Labrador. Gnadau 1831. — **533.**

KÖNIG, H. Das Recht der Polarvölker. Anthropos, Vol. XXIII—XXIV. Mödling 1928 —29. — **533 a.**

—»— Im Hundeschlitten durch Labrador. Gelbe Hefte, Jahrg. 10, p. 432. München 1934. — **534.**

KÖPPEN, W. Klassifikation der Klimate nach Temperatur, Niederschlag und Jahreslauf. Dr. A. Petermanns Geogr. Mitteil., Bd. 64. Gotha 1918. — **534 a.**

—»— Baumgrenze und Lufttemperatur. Dr A. Petermanns Mitteil., Bd. 65. Gotha 1919. — **534 b.**

—»— Die jährliche Temperaturgang in den gemässigten Zonen und die Vegetationsperiode. Meteorolog. Zeitschr., Bd. 43, p. 167. Wien 1926. — **534 c.**

KRANCK, E. H. Bedrock Geology of the Coastal Region of Newfoundland-Labrador. Newfoundland Geol. Survey Bull., No. 19. St. John's 1939. — **535.**

—»— The Rock-Ground of the Coast of Labrador and the Connection between the Pre-Cambrian of Greenland and North-America. Comptes Rendus d. l. Soc. géolog. de Finlande, No. XIII, p. 1. Helsingfors 1939. — **536.**

—»— Med motor och hammare längs Labradorkusten. Helsingfors 1940. — **536 a.**

—»— Urbergsforskning på Labradorkusten. En sammanfattning av petrografisk-geologiska observationer 1939. Soc. Scient. Fennica, Årsbok XIX B, No. 2. Helsingfors 1941. — **536 b.**

—»— Geologische Beobachtungen während einer Forschungsfahrt nach Labrador im Sommer 1939. Erste vorläufige Mitteilung. Mitt. d. Naturforschenden Gesellsch. Schaffhausen (Schweiz), — **536 c.**

KRAUSE, F. Über die Entstehung der nordamerikanischen Kulturprovinzen. Tagungsber. d. Deutsch. Anthropol. Gesellsch. Augsburg 1926. — **537.**

KRETSCHMER, KONRAD. Die Entdeckung Amerika's in ihrer Bedeutung für die Geschichte des Weltbildes. Mit Atlas. Berlin 1892. — **538.**

KRICKEBERG, WALTER. Das Schwitzbad der Indianer. Ciba [Zeitschrift d. Gesellsch. f. Chem. Industrie in Basel]. Basel 1934. — **539.**

KROEBER, A. L. Native American population. Amer. Anthropolog., Vol. 36, p. 1. Menasha 1934. — **539 a.**

KROGMAN, W. M. Are the North American Indians increasing in numbers? Zeitschr. f. Rassenkunde, Bd. IV, p. 203. Stuttgart 1936. — **539 b.**

KUNSTMANN, Fr. Über einige den ältesten Karten Amerika's. Anh, z. Entdeckungsgeschichte Amerikas. München 1859. — **540.**

—»— Die Entdeckung Americas. Monumenta Secularia. Herausgg. v.d. Kgl. Bayrischen Akademie der Wissenschaften zur Feier ihres hundertjährigen Bestehens. München 1859. — **541.**

LACASSE, ZACHARIE. O. M. I. Lettre, Betsiamits 5 Janv. 1874. See STREIT, — **543.**

LACEY, AMY. See AMY LACEY.

LAGUNA, F. DE. See DE LAGUNA, F.

LAFLAMME, J. C. K. Essai de géographie physique, Le Saguenay. Bull. Soc. Géogr. Québec, Vol. 1, p. 47. Québec 1885. — **543 a.**

LAGDEN, SIR GODFREY. The Native Races of the Empire. London 1924. — **544.**

LAMBE, LAWRENCE M. Notes on Fossil Corals collected by Mr. A. P. Low at Beechley Island, Southampton Island and Cape Chidley in 1904. Geol. Surv. Canada. Ottawa 1907. — **545.**

LANE, MICHAEL. See COOK, JAMES.

LANGLOIS, LT.-COLONEL. La découverte de l'Amérique par les Normands vers l'an 1000. Deux Sagas Islandaises. Soc. Edit. Géogr., Marit. et Colon. Paris 1924. — **546.**

LA RONCIÈRE, CHARLES DE. La traversée de l'Océan Atlantique des Vikings à Christophe Colomb. Revue Scientifique, 15 mars 1937, p. 81. Paris 1937. — **547.**

LARSEN, SOFUS. Danmark og Portugal i det 15de Aarhundrede. Aarbog f. nord. Old-kynd. og. Hist, III R., 9. Bd., p. 236. København 1919. — **548.**

—»— La découverte du continent de l'Amérique septentrionale en 1472—1473, par les Danois et les Portugais. Resumé d'un mémoire. Acad. das sci. de Lisboa. Coimbra 1922. — **549.**

—»— Grønlands Genopdagelse. Svar til d'Herrer Direktör P. R. Solbed og Professor, Dr O. Solberg i Norsk Geogr. Aarbok. Det grönlandske Selskabs Aarsskrift 1922 —23, p. 72. København 1923. — **550.**

—»— Discovery of the North-American Mainland twenty years before Columbus. Con-férence faite au XXe Congr. Int. d. Américanistes, Sess. de Göteborg, 20—26 Août, p. 285. Göteborg 1924. — **551.**

—»— The Discovery of North America twenty years before Columbus. Copenhagen 1925. — **552.**

LARSON, LAURENCE M. Did John Scolvus visit Labrador and Newfoundland in or about 1476? Soc. for the Advancement of Scandinav. Stud., Vol. 7, p. 81. Menasha, Wis. 1922. — **553.**

LATROBE, B [DE LA TROBE]. A succinct view of the missions established among the heathen by the church of the Brethren, or Unitas fratrum. London 1771. — **554.**

—»— A brief account of the missions established among the Esquimaux Indians on the Labrador coast, established by the Church of the Brethren's society for the fur-therance of the gospel. [London?] 1774. — **555.**

LAURE, R. P. PIERRE. Relation du Saguenay, 1720 à 1730, par le R. P. Pierre Laure. Se Thwaites: The Jesuit Relations and Allied Documents. Vol. LXVIII. — **556.**

LAVERDIÈRE, C.-H. OEuvres de Champlain. T. I & II. Québec 1870. — **557.**

LAWSON, GEORGE. Monograph of Ranunculaceæ of the Dominion of Canada and the adjacent parts of British America. Proceed. and Transact. Nova Scotia Inst. Nat. Sci., Vol. II, p. 17. Halifax 1870. — **558.**

LAVRITZ. Ist bei den Eskimos in Labrador gewesen. Büschings Nachrichten, Bd. II, pp. 72, 87. Berlin 1774. — **559.**

LEE, CUTHBERT. With Dr Grenfell in Labrador. New York 1914. — **560.**

LEE, SIDNEY. See STEPHEN, LESLIE.

LEECHMAN, DOUGLAS. Whence came the Eskimo? The Beaver, p. 38. Winnipeg 1935. — **561.**

LEITH, C. H. and A. T. A summer and winter in Hudson Bay. Madison 1912. — **562.**

LELAND, WALDO G. Guide to Materials for American History in the Libraries and Archives of Paris. Washington 1932. — **563.**

LEMOINE, GEORGE. Dictionnaire Français-Montagnais. Boston 1901. — **564.**

LESCARBOT, MARC. The History of New France. Publications of the Champlain Society. Toronto 1914. — **564 a.**

LESLIE, LIONEL A. Wilderness trails in three continents, an account of travel, big game hunting and exploration in India, Burma, China, East Africa and Labrador. London 1931. — **565.**

LEVESON-GOWER, R. H. G. H. B. C. and the Royal Society. The Beaver, Outfit 265, No. 2, p. 29. Winnipeg 1934. — **566.**

LEVIN. [TH. REICHEL, Mitglied der Direction der Brüder-Unität]. Labrador, Bemerkungen über Land und Leute. Mit zwei Original-Karten. Petermann's Geogr. Mitth., Bd. IX, p. 121. Gotha 1863. — **567.**

LÉVY-BRUHL. L. Un ouvrage sur les Naskapi. Journ. d. l. Soc. d. Américanistes, T. XXIX, p. 219. Paris 1937. — **568.**

LEWIS, ARTHUR. The Life and Work of the Rev. E. J. Peck among the Eskimos. London 1905. — **569.**

LEWIS, H. F. Notes on Birds of the Labrador Peninsula in 1923. The Auk, Vol. 42, pp. 74, 278. New York 1925. — **570.**

—»— Birds of Labrador Peninsula in 1925 and 1926. Ibid., Vol. 44, p. 59. New York 1927. — **571.**

—»— An annotated List of Vascular Plants collected on the North Shore of the Gulf of St. Lawrence. Can. Nat., Vol. XLV, pp. 129, 174, 199, 225, Vol. XLVI, pp. 12, 36, 64, 89. Ottawa 1931 & 1932. — **572.**

LEWIS, HARRISON F. Notes on Some Details of the Explorations by Jacques Cartier in the Gulf of St. Lawrence. Trans. Roy. Soc. Canada, 3rd Ser., Vol. XXVIII, p. 117. Ottawa 1934. — **573.**

LIBBEY, W. The Relations of the Gulf Stream and the Labrador Current. Nat. Geogr. Magz., Vol. 5, p. 161. Washington 1893. — **574.**

LIEBER, OSKAR M. Sketch showing the Geology of the Coast of Labrador. Rep. U. S. Coast Survey for 1860, App. No. 42, Sketch No. 38. Washington 1861. — **575.**

—»— Die amerikanische astronomische Expedition nach Labrador im Juli 1860. Dr A. Petermanns Geogr. Mitteil., Bd. VII, p. 213. Gotha 1861. — **576.**

LINDOW, HARALD. Labrador och dess eskimåer. Terra, Bd 35, s. 4. Helsingfors 1923. — **577.**

—»— Blandt Eskimoerne i Labrador. Det grønlandske Selsk. Aarsskrift København 1924. — **578.**

—»— Sekinermiut. Labradorime inuit akornàne angalanermik nalunaerut. Nungme 1923. — **578 a.**

LIPS, JULIUS E. Trap Systems among the Montagnais-Naskapi Indians of Labrador Peninsula. The Ethnograph. Mus. of Sweden, Smärre Med., No. 13. Stockholm 1936. — **579.**

—»— Public Opinion and Mutual Assistance among the Montagnais-Naskapi. Amer. Anthropolog., Vol. 39, p. 222. Menasha 1937. — **580.**

LITTLE, HOMER P. Ordovician Fossils from Labrador. Science, Vol. 84, No. 2177, p. 268. New York 1936. — **581.**

LLOYD, F. E. L. Two years in the region of ice-bergs. London 1886. — **582.**

LLOYD, T. G. B. On the Beothuks, a Tribe of Red Indians. J. R. A. I. 4. London 1874. — **583.**

LLOYD, T. G. B. Notes on Indian Remains Found on the Coast of Labrador. Journ. of Anthropolog. Instit. of Great Britain and Ireland, T 4: 39—44. London 1874. —**584.**

LODGE, T. Dictatorship in Newfoundland. Are we mismanaging our Colonies? Pp. 136, 501, 52, 91, 101 ff. London 1939. — **585.**

LOGAN, W. E. Geology of Canada, Ottawa 1863. — **585 a.**

LONG, WILLIAM J. Northern trails. Some studies of animal life in the Far North. Boston 1905. — **586.**

LOW, A. P. The recent exploration of the Labrador Peninsula. Canad. Record of Sc., 1894. Montreal 1894. — **587.**

—»— Reports on Explorations in the Peninsula along the East Main, Koksoak, Hamilton, Manicuagan and portions of other Rivers in 1892—93—94—95. Geological Survey of Canada. Ann. Rep. Vol. VIII, N. S., Part L, pp. 1 L—387 L. Ottawa 1896. —**588.**

—»— List of Mammalia of the Labrador Peninsula, with Short Notes on their Distribution, etc. Appendix I to Low's Report (588) Ottawa 1896. — **588 a.**

—»— List of Birds of the Interior of the Labrador Peninsula. Ibid., Appendix II. —**588 b.**

—»— List of the principal Food Fishes of the Labrador Peninsula with Short Notes on their Distribution. Ibid., Appendix III. — **588 c.**

—»— Report of the Mistassini expedition. Geol. Surv. Canada, Ann. Rep. 1, Rep. D. Ottawa 1885. — **588 d.**

—»— Southern part of Portneuf, Québec, and Montmorency Countries, Québec. Ibid., Rep. 5, Part 1, Rep. 2. Ottawa 1890—91. — **588 e.**

—»— A traverse of the northern part of the Labrador peninsula from Richmond Gulf to Ungava Bay. Geol. Surv. Canada, Ann. Rep. IX, N. S., part L. Ottawa 1896. — **589.**

—»— The Labrador Peninsula. In Grinnell, George Bird & Roosevelt, Theodore: Trail and Camp Fire. The Book of the Boone and Crocket Club, p. 15. New York 1897. — **589 a.**

—»— The Labrador Area. Ottawa Naturalist, Vol. X, p. 208. Ottawa 1897. — **590.**

—»— Report on an Exploration of Part of the South Shore of Hudson Strait and of Ungava Bay. Geol. Survey of Canada, Vol. XI. Ottawa 1899. — **591.**

—»— Exploration of the east coast of Hudson Bay. Geol. Surv. Canada, Ann. Rep., Vol. XIII, N. S., Part D. Ottawa 1900. — **591 a.**

—»— Iron ores of Labrador Peninsula. Engin. Mag. 1900, p. 205. New York 1900. — **592.**

—»— An exploration of the east coast of Hudson Bay from Cape Wolstenholme to the south end of James Bay. Geol. Surv. Canada, Ann. Report 13, N. S., Pt D. Ottawa 1900. — **592 a.**

—»— The Cruise of the Neptune. Report on the Dominion government expedition to Hudson Bay and the Arctic islands on board the D.G.S. Neptune, 1903—1904. Ottawa 1906. — **593.**

—»— The Hamilton River and the Grand Falls. In Grenfell's Labrador. The Country and the People, p. 140. New York 1922. — **594.**

LUCAS, FREDERIC A. The expedition to the Funk Island, with observations upon the history and anatomy of the great auk. Rept. U. S. National Museum, 1887—88, p. 493. Washington 1890. — **595.**

—»— Explorations in Newfoundland and Labrador in 1887 made in connection with the cruise of the U. S. Fish Commission Schooner Grampus. Rep. of the National Mus. 1889, p. 709. Washington 1891. — **596.**

LYON, G. F. The Private Journal. London 1824. — **597.**

LØFFLER, M. The Vineland-excursions of the ancient Scandinavians. Congr. Int. d. Américanistes. Copenhague 1883, p. 64. Copenhague 1884. — **598.**

MACHATSCHEK, F. see DECKERT, E.

MC DONALD, A. A. The Trapping of Furs. The Beaver, Vol. IV, No. 4, p. 130. Winnipeg 1924. — **599.**

MC DUNNOUGH, J. Notes on a collection of Labrador Lepidoptera. Canad. Entomologist, Vol. 53, p. 81. London Ont. 1921. — **600.**

MC FEE, WM. Sir Martin Frobisher. London 1928. — **601.**

MC GRATH, Patrik T. The Labrador Boundary Decision. Geograph. Review, Vol. 17. p. 643. New York 1927. — **602.**

MAC GREGOR, W. M. Report of an Official Visit to the Coast of Labrador by His Excellency the Governor of Newfoundland, during the month of August 1905. Dat. 5th Febr. 1906. St. John's 1906. — **603.**

MAC GREGOR, SIR WM. Report of His Excellency Sir Wm. MacGregor . . . of official visits to Labrador, 1905 & 1908. St. John's[?] 1909. — **604.**

MC GRIGOR, G. M. The Labrador boundary. The Spectator 1927, p. 596. London 1927. — **605.**

MC ILWRAITH, T. F. The Progress of Anthropology in Canada. Canadian Histor. Review, 1930, p. 132. Toronto 1930. — **606.**

MAC KAY, A. H. Labrador Plants [Collected by W. H. Prest on the Labrador Coast north of Hamilton Inlet, from the 25th of June to the 12th of August 1901]. Proceed. and Transact. of the Nova Scotian Institute of Science, Halifax, Nova Scotia, Vol. X, p. 507. Halifax 1903. — **607.**

MACKAY, DOUGLAS. The honourable company: a history of the Hudson's Bay Company. Maps by R.H.H. Maculay. Toronto 1936. — **608.**

MC KENZIE, JAMES. The King's posts in 1808. See Masson, L. R. Les bourgeois . . . (**634**), Vol. II. — **609.**

MC LEAN, JOHN. Notes of a Twenty-five Years' Service in the Hudson's Bay Territory. London 1849. — **610.**

—»— John McLean's Notes of a twenty-five year's service in the Hudson's Bay Territory. Edited by W. S. Wallace. The Champlain Society, Vol. XIX. Toronto 1932. — **611.**

MAC LEAN, J. P. A critical examination of the evidences adduced to establish the theory of the Norse discovery of America. Chicago Ill. 1892. — **612.**

MC LEAN, N. B. Report of the Hudson Strait Expedition 1927—28 (1929). Ottawa 1932. — **613.**

MAC MILLAN, DONALD B. The MacMillan Arctic Expedition returns. Nat. Geogr. Soc. Mag., Vol. XLVIII, p. 477. Washington 1925. — **614.**

—»— Tribute to Grenfell. Among the Deep Sea Fishers, Vol. 28, No. 1, p. 3. New York 1930. — **615.**

MAC RITCHIE, DAVID. Kayaks of the North Sea. Scot. Geogr. Mag., Vol. XXVIII, p. 126. Edinburgh 1912. — **615 a.**

MACOUN, JOHN. Additional Plants from Labrador and Hudson Strait. Geol. Surv. Canada, Ann. Rept.. N. S., Vol I, Pt. DD. p. 25. Ottawa 1886. — **615 b.**

MACOUN, JAMES MELVILLE. List of plants collected on the Rupert and Moose Rivers, along the shore of James' Bay, and on the islands in James' Bay, during the summers of 1885 and 1887. Rept. Geolog. and Nat. Hist. Survey of Canada for 1887—88, Vol. III, pt. 2, Appendix I. Ottawa. 1889. — **616.**

MACOUN, JAMES MELWILLE. Catalogue of the plants reported by various travellers as growing on the coast of Labrador. Chap. XVI, p. 448 in Packard: The Labrador coast (716). 1891. — 617.

—»— The forests of Canada and their distribution, with notes on the more interesting species. Mém. et comptes rendus d.l. Soc. Roy. du Canada. Montréal 1894. — 618.

—»— List of the plants known to occur on the coast and in the interior of the Labrador peninsula. Geol. Surv. Canada. Ann. Rept., N. S., Vol. VIII, App. VI, p. 353 L. Ottawa 1895. — 619.

—»— Catalogue of Canadian Birds. Geolog. Surv. Canada. Ottawa 1900, 1903 and 1904. — 620.

MACOUN, J. M. and MALTE, M. O. The Flora of Canada. Can. Dept. Mines, Geol. Surv., Mus. Bull, No. 26. Ottawa 1917. — 621.

MACULAY, R. H. H. Trading into Hudson's Bay. A narrative of the visit of Patrick Ashley Cooper, thirtieth governor of the Hudson's Bay Company, to Labrador, Hudson Strait and Hudson Bay in the year 1934. Winnipeg 1934. — 622.

MAGNAGHI, ALBERTO. Precursori di Colombo? Il tentativo di viaggio transoceanico dei genovesi fratelli Vivaldi nel 1291. Mem. della R. Soc. Geogr. Italiana, T. XVIII, p. XIV. Roma 1935. — 623.

MALDONADO, LORENZO FERRER. A Relation of the Discovery of the Strait of Assian made by me, Capt. Lorenzo Ferrer Maldonado, in the Year 1588; in which is given the Course of the Voyage, the situation of the Strait, the Manner in which it ought to be fortified, and also, the Advantages of this Navigation, and the Loss which will arise from not prosecuting it. — See Barrow 1818 appendix 2. — 624.

MALLALIEU, WANDA S. Whaling. Among the Deep Sea Fishers, Vol. 27, No. 1, p. 26. New York 1929. — 625.

MALLET, THIERRY. Glimpses of the Barren Lands. The Beaver, No. 2., Outfit 262, p. 278. Winnipeg 1931. — 626.

MALTE, M. O. See MACOUN, J. M.

MALTE-BRUN, M. Proposition sur les moyens de donner une direction méthodique aux travaux géographiques en général, et à ceux de la Société de Géographie en particulier, lue dans la Séance du 15 février. Bull. de la Soc. de Géographie, Sér. 1, T. I, p. 50. Paris 1822. — 627.

MARCHANT, IANES. The first voyage of Master Iohn Dauis, undertaken in June 1585. for the Discouerie of the North-west Passage. Hakluyt, p. 776. London 1589. — 628.

MARCOLINI, F. Discoverie of Estotiland, Drageo, and Icaria, by Nicolas Zeno and Antonio, his brother. Gathered out of their letters. (Hakluyt: Voyages, Vol. III) London 1811. — 628 a.

MARIE-VICTORIN, FRÈRE. Flore Laurentienne. Montréral 1935. — 628 b.

MARKHAM, ALBERT HASTINGS. See DAVIS, JOHN.

MARKHAM, CLEMENTS R. See BAFFIN, WILLIAM.

MARMER, H. A. Arctic Ice and Its Distribution in the North Atlantic Ocean. Geogr. Review, Vol. 22, p. 157. New York 1932. — 629.

MARTENS, J. H. C. The mineral composition of some sands from Quebec, Labrador and Greenland. Field Mus. Nat. Hist., Geol. ser. Vol. 5, No. 2, p. 17. Chicago 1929. — 630.

MARTIN, A. AND OTHERS. Aglait illunainortut. Okiok 1909. Nain 1909. — 631.

MASON, M. H. The arctic forests. London 1924. — 632.

MASON, OTIS T. The ulu or woman's knife of the Eskimo. Report of the United States National Museum. Washington 1890. — **638.**

MASSON, L. R. Les Bourgeois de la Compagnie du Nordouest. Québec 1890. — **634.**

MATHIASSEN, THERKEL. Archeology of the Central Eskimos. Rept. of the 5th Thule Expedition. Vol. IV. København 1927. — **635.**

—»— Nordboruiner i Labrador? Dansk Geogr. Tidskr., Bd. 31, p. 75. København 1928. — **636.**

—»— Norse Ruins in Labrador? American Anthropologist, Vol. 30, No. 4, p. 569. Menasha 1928. — **637.**

—»— Eskimoerne i Nutid og Fortid. København 1929. — **638.**

—»— The question of the origin of Eskimo culture. The Americ. Anthropologist, N. S., Vol. XXXII. Menasha 1930. — **639.**

—»— The Question of the Origin of Eskimo Culture. Ibid., Vol. 32, No. 4, p. 591. Menasha 1930. — **640.**

—»— The Eskimo archaeology of Julianehaab District; With a brief summary of the prehistory of the Greenlanders. Medd. om Grønland, Bd. 118, No. 1. København 1936. — **640 a.**

MATTHEW, G. F. Fossil Chrysochloridae in North America. Science, N. S., Vol. 24, p. 786. Cambridge Mass. 1906. — **641.**

MATTHEWS, D. J. Report on the work carried out by the S.S. *Scotia*, 1913. Ice observations, meteorology and oceanography in the North Atlantic Ocean, p. 4. London 1914. — **642.**

MAWDSLEY, J. B. St Urbain area, Charlevoix district, Quebec. Geol. Surv. Canada, Mem. 152. Ottawa 1927. — **642 a.**

MECKING, LUDWIG. The geography of the polar regions. Amer. Geogr. Soc., No. 8. New York 1928. — **643.**

—»— Das Bodenrelief der Meere und seine Beziehungen zum Bau der Erde. Nova Acta Leopoldina, N. F., Bd. 7. Halle 1939. — **643 a.**

—»— Ozeanischen Bodenformen und ihre Beziehungen zum Bau der Erde. Petermanns Geogr. Mitt., Jahrg. 86. H. 1. Gotha 1940. — **643 b.**

MERRICK, ELLIOT. Indian Harbor. Among the Deep-Sea Fishers, Vol. 31, No. 3, p. 115. New York 1933. — **644.**

—»-- True North. New York 1935. — **645.**

MEYER, ERNST. De Plantis Labradoricis libri tres. Lipsiae 1830. — **646.**

MEYLAN, A. Histoire de l'évangelisation des Lapons, et l'Evangile au Labrador. Paris 1863. — **647.**

MIGNARD, JACQUES. Quelques Escrocs Anglais demasqués, ou les Déserts de l'Amérique du Nord présentés tels qu'ils sont. Paris 1798. — **648.**

MILLER, O. M. Making a Map of Northern Labrador. Among the Deep Sea Fishers, Vol. 33, No. 1, p. 16. New York 1935. — **649.**

—»— Notes on the construction of the Cape Chidley sheet. Geogr. Review, Vol. 26, p. 53. New York 1936. — **650.**

MITCHELL, HELEN S. Food Problems on the Labrador. Among the Deep Sea Fishers, Vol. 27, No. 3, p. 99. New York 1929. — **651.**

MORANT, G. M. A contribution to Eskimo craniology based on previously published measurements. Biometrika, Vol. 29, Pts I and II. Cambridge 1937. — **651 a.**

MONTGON, A. DE. See QUINEL, CH.

MORAVIAN MISSION. Voyages to Labrador in the years 1765—1770. I. Extract from the Journals of John Hill, Jens Haven, Chr. Danchard and A. Schloezer of their Voyage to the Coast of Labrador in the year 1765. II. Extract from Jens Haven's & Chr. Danchard's accounts of the Voyage of the Jersey Packet in the year 1770. III. Journal of the Voyage of Jersey Packet to Labrador and Newfoundland taken from the Papers of Jens Haven and Chr. Danchard 1770. Manuscr. in the Moravian Miss. Office in London. Sign XLIV a. — **652.**

—»— Diaries and Voyages, Labrador 1765—1800. 14 Vols in the Society's Office, London. Signum XLIV a. — **653.**

—-»— Missionen der evangelischen Brüder in Grönland und Labrador. Gnadau 1821. — **654.**

—»— Übersicht der Brüder Missionsgeschichte. Gnadau 1833. — **655.**

—»— Log Books of the Ship Harmony 1821—3, 1825, 1834—36, 1837, 1837—39, 1840, 1843, 1847. Mor. Miss. Office London. Signum XIV a. — **655 a.**

—»— Nachrichten aus der Brüdergemeinde 1876—94. Moravian Soc. Office, London. Sign. VI, 1—18. — **656.**

—»— Missionsatlas der Brüderunität, 16 Karten und Text. Herrnhut 1895. — **657.**

—»— Labrador. Extracts from the Station Diaries, July 1, 1903 — July 1, 1904. Period. Accounts Morav. Miss., T. 5, p. 608. — **658.**

—»— Monatliche Nachrichten aus der Brüdergemeinde 1865—98. 3 Vols. Ibid. Sign. XIII. — **659.**

—»— Newfoundland & Labrador. Map & List of Missions. Ibid. Sign. XLVI, 7. — **660.**

—»- - Life and Work of E. J. Peck among the Eskimos of Labrador. Ibid. Sign. LXI, 76. — **661.**

—»— Diaries, Labrador 1870—1908. 23 Vols, Sign. XLIV. — **662.**

—»— The Moravian Missionary Atlas containing An Account of the Various Countries in which the Missions of the Moravian Church are carried on, and of Its Missionary Operations. New Edition. With Eighteen Maps and a Mission Field and Station Index. London 1908. — **663.**

—»— Labrador. Cuttings. Periodical accounts. In the Morav. Miss. Office in London. Signum LV, 1. — **664.**

—»— List of Missions. Newfoundland and Labrador. Ibid., Sign. XLVI, 7. — **665.**

—»— Meteorology. Ibid., Sign. LIV, 17. — **666.**

—»— Brüdergemeinde 1876—1908. Nachrichten und Mitteilungen. Vols 1—32. Gnadau 1876—1930. —**667.**

MORTON B. R. The native trees of Canada. Dept. of the Interior, Forestry Branch, Bull. No. 61. Ottawa 1921. — **667 a.**

MOSBY, OLAV. From the Northwestern North Atlantic. Norsk Geogr. Tidskr., Bd. VII, h. 5—8, p. 253. Oslo 1939. — **668.**

—»— See SMITH, EDWARD H.

MOSDELL, H. M. Newfoundland: its manifold attractions for the capitalist, the settler, and the tourist. St. John's 1920. — **668 a.**

MUDDIMAN, B. A lonely fur factor: a day at Wakeham Bay on the Labrador coast. Rod and Gun, Vol. 16, p. 679. 1913. — **668 b.**

MUIR, ETHEL GORDON. The People on the Labrador and Their Needs. Among the Deep Sea Fishers, Vol. 8, No. 2, p. 35. Toronto 1910. — **669.**

MUNN, W. A. Wineland voyages. Location of Helluland-Markland and Vinland. St. John's 1929. — **670.**

MURPHY, ROBERT CUSHMAN. Progress in Oceanographic Research. Geogr. Review, Vol. 27, p. 500. New York 1937. — **671.**

MURRAY, A. Glaciation of Newfoundland. Mém. et comptes rendus d. l. Soc. Roy. du Canada, Vol. I, sect. V (1882), p. 55. Montréal 1883. — **672.**

MÜLLER, PETER ERASMUS. Sagabibliothek med Anmærkninger og inledende Afhandlinger. Kjøbenhavn 1817—20. — **672 a.**

NANSEN, FRIDTJOF. In northern mists. Arctic exploration in early times. London 1911. — **673.**

—»— Nord i Tåkeheimen. Utforskningen av jordens nordlige strøk i tidlige tider. Kristiania 1911. — **674.**

—»— Nebelheim. Entdeckung und Erforschung der nördlichen Länder und Meere. Leipzig 1911. — **674 a.**

—»— The Norsemen in America. Geogr. Journ., Vol. XXXVIII. London 1911. — **675.**

NECKEL, GUSTAV. Die erste Entdeckung Amerikas im Jahre 1000 n. Chr. Voigtländers Quellbücher, Bd. 43. Leipzig 1913. — **675 a.**

NEWFOUNDLAND COLONIAL SECRETARY's office. Census of Newfoundland and Labrador, 1891. St. John's 1893. — **675 b.**

—»— Census of Newfoundland and Labrador, 1911. St. John's 1914. — **676.**

NEWFOUNDLAND GOVERNMENT: List of Newfoundland, Canadian and French Lights and Fog Alarms on the coasts of Newfoundland and Labrador. St. John's 1939. — **677.**

NEWFOUNDLAND HIGH COMMISSIONER IN LONDON. Dominion of Newfoundland and Labrador. Some information about the resources of the ancient colony. London 1921. — **678.**

NEWFOUNDLAND ROYAL COMMISSION 1933. London 1934. — **679.**

NEWFOUNDLAND AND LABRADOR PILOT. The Coast of Newfoundland, the Strait of Belle Isle and the Southeast and East Coasts of Labrador from Blanc Sablon to Cape Chidley. Sixth Edition. London 1929. — **680.**

NIELSEN, YNGVAR. Nordmænd og Skrælinger i Vinland. Norsk Geogr. Selsk. Aarbog 1904—05, p. 1. Kristiania 1905. — **681.**

—»— Die ältesten Verbindungen zwischen Norwegen und Amerika. Congr. Int. d. Américanistes. T. XIV. Stuttgart 1906. — **682.**

NOBLE, LOUIS L. After icebergs with a painter: a summer voyage to Labrador and around Newfoundland. New York 1861. — **683.**

NORAKOVSKY, STANISLAUS. Arctic or Siberian hysteria as a reflex of the geographical environment. Ecology, Vol. V, No. 2, p. 113. Brooklyn, N. Y. 1924. — **684.**

NORTON, ARTHUR H. Bodwoin College Exploration, 1891. — **685.**

—»— Birth of the Bodwoin College expedition to Labrador in 1891. Proceed. Portland Soc. of Nat. History, Vol. 2: 5. Portland 1891. — **685 a.**

—»— Birds of the Bodwoin College Expedition to Labrador in 1891. Proceed. Portland Soc. Nat. Hist., Vol. II. Portland, Main 1901. — **686.**

NUTTALL, THOMAS. A manual of the ornithology of the United States and Canada 1832 —34. Vol. I. The Land Birds. Cambridge 1832. Vol. II. The Water Birds. Boston 1834. — **687.**

NØRLUND, POUL. De gamle Nordbobygder ved Verdens Ende. Skildringer fra Grøn-
lands Middelalder. København 1934. — **688.**
—»— Viking settlers in Greenland, and their descendants during five hundred years.
London & Copenhagen 1936. — **689.**
ODELL, N. E. Explorations in the Mountains of Northern Labrador. Canadian Alpine
Journ., 1932, p. 67. Winnipeg 1932. — **690.**
—»— Surveying in Northern Labrador. Discovery, Vol. XIII, p. 47. London 1932. — **691.**
—»— The Mountains of Northern Labrador. Geogr. Journ., Vol. LXXXII, pp. 193,
315. London 1933. — **692.**
—»— Within and without the Arctic circle. The Alpine Journ., Vol. XLVI, p. 27. Lon-
don 1934. — **693.**
—»— The geology and physiography of northernmost Labrador. In Forbes, Alex., Amer.
Geogr. Soc., Spec. Publ. No. 22. 1938. — **694.**
OETTEKING, BRUNO. Ein Beitrag zur Kraniologie der Eskimo. Abh. u. Ber. d. Kön. Zool.
u. Anthrop.-Etnogr. Mus. zu Dresden, Bd 12, No. 3. Dresden 1908. — **694 a.**
O'HARA. Reise nach dem Süden von Hoffenthal in Labrador. Missionsblatt aus der
Brüdergemeinde, August 1871, p. 211. Gnadau 1871. — **695.**
OLDMIXON, J. The British Empire in America. London 1741. — **696.**
PACKARD, ALPHAEUS SPRING. A List of Animals Dredged Near Caribou Island, Southern
Labrador, during July and August, 1860. Canad. Nat. and Geolog., Vol. VIII,
p. 401. Montreal 1863. — **697.**
—»— Results of observations on the drift phenomena of Labrador and the Atlantic
coast southward. Amer. Journ. of Sc. and Arts, 2nd Ser., Vol. XLI, p. 30. New
Haven 1866. — **698.**
—»— List of vertebrates observed at Okkak, Labrador, by Rev. Samuel Weiz, with
annotations by A. S. Packard, Jr., M.D. Proc. Boston Soc. Nat. Hist, Vol. X,
p. 264. Boston 1866. — **699.**
—»— Observations on the Glacial Phenomena of Labrador and Maine with a View of
the Recent Invertebrate Fauna of Labrador. Boston Soc. Nat. Hist., Mem., Vol. 1,
p. 210. Boston 1867. — **700.**
—»— View of the lepidopterous fauna of Labrador. Ibid., Vol. XI, p. 32. Boston 1867.
— **701.**
—»— List of the Coleoptera collected in Labrador. Ann. Rep. Peab. Acad. Sc., 1871,
p. 92. Salem Mass. 1872. — **702.**
—»— The Esquimaux of Labrador. Appelton's Journal, 1871, p. 65. New York 1871.
Repr. in Beach's Indian Miscellany, Albany 1877. — **703.**
—»— Pan-ice work and glacial marks in Labrador. Amer. Natur. Vol. 11, p. 674. Salem
Mass. 1877. — **704.**
—»— Glacial marks in Labrador. American Naturalist, Vol. XVI, p. 30. New York
1882. — **705.**
- »— Glacial Marks in Labrador. Ibid., Vol. 16, p. 30. Salem Mass. 1882. — **706**
—»— Do Labrador dogs bark? Ibid., Vol. XVIII, p. 1063, 1884. — **707.**
—»— The bees, wasps, etc., of Labrador. Ibid., p. 1267. — **708.**
—»— Life and nature in southern Labrador. Ibid., Vol. XIX, pp. 269, 365. 1885.
— **709.**
—»— Notes on the Labrador Eskimo and their former range southward. Ibid., p. 471.
— **710.**

PACKARD, ALPHAEUS SPRING. Notes on the Physical Geography of Labrador. Bull. Amer.
    Geogr. Soc., Vol. XIX p. 403. New York 1887. — **711.**

—»— Who First Saw the Labrador Coasts? Ibid., Vol. XX, p. 197. 1888. — **712.**

—»— The Geographical Evolution of Labrador. Ibid., Vol. XX, p. 208. 1888. — **713.**

—»— List of the spiders, myriopods, and insects of Labrador. Canadian Naturalist.
    p. 141. Ottawa 1888. — **714.**

—»— A summer's cruise to northern Labrador. I. From Boston to Square Island.
    II. From Henley Harbor to Cape St. Michael. III. From Cape St. Michael to
    Hopedale. Bull. Amer. Geogr. Soc., Vol. XX, p. 337. New York 1888. — **715.**

—»— The Labrador Coast. A journal of two summer cruises to that region. With notes
    on its early discovery, on the Eskimo, on its physical geography, geology, and
    natural history. New York 1891. — **716.**

—»— See WEIZ, SAMUEL.

PADDON, HARRY L. Mud Lake Winter Hospital, Among the Deep Sea Fishers, Vol. 12,
    No. 2, p. 50. New York 1914. — **717.**

—»— People Worth Knowing in Hamilton Inlet. Ibid. Vol. 18, No. 1, **pp.** 15 & 51. New
    York 1920. — **718.**

—»— Winter Season at Northwest River. Ibid., Vol. 21, No. 3, p. 92. New York 1923.
    — **719.**

—»— People and Things and the Future of Labrador. Ibid., Vol. 23, No. 2, p. 66. New
    York 1925. — **720.**

—»— The Mission's Most Northerly Outposts. Ibid., Vol. 23, No. 4, p. 152. New York
    1926. — **721.**

—»— Events in Dr. Paddon's Districts. Ibid., Vol. 24, No. 3, p. 95. New York 1926. — **722.**

—»— Why Work in Labrador is Worth While. Ibid., Vol. 25, No. 1, p. 8. New York 1927.
    — **723.**

—»— Ye Goode Olde Days. I & II. Ibid., Vol. 27, No. 2 & 4, pp. 151 & 155. New York
    1929 & 1930. — **724.**

—»— The Log of a Winter Cruise. Ibid., Vol. 28, No. 2, p. 69. New York 1930. — **725.**

—»— Log of a Komatik Trip. Ibid., Vol. 29, No. 2, p. 64. New York 1931. — **726.**

—»— Cruise of the »Jessie Goldthwait» Summer of 1933. Ibid., Vol. 31, No. 3, p. 130.
    New York 1933. — **727.**

—»— Early Winter on Hamilton Inlet. Ibid., Vol. 32, No. 1, p. 19. New York 1934. — **728.**

—»— With the Trappers to Muskrat Falls. Ibid., Vol. 33, No. 4, p. 145. New York 1936.
    — **729.**

—»— Northwest River Comes of Age. Among the Deep Sea Fishers, Vol. XXXV, p. 61.
    New York 1937. — **730.**

—»— Labrador: its People and its Problems. A viewpoint twenty-six years long. The
    Daily News Labrador Souvenir supplement, October 19. St. John's 1938. — **731.**

—»— Change and . . .? A retrospective view. Ungava. Mag. of Labrador, Vol. 4, No. 1,
    p. 1. Cartwright 1939. — **732.**

—»— Of Northwest River. Among the Deep Sea Fishers, Vol. XXXVII, No. 4, p. 136
    New York 1940. — **733.**

PAIN, B. H. see DUCKWORTH, W. J.

PALMER, DOUGLASS. The Doctor of the Labrador. [Separate in the library of Nat. Geogr.
    Soc., Washington] — —. — **734.**

PALMER, HENRY WEBSTER. Coasting along Labrador. Outing, Vol. 39, p. 519. New York 1902. — **735.**

PARKER, DAVID W. Guide to the materials for United States History in Canadian Archives Washington 1913. — **736.**

—♦— A Guide to the Documents in the Manuscript Room at the Public Archives of Canada. Vol. I. Ottawa 1914. — **737.**

PARKMAN, FRANCIS. Pioneers of France in the New World. Boston 1914. — **738.**

PARKS, HESTER. A Newcomer Along the Coast. Among the Deep Sea Fishers, Vol. 21, No. 2, Page 1. New York 1923. — **739.**

PATTERSON, GEORGE. The Portuguese on the North-East coast of America, and the first European attempt at Colonization there. A lost chapter in American History. Mém. et comptes rendus d. l. Soc. Roy. du Canada. T. VIII, Sect. II, p. 127. Montréal 1891. — **740.**

PAULLIN, CHARLES O. and PAXSON, FREDERIC L. Guide to the Materials in London Archives for the History of the United States since 1783. Washington 1914. — **741.**

PAXSON, FREDERIC L. See PAULLIN, CHARLES O.

PAYNE, F. F. Eskimo of Hudson Strait. Proc. Canadian Institute. Toronto 1889. — **742.**

PEATTIE, RODERICK. The Isolation of the Lower St. Lawrence Valley. Geogr. Review, Vol. V, No. 2. New York 1918. — **743.**

PECK, EDMUND JAMES. Journey from Richmond Bay to Ungava Bay. — 1887. — **744.**

PERRET, ROBERT. La géographie de Terre-Neuve. Paris 1913. — **745.**

PETTERSSON, OTTO. Changes in the Oceanic circulation and their climatic consequences. Geogr. Review, Vol. 19, p. 121. New York 1929. — **746.**

PHINNEY, FRANK D. The Eye Doctor. Among the Deep Sea Fishers, Vol. 26, No. 4, p. 152. New York 1929. — **747.**

PICKERSGILL, RICHARD. Track of His Majesty's armed Brig Lion, from England to Davis' Streights and Labrador; with Observations for determining the Longitude by Sun and Moon, and Error of Common Reckoning; also of Variation of the Compass and Dip of the Needle, as observed during the said Voyage in 1776. R. S. Phil. Trans., Part II, Vol. LXVIII, p. 1057. London 1778. — **748.**

PIKE, W. The Barren Grounds of northern Canada. London 1892. — **749.**

PILLING, JAMES CONSTANTINE. Bibliography of the Eskimo Language. Smithsonian Instit., Bureau of Ethnology. Bull. No 1. Washington 1887. — **750.**

PINKERTON, JOHN. Modern Geography digested on a new plan. Vol. II. London 1802. — **751.**

PINKERTON, ROBERT E. The Gentlemen Adventurers. Introduction by STEWART EDWARD WHITE. Toronto 1931. — **751 a.**

PITTARD, EUGENE. Contribution à l'étude anthropologique des esquimaux du Labrador et de la Baie d'Hudson. Bull. d. l. Soc. Neuchâteloise de Géographie, T. XIII, p. 158. Neuchâtel 1901. — **752.**

[POLARFORSCHUNG.] Die Internationale Polarforschung 1882—1883. Die Beobachtungs-Ergebnisse der Stationen. Bd. I. Kingua-Fjord und die meteorologischen Statio-nen II. Ordnung in Labrador, S. 123. Beobachtungen: VIII, 1882—IX 1883. Ber-lin 1901. — **752 a.**

♦POLARIS♦ FLOE PARTY. Drift of the ♦Polaris♦ floe Party 1871—73. Monthly Meteorol. Chart of the North Atlantic, Febr. 1923. Meteorol. Office. London 1923. — **753.**

POLUNIN, NICHOLAS. The Isle of the Auks. London 1932. — **754.**

—»— Some Notes on the Vegetation of Akpatok Island: Appendix III, Botany. Geogr. Journ., Vol. LXXX, p. 229. London 1932. — **754 a.**

—»— The Flora of Akpatok Island, Hudson Strait. Journ. of Botany, Vol. LXXII, p. 197. 1934. — **754 b.**

—»— The Vegetat. on of Akpatok Island. Parts I & II. Journ. of Ecology, Vols XXII, p. 337; XXIII, p. 161. Brooklyn N.Y. 1934, 1935. — **754 c.**

—»— The Birch 'forests' of Greenland. Nature, Vol. CXL, p. 939. 1937. — **754 d.**

—»— The Vegetation of Akpatok Island. (Corrections of the Determination of Species) Journ. of Ecology, Vol. XXV, p. 570. 1937. — **754 e.**

—»— Botany of the Canadian Eastern Arctic. Part I. Nat. Mus. of Canada, Bull. no 92. Ottawa 1940. — **754 f.**

PORSILD, MORTEN. P. Stray Contributions to the Flora of Greenland. XII. On some Herbaria from Greenland and Labrador collected by the Moravian Brethren. Medd. om Grønland, Bd. XCIII, No 3, p. 84. København 1935. — **755.**

PORTER, RUSSEL W. Report on Method of Survey [of St. Augustine River]. See HENRY G. BRYANT 1913. — **756.**

PREST, W. H. On Drift Ice as an Eroding and Transporting Agent. Proceed. and Transactions of the Nova Scotia Institute of Sc., Sess. 1900—01, Pt. III, p. 333. Supplementary Notes on the Above. Ibid., Sess. 1901—02, Pt. VI, p. 455. Halifax 1902. — **757.**

PRICHARD, H. HESKET. On the Labrador. Cornhill Mag., Vol. 27, p. 176. London 1909. — **758.**

- -»— Across Labrador from Nain to the George or Barren Grounds River. Geogr. Journ., Vol. XXXVI, p. 691. London 1910. — **759.**

—»— Across unknown Labrador. Fry's Mag., Vol. 14, pp. 3, 99, 195, 307, 423, 494. London 1910. — **760.**

—»— Through trackless Labrador. With a chapter on fishing by G. M. Gathorne-Hardy. London 1911. — **761.**

- »— Across unknown Labrador. Wide World Mag. Vol. 27, p. 55. London 1911. — **762.**

—»— The long Labrador trail. Travel & Exploration, Vol. V, p. 120. London 1911. — **763.**

—»— See GATHORNE-HARDY, G. M.

PRIVY COUNCIL [GREAT BRITAIN]. Documents descriptive of physiography, geology and ethnography of the Labrador Peninsula, Canada. Privy Council. In the matter of the boundary between the Dominion of Canada and the Colony of Newfoundland. Vol. V of Joint Appendix, p. 2590. London 1926. — **764.**

—»— Judicial committee. In the matter of the boundary between the Dominion of Canada and the colony of Newfoundland in the Labrador Peninsula, between the Dominion of Canada of the one part and the colony of Newfoundland on the other part. Report of the Lords of the Judicial Committee of the Privy Council, delivered the 1st March 1927. 12 volumes, 3 maps. Labrador. London 1927. — **765.**

PRIVY COUNCIL REPORT: Boundary between Canada & Newfoundland Labrador. Vols. I—XII. London 1927. — **766.**

PROWSE, DANIEL W. A history of Newfoundland from the English colonial and foreign records. With a prefatory note by Edmund Gosse ... London 1895. — **767.**

PROWSE, DANIEL, W. A history of the churches in Newfoundland. By various writers
    . . . London 1895. — **768.**
—»— The Newfoundland Guide Book, including Labrador. London 1905. —**769.**
PROWSE, G. R. F. Cabot's Bacalhaos. Transact. Roy. Soc. Canada, Ser. 3rd, Vol. XXXII,
    Sect II, p. 139. Ottawa 1938. — **770.**
PUMPHREY, BEVAN. Experiences of Two Outdoor Workers. Among the Deep Sea Fishers,
    Vol. 29, p. 166. New York 1932. — **770 a.**
PURCHAS, SAMUEL. Purchas his Pilgrimage. London 1626. — **771.**
PUTNAM, G. P. The Putnam Baffin Island Expedition. Geogr. Rev., Vol. 18, p. 1. New
    York 1928. — **772.**
PUYALON, HENRY DE. Report on the Copper, etc. found to exist on the North Shore
    of the Gulf of St. Lawrence. Report of Com. of Crown Lands, Province of Quebec.
    1883. — **773.**
—»— Report of Exploration for minerals on North Shore of the Gulf of St. Lawrence.
    Ibid. Quebec 1884. — **773 a.**
—»— Récits du Labrador. Montréal 1894. — **773 b.**
QUINEL, CH. ET MONTGON, A. DE. Jacques Cartier. Le découvreur du Canada. Paris 1936.
    — **774.**
RAFN, CARL CHR. Antiquitates Americanae. Boston 1837. — **775.**
—»— Americas arctiske Landes gamle Geographie efter de nordiske Oldskrifter. Særsk.
    Aftr. af Grönlands historiske Mindesmærker. Kjøbenhavn 1845. — **776.**
RAMUSIO, GIO. BATTISTE. Delle navigatione et viaggi. Discurso supra la Terra Ferma
    dell' Indie Occidentali, del Labrador de los Baccalaos, et Della nuova Francia.
    Vol. 3. Venezia 1606. — **777.**
—»— Navigationi et viaggi. Vol. II. Venezia 1688. — **778.**
RASMUSSEN, KNUD. Adjustment of the Eskimos to European civilization with special
    emphasis on the Alaskan Eskimos. Proceed. of the 5th Pacific Sc. Congr., B 2.
    23, p. 2889. Toronto 1935. — **779.**
—»— Eskimos and stone age peoples. Ibid., B2. 8, p. 2767, 1935. — **780.**
—»— Fra Grønland til Stille Havet. Bd. I—II. København 1925—26. — **781.**
RATHBUN, MARY J. The Marine Crustacea. In Grenfell's Labrador. The Country and
    the People, p. 473. New York 1922. — **782.**
—»— List of Crustacea of the Labrador Coast. In Grenfell's Labrador. The Country
    and the People, p. 506. New York 1922. — **783.**
RECLUS, ELISÉE. Nouvelle géographie universelle. La Terre et les hommes. T. XV,
    Amérique boréale, pt. VII, p. 618. Labrador. Paris 1890. — **783 a.**
REEVES, A. M. The Finding of Wineland the Good. The History of the Icelandic Dis-
    covery of America. London 1890. — **784.**
REGAN, C. T. Observations on the fauna and flora. Labrador Appeal, Vol. 3, p. 2509.
    Cfr **764.** London 1926. — **785.**
REICHEL, EDWARD H. Historical sketch of the church and missions of the United Brethren,
    commonly called Moravians. Bethlehem, Pa. 1848. — **786.**
REICHEL, LEVIN THEODOR. Visitationsreise nach Labrador von Mai bis November 1861.
    Budissin. Annex. Manuskr. »Labrador. Letters to my Children, 1861». Moravian
    Miss. Soc. Office London. Sign. LXI, 295. — **787.**
—»— The Moravian Missions in Labrador, by Brother L. T. Reichel. Append. No. VII,
    p. 262, in HIND (**434**) 1863. — **787 a.**

REICHEL, LEVIN THEODOR. Labrador, Bemerkungen über Land und Leute. Petermanns Geogr. Mitteil., Vol. X. Gotha 1863. — **788.**

—»— Missionsatlas der Brüder-Unität. Herrnhut 1861. — **789.**

—»— Die Missionen der Brüder-Unität. I. Labrador. Gnadau 1873. — **790.**

—»— Labrador. Aivektök oder Eskimo Bay. Mit Karte 1 : 2,300,000. Missionsblatt der Brüdergemeinde. Gnadau 1873. — **791.**

—»— The early history of the Church of the United Brethren (Unitas Fratrum) commonly called Moravians, in North America, A.D. 1734—1748. Nazareth Pa 1888. — **792.**

REICHEL, TH. See LEVIN.

REID NEWFOUNDLAND COMPANY. Newfoundland and Labrador, unique nature attractions, excellent transportation facilities, guide for sportsman, tourist, and health seeker. Chicago circ. 1910. — **792 a.**

RELIGIOUS TRACT SOCIETY. British North America. Comprising Canada, British Central North America, .British Columbia, Vancouver's Island, Nova Scotia and Cape Breton, New Brunswick, Prince Edward's Island, Newfoundland and Labrador. London 1864. — **793.**

—»— Bericht des Eskimobruders Daniel. Brief von Dr. Ribbach in Hoffenthal. Mit Karte von Samuel Weiz. Missionsblatt aus der Brüdergemeinde. Dec. 1868, Jan. 1869. Gnadau 1869. — **794.**

RIBBACH, C. A. Labrador vertaald door J. H. Van Lennep. Tijdschr. van het aardrijks-kund. Genootsch., Bd. I, p. 284. Amsterdam 1875. — **795.**

RICHARDS, G. M. Geological notes and microscopical features of rock specimens from Labrador. See WALLACE, DILLON. The long Labrador Trail (995). — **796.**

RICHARDSON, JAMES. List of plants collected on the island of Anticosti and coast of Labrador in 1860. Canada Bot. Soc. Ann., Vol. 1, p. 58. Kingston 1861—62. — **797.**

—»— Fauna Boreali-Americana. Vol. I—IV. London 1829—1837. — **797 a.**

RICHARDSON, WILLIAM. Journal of William Richardson who visited Labrador in 1771. The Canad. Histor. Rev., N. S., Vol. XVI, p. 54. Toronto 1935. — **798.**

RICHET, ETIENNE. Rapport sur un projet d'expédition au Labrador. Bull. d. l. Soc. Roy. de Géogr. d'Anvers, T. XXII, p. 283. Anvers 1898. — **799.**

RICKARD, T. A. Drift Iron. A fortuitous factor in primitive culture. Geogr. Review, Vol. 24, p. 525. New York 1934. — **800.**

RICKETTS, NOBLE G. and TRASK, PARKER D. The bathymetry and sediments of Davis Strait. The *Marion* Expedition to Davis Strait and Baffin Bay under the direction of the U. S. Coast Guard, 1928. U. S. Treasury Department, Coast Guard Bull. No. 19, Part 1. Scientific Results. Washington 1932. — **800 a.**

RIIS-CARSTENSEN, EIGIL. The *Godthaab* Expedition, 1928. Report on the Expedition Medd. om Grønland, Bd 78, No. 1. København 1931. — **800 b.**

RINK, H. Eskimoiske Eventyr og Sagn. København 1866. — **801.**

—»— Om Eskimoernes Herkomst. Aarb. f. Nord. Oldkynd. og Hist., 1871, p. 269 & 1890, p. 189. Kjøbenhavn 1871, 1890. — **802.**

—»— Tales and traditions of the Eskimo. London 1875. — **803.**

RIVERS. W. H. R. The colour vision of the Eskimo. Proceed. of the Cambridge Philos. Soc., Vol. XI, p. 143. Cambridge 1902. — **804.**

ROBERTSON, SAMUEL. Notes on the Coast of Labrador. Transact. Litt. and Hist. Soc. of Quebec, T. IV, pt. 1, p. 27. Quebec 1843. — **805.**

ROBINSON, CHARLES BUDD. Plant studies on the northern coast of the gulf of St. Lawrence. Torreya, Vol. VII, p. 222. Lancaster. Pa 1907. — 806.

ROBINSON, E. C. In an unknown Land. London 1909. — 807.

ROBINSON, H. Private journal kept on board H. M. S. Favorite on the Newfoundland
station. By Capt. H. Robinson, R. N., 1820. [Mealy Mts.]. Journ. R. Geogr. Soc.,
Vol. IV. London 1843. — 808.

ROCHEMONTEIX, C. DE. Les Jésuites et la Nouvelle France. Paris 1905. — 809.

ROCHEMONTEIX, P. C. DE. see SILVY, A.

ROCHETTE, EDGAR. Notes sur la Côte Nord du Bas Saint-Laurent et le Labrador canadien.
Québec 1926. — 810.

ROE, F. G. The Hudson's Bay Company and the Indians. The Beaver, 1936, pp. 8,
64. New York 1936. — 811.

ROGERS, J. D. A historical geography of the British Colonies. Vol. V, Part IV. Newfoundland. Oxford 1931. — 812.

LA RONCIÈRE, CHARLES DE. Les navigations françaises au quinziéme siècle. Bull. d.
Géogr. historique et descriptive. Paris 1895. — 813.

ROSS, A. H. D. The Forest Resources of the Labrador Peninsula. Canad. Forestry
Journ., 1905, p. 28. Ottawa 1905. — 814.

ROSS, COLIN. Zwischen U.S.A. und dem Pol. Durch Kanada, Neufundland, Labrador
und die Arktis. Leipzig 1934. — 815.

ROTH, J. Ueber das Vorkommen von Labrador. Sitzungsber. Akd. d. Wissensch., Bd.
XXVIII, p. 697. Berlin 1883. — 815 a.

ROTHNEY, GORDON O. The case of Bayne and Brymer. An incident in the early history
of Labrador. Canadian Histor. Rev., Vol. 15, p. 264. Toronto 1934. — 816.

ROUILLARD, E. La Cote Nord du Saint-Laurent et le Labrador canadien. Québec 19c8.
— 817.

—›— Une tribu sauvage au Labrador. Bull. Soc. Géogr. de Québec, p. 54. Québec 1908.
— 818.

ROUILLARD, O. E. Les premiers sauvages au Labrador au temps de Cartier. Soc. Géogr.
de Québec, Vol. 4, p. 314. Québec 1910. — 819.

ROUSSEAU, V. The Liveyeres of Labrador. Chamb. Journ., Vol. 10, p. 588. London 1920.
— 820.

ROY, S. K. Upper Canadian (Beekman-town) drift fossils from Labrador. Field Mus.
Nat. Hist. Publ. 307, Geol. Ser., Vol. 6, No. 2, p. 29. Chicago 1932. — 821.

RUBINSTEIN, E. Beziehungen zwischen dem Klima und dem Pflanzenreich. Meteorolog.
Zeitschr., Bd. 41, p. 15. Wien 1924. — 822.

RUEDEMANN, RUDOLPH. On some Fundamentals of Pre-cambrian Palaeogeography. Proceed. Nat. Acad. Sci, Vol. 5, p. 1. Baltimore 1919. — 822 a.

RUGE, S. Kartographie von Nord-Amerika. Petermanns Geogr. Mitteil., Ergänzh. 106,
p. 54. Gotha 1892. — 823.

RUSSELL, FRANK and HUXLEY, H. M. A comparative study of the physical structure of
the Labrador Eskimos and the New England Indians. Proceed. of the Amer. Ass.
for the Advancement of Sci. 48th meeting, held at Columbus, Ohio, p. 365. 1899.
— 823 a.

RUTHERFORD, H. M. See EWING, MAURICE.

RYERSON, KNOWLES. Health and Happiness in Labrador. Among the Deep Sea Fishers,
Vol. 27, No 1, p. 14. New York 1929. — 824.

SAINT-CYR, DOMINIQUE NAPOLEON. List of plants gathered on the North Shore, from
St. Paul Bay to Ouatchechou, and in the islands of Mingan, Anticosti, and Grand
Mécatina, during the summer of 1882 and the month of July, 1885, during the leisure
hours of his two trips to the Lower St. Lawrence and the Gulf. Return (17 B)
to an address of the Legislative Assembly, Dept. of Publ. Instr. April 19th 1886.
Québec 1886. — 825.

—»— Catalogue of plants in the Museum of the Dept. of Publ. Instr. gathered by D. N.
Saint-Cyr, up to 1885, or acquired by exchange or purchase. Bound with the
preceding. — 826.

ST. JOHN, HAROLD. A Botanical Exploration of the North Shore of the Gulf of St. Law-
rence, Including an Annotated List of the Species of Vascular Plants. Canada
Department of Mines, Victoria Memorial Museum, Biological Series, Mem. 126,
No. 4. Ottawa 1922. — 827.

SANDEMOSE, AKSEL. Fortællinger fra Labrador. Kjøbenhavn 1932. — 828.

SAUVAGEST, A. Eskimo et ouralien. Journ. d. l. Soc. d. Américanistes, N. S., Vol. XVI.
Paris 1924 — 829.

SCHENK, ALEXANDRE. Note sur deux cranes d'Esquimaux du Labrador. Bull. Soc. Neuchâteloise de Géogr., T. 11, p. 166. Neuchâtel. 1899. — 829 a.

SCHMIDT, P. WILHELM. Das Eigentum auf den ältesten Stufen der Menschheit. Bd. I.
Münster 1937. — 829 b.

SCHERHAG, R. Die aerologischen Entwicklungsbedingungen einer Labrador-Sturmzyklone.
Ann. d. Hydrogr.. Bd. 65, p. 40. 1937. — 829 b 1.

SCHOOLCRAFT, HENRY R. Information respecting the history, condition and prospects
of the Indian Tribes of the United States... Vol. II. Philadelphia 1852. — 830.

SCHOTT, CHARLES A. Tables of Atmospheric Temperature in the United States. Smithsonian
Contributions No. 277. Washington. — 830 a.

SCHOTT, GERHARD. Die Nebel der Neufundland-Bänke. Ann. d. Hydrogr. u. Marit.
Meteorologie. Hamburg 1897. — 831.

—»— Beiträge zur Hydrographie des St. Lorenz-Golfes. Ann. d. Hydrographie u.
Marit. Meteorologie, Jahrg. XXIV, p. 221, Jahrg. XXV, p. 116. Berlin 1896, 1897.
— 831 a.

SCHOW. Essai sur la Géographie botanique. Bull. de la Soc. de Géographie, Sér. 1, T. VI,
p. 132. Paris 1826. —· 832.

SCHUCHERT, C. and TWENHOFEL, W. H. Ordovicic-Siluric section of the Mingan and
Anticosti islands, Gulf of St. Lawrence. Bull. Geol. Soc. Am., Vol. 21, p. 677.
New York 1910. — 832 a.

SCHRANCK, R. R. VON. Aufzählung einiger Pflanzen aus Labrador, mit Anmerkungen.
Denkschr. d. Königl. Bayr. Bot. Gesellsch., II. Abt. Regensburg 1815. — 832 a1.

SCHUCHERT, CHARLES AND DUNBAR, CARL O. Stratigraphy of Western Newfoundland.
Geol. Soc. of America, Memoir 1. Washington 1934. — 833.

SCHULZE, AD.· Abriss einer Geschichte der Brüder Mission. Gnadau [?] 1901. — 834.

SCHULTZE, ERNST. Einführung zahmer Renntiere nach Labrador. Deutsch. Rundschr.
für Geogr. u. Statist., Jahrg. XXX, p. 399. Wien & Leipzig 1908. — 835.

SCHÜBLER, M. Om den Hvede som Nordmændene i Aaret 1000 fandt vildvoksende i
Viinland. Forhandl. i Vidensk. Selsk. 1858, p. 21. Christiania 1859. — 836.

SCOFIELD, EDNA. Migrations and Changes of Culture on the Labrador Peninsula. Geogr.
Review, Vol. XXX. p. 487.. New York 1940. — 836 a.

SCOTT, J. M. The Land that God gave Cain. An Account of H. G. Watkin's Expedition to Labrador, 1928—1929. London 1933. — **887.**

SCUDDER, SAMUEL HUBBARD. Description of some Labradorian Butterflies. Proc. Soc. Nat. Hist. of Boston, Vol. XVII, p. 294. Boston 1874. — **888.**

—»— A revised list of the butterflies obtained in Labrador by Dr. A. S. Packard. Canadian Entomologist, 1888, p. 148. Orillo 1888. — **889.**

SEARS, FRED C. The Agricultural Possibilities in Labrador. Among the Deep Sea Fishers Vol. 26, No. 4, p. 144. New York 1929. — **840.**

—»— The Photographic Lure of Labrador. Ibid., Vol. 27, No. 1, p. 3. New York 1929. — **841.**

—»— The Labrador Garden Campaign. Ibid., Vol. 30, No. 4, p. 167. New York 1933. — **842.**

—»— Some Interesting Characters of the Labrador. Ibid., Vol. 31, No. 1, p. 6. New York 1933. — **843.**

—»— The 1933 Labrador Garden Campaign. Ibid., Vol. 32, No. 1, p. 8. New York 1934. — **844.**

—»— The Nupuktulik Island Project. Ibid., Vol. 32, No. 4, p. 145. New York 1935. — **845.**

SEARS, P. B. Post-glacial Climate in eastern North America. Ecology, Vol. XIII, p. 1. Brooklyn. N. Y. 1932. — **846.**

SEARS, PAUL B. Types of North American Pollen Profiles. Ecology, Vol. 16, No. 3. Brooklyn 1935. — **846 a.**

—»— Glacial and Post-glacial Vegetation. Bot. Rev., Vol. I, p. 37. Lancaster 1935. — **846 b.**

—»— Climatic Interpretation of Postglacial Pollen Deposits in North America. Bull. Amer. Meteor. Soc., Vol. 19, p. 177. Milton, Mass. 1938. — **846 c.**

SEITZ, D. C. Newfoundland. London 1927. — **847.**

SETTLE, DIONISE. The Second voyage of Master *Martin Frobisher*, made to the West and Northwest Regions, in the yeare 1577, with a description of the Country, and People: Written by Master D. S. Hakluyt, III, p. 32. London 1600. — **848.**

SEWALL, K. W. Blood, taste, digital hair and color of eyes in eastern Eskimo. Amer. Journ. of Phys. Anthropol., Vol. XXV, p. 93. Philadelphia 1939. — **849.**

SEWARD, WILLIAM H. A cruise to Labrador. Log of the schooner Emerald. Albany Evening Journal. Albany 1857. — **850.**

SHAW, E. B. The Newfoundland Forest Fire of August 1935. Monthly Weather Rev., Vol. 64, p. 171. 1936. — **851.**

SHERMAN, JR, JOHN D. The Beetles. In Grenfell's Labrador, 1922, Appendix 1. New York 1922. — **851 a.**

[SILVY, A.?] Relation par lettres de l'Amérique septentrionale 1709—10. — Ed. P. C. de Rochemonteix. Paris 1904. — **852.**

SIMMONS, H. G. Eskimåernas forna och nutida utbredning samt deras vandringsvägar. Ymer, Bd. XXV. Stockholm 1905. — **853.**

SIMPSON, G. C. World Climate during the Quaternary Period. Quart. Journ. Roy. Meteorol. Soc., Vol. LX, p. 425. London 1934. — **854.**

SKINNER, ALANSON. Notes on the Eastern Cree and Northern Saulteaux. Anthropolog. Pap. of the Amer. Mus. of Natural Hist., Vol. IX, Pt 1. New York 1911. — **855.**

SMALLWOOD, J. R. The Book of Newfoundland, Vol. I. St. John's 1937. — **856.**

SMILLEY, WILSON G. Trout Fishing at Northwest River. Among the Deep Sea Fishers, Vol. 26, No. 4, p. 158. New York 1929. — **857.**

SMITH, EDWARD H. Expedition of U. S. Coast Guard cutter »Marion» to the region of Davis Strait in 1928. Science, Vol. LXVII, No. 1768, Nov. 16. Cambridge 1928. — **858.**

—»— Arctic ice; with special reference to its distribution in the North Atlantic Ocean. The *Marion* Expedition to Davis Strait and Baffin Bay under the direction of the U. S. Coast Guard, 1928. U. S. Treasury Department. Coast Guard Bull. No. 19, Part 3. Scientific Results. Washington 1931. — **858 a.**

SMITH, EDWARD H., SOULE, FLOYD M. and MOSBY, OLAV. The Marion and General Greene Expeditions to Davis Strait and Labrador Sea under the direction of the United States Coast Guard 1928—1931—1933—1934—1935. Scientific Results. Part 2. Physical oceanography. U. S. Treasury Department. Coast Guard. Bull. No. 19. Washington 1937. — **8.9.**

SMITH, HARLAM I. Notes on Eskimo Traditions (Labrador Eskimo). Journ. of Amer. Folklore, Vol. VII, p. 209. New York 1894. — **860.**

—»— An Album of Prehistoric Canadian Art. Canadian Depart. of Mines, Anthropology Series No. 8. Bull. No. 37. Ottawa 1923. — **861.**

SMITH, NICHOLAS. Fifty-two years at the Labrador Fishery. London 1936. — **862.**

SMITH, PAUL A. Atlantic submarine valleys of the United States. Geogr. Review, Vol. XXIX, p. 648. New York 1939. — **862 a.**

SMITH, PAUL A. See VEATCH, A. C.

SMITH, WILLIAM. Labrador boundary case. Queen's Quart., Vol. 36, p. 267. Kingston 1929. — **863.**

SNELGROVE, A. K. Mines and Mineral Resources of Newfoundland. Newfoundland Geol. Survey, Information Circular No. 4. St. John's 1938. — **864.**

SNIFFEN, STEWART B. The Travelling Labrador Health Unit. Among the Deep Sea Fishers, Vol. 20, Nos 3 & 4 combined, p. 109. New York 1923. — **865.**

SORNBORGER, J. D. See FERNALD, M. L.

SOULE, FLOYD M. Oceanography. International Ice Observation and Ice Patrol Service in the North Atlantic Ocean. Season 1935. U. S. Coast Guard Bull., No. 25. Washington 1936. — **865 a.**

—«— See SMITH, EDWARD H.

SPECK, FRANK G. The Family Hunting Band as the Basis of Algonkian Social Organisation. Amer. Anthropologist, Vol. 17, p. 289. Menasha 1915. — **866.**

SPECK, FRANK G. and HEYE, GEORGE G. Hunting charms of the Montaignais and the Mistassini. Indian Notes and Monographs by F. W. Hodge, A series of publ. relat. to the Amer. aborigines. Mus. of the Amer. Indian Heye Foundation. New York 1921. — **867.**

SPECK, FRANK G., Mistassini Hunting Territories in the Labrador Peninsula. American Anthropologist, Vol. 25, p. 452. Menasha 1923. — **868.**

—»— Montagnais and Naskapi Tales from the Labrador Peninsula. Journ. Amer. Folk-Lore, Vol. 38, p. 1. New York 1925. — **869.**

—»— Dogs of the Labrador Indians. Nat. Hist., Jan.—Febr. 1925, p. 58. New York 1925. — **870.**

—»— Culture Problems in Northeastern North America. Proc. Amer. Philos. Soc., Vol. 65, No. 4, p. 272. Philadelphia 1926. — **871.**

—»— Family Hunting Territories of the Lake St. John, Montagnais and Neighboring Bands. Anthropos, Bd. 22, p. 387. Mödling 1927. — **872.**

SPECK, FRANK G. Montagnais-Naskapi Bands and Early Eskimo Distribution in the Labrador Peninsula. Amer. Anthropologist, Vol. 33, p. 457. Menasha 1931. — 873.
—»— Collections from Labrador Eskimo. Indian Notes, Vol. I. New York 1924. — 874.
—»— The Montagnais of Labrador. Home Geogr. Monthly, Vol. 2, p. 7. — Worcester, Mass. 1932. — 875.
—»— Labrador Eskimo mask and clown. The General Magz., Univ. of Pennsylv., Vol. XXXVII. Philadelphia 1935. — 875 a.
—»— Ethical attributes of the Labrador Indians. Amer. Anthropologist, Vol. 35, No. 4, p. 559. Menasha 1933. — 876.
—»— Montagnais-Naskapi. The Savage Hunters of the Labrador Peninsula. Oklahoma 1935. — 877.
—»— Inland Eskimo bands of Labrador. In Essays in Anthropology, presented to A. L. Kroeber, p. 313. Berkeley 1936. — 878.
—»— Land Ownership Among Primitive People. Int. Congr. of Americanists 1926. T. XXII, p. 325. — 878 a.
SPECK, FRANK G. and EISELEY, LOREN C. Significance of hunting territory systems of the Algonkian in Social Theory. Amer. Anthropolog., Vol. 41, p. 269. Menasha 1939. — 879.
SPENCER, J. W. W. See HULL, EDWARD.
SPENCER, MILES. Notes on the breeding habits of certain mammals, from personal observations and enquiries from Indians. Rept. Geolog. and Nat. Hist. Survey of Canada for 1887—88, Vol. III, p. 82, Appendix III. Ottawa 1889. — 880.
SPIESS, F. Meteorologische Beobachtungen in Labrador während des 2. Internationalen Polarjahrs 1932—33 und der anschliessenden Jahre. Aus d. Arch. d. Deutschen Seewarte n. d. Marineobservat., Bd. 60, Nr. 8. Hamburg 1940. — 880 a.
STAINER, JOHN. A canoe journey in Canadian Labrador. Geogr. Journ., Vol. 92, p. 153. London 1938. — 881.
STEARNS, WINFRID ALDEN. Notes on the Natural History of Labrador. Proceed. U. S. Nat. Mus., Vol. VI, p. 111. Washington 1883. — 882.
—»— Labrador. A sketch of its peoples, its industries and its natural history. Boston 1884. — 883.
—»— Wrecked in Labrador. Boston 1888. — 884.
—»— Bird Life in Labrador. [also »American Field», Apr. 26, 1890—Oct. 11. 1890]. Amherst 1886. — 885.
STEENSBY, H. P. Om Eskimokulturens Oprindelse. København 1905. — 886.
—»— An anthropological study of the origin of the Eskimo culture. Medd. om Grønland, Vol. LIII. København 1917. — 887.
—»— The Norsemen's Route from Greenland to Wineland. Medd. om Grønland, Bd LVI. København 1917. — 888.
—»— See JENSEN.
STEFANSSON, V. The Three Voyages of Martin Frobisher. I & II. London 1938. — 889.
STEIN, R. Bernh. Hantsch's Notes on Northeastern Labrador. Bull. Amer. Geogr. Soc., Vol. 41, p. 566. New York 1909. — 890.
STEINERT, WALTHER. Die Wirkung des Landschaftszwanges auf die materielle Kultur des Eskimo. Hamburg 1935. — 891.
STEINHAUR, HENRY. Notice relative to the geology of the Labrador coast. Trans. Geolog. Soc., Vol. II, p. 488. London 1814. — 892.

STEPHEN, LESLIE and LEE, SIDNEY. The Dictionary of National Biography founded in 1882 by George Smith. Vol. XI. p. 254. Oxford 1921—22. — **893.**

STEPHENS, CHARLES ASHBURY. Left on Labrador, or the cruise of the schooner-yacht Curlew. Boston 1872. — **894.**

—»— The Knock-about Club along shore. The adventures of a party of young men on a trip from Boston to the land of the midnight sun. Boston 1883. — **895.**

STETSON, H. C. Geology and Paleontology of the Georges Bank Canyon. Bull. Geol. Soc. of America, Vol. 47, p. 339. Washington 1936. — **895 a.**

STEWARD, JULIAN H. The Economic and Social Basis of Primitive Bands. Essays in Anthropology presented to A. L. Kroeber, Univ. of California, p. 331. Berkeley 1936. — **895 b.**

STEVENSON, EDWARD LUTHER. Maps illustrating early Discoveries and Exploration in America. New Brunswick. N. J. 1903. — **896.**

STEWART, NORMAN B. Indian Harbour Items. Among the Deep Sea Fishers, Vol. 8, No. 4, p. 11. Toronto 1911. — **897.**

STEWART, T. D. Anthropometric Observations on the Eskimos and Indians of Labrador. Field Museum of Natural History, Anthropolog. Series, Vol. 31, No. 1. Chicago 1939. — **897 a.**

—»— New measurements on the Eskimos and Indians of Labrador. Abstract No. 15, Proceed. of the Seventh Annual Meeting of the Amer. Ass. of Physical Anthropologists. Amer. Journ. of Phys. Anthropology, Vol. 21, No. 2, and Supplement, p. 10. Washington 1936. — **897 b.**

—»— Change in physical type of the Eskimos of Labrador since the 18th century. Abstr. No. 9, Proceed. of the Ninth Ann. Meet. of the Amer. Ass. of Phys. Anthrop. Ibid., Vol. 23, No. 4, p. 493. 1938. — **897 c.**

STINSON, BEN A. The two outstanding facts in the physical environment of Newfoundland. The Journ. of Geogr. Magaz. for Teachers, Vol. XXIII, p. 37. Chicago 1924. — **898.**

STOCKS, THEODOR und WÜST, GEORG. Morphologie des Atlantischen Ozeans. Erste Lieferung. Die Tiefenverhältnisse des offenen Atlantischen Ozeans. Begleitworte zur Übersichtskarte 1 : 20 mill. Wissenschaftliche Ergebnisse der Deutschen Atlantischen Expedition aus dem Forschungs- und Vermessungsschiffe *Meteor* 1925 —1927. Bd. III, Erste Teil. Berlin 1935. — **898 a.**

STORER, HORATIO ROBINSON. Observations on the fishes of Nova Scotia and Labrador, with description of new species. The Boston Journ. of Natural Hist., Oct. Boston 1850. — **898 b.**

STORM, G. Søfareren Johannes Scolvus og hans reise til Labrador eller Grønland. Norsk Hist. Tidskr., 2. ser., Vol. 5, p. 385. Kristiania 1886. — **899.**

—»— Studies on the Vineland Voyages. Mém. d. l. Soc. Roy. des Antiquaires du Nord, Copenhagen 1884—89. — **900.**

—»— Studier over Vinlandsreisene. Aarb. f. nord. Oldkyndighed. København 1887. — **901.**

—»— EIRIKS SAGA RAUDA og FLATØBOGENS GROENLENDINGADATTR, samt Uddrag af Olafssaga Tryggvasonar. Udg. for Samfund til udg. af gammel nordisk Litteratur ved Dr GUSTAV STORM. Kjøbenhavn 1891. — **902.**

STOSCH, *Oberleutnant zur See*. In den Gewässern von Labrador. Gaea, Natur und Leben, Bd 45, p. 257. Leipzig [Stuttgart] 1909. — **903.**

STOUGHTON, See HODDER.

STREIT, ROBERT, O. M. I. Bibliotheca Missionum. Zweiter Band (1924). Amerikanische Missionslitteratur 1493—1699. Dritter Band (1927). Amerikanische Missionslitteratur 1700—1909. Aachen 1924, 1927. — 904.

STRONG, WILLIAM DUNCAN. Notes on the Mammals of the Labrador Interior. Journ. of Mammalogy, Vol. XI, No. 1. Ann Arbor 1930. — 905.

—»— Under »Anthropological notes and news». American Anthropologist, Vol. XXX, p. 173. Menasha 1928. —· 906.

—»— A Stone Culture from Northern Labrador and its Relation to the Eskimo-like Cultures of the Northeast. The American Anthropologist, N. S., Vol. XXXII, p. 126. Menasha 1930. — 907.

—»— Cross-cousin marriage and the culture of the northeastern Algonkian. Amer. Anthropologist, N. S., Vol. XXXI, p. 277. Menasha 1929. — 908.

STRÖMBÄCK, DAG. The Arna-Magnæan Manuscript 557 4-to containing inter alia the History of the first Discovery of America. With an introduction by Dag Strömbäck. Copenhagen 1940. — 908 a.

SUK, V. On the occurrence of syphilis and tuberculosis amongst Eskimos and mixed breeds of the North Coast of Labrador. A contribution to the question of the extermination of aboriginal races. — Spisy vydávané přirodovědeckou fakultou Masarykovy university. Čislo 84. Brno 1927. — 909.

—»— The occurrence of evanescent congenital spots (the »Mongol Spot») among Eskimo children in N. Labrador. — Anthropologie. Vydává Anthropolický ústav Karlovy university v Praze. Vol. 6 (1928), pp. 28—34. Prague 1928. — 910.

SUPAN, A. Die Temperaturverhältnisse an der Nordküste von Labrador. Petermanns Geogr. Mitteil. Gotha 1889. — 911.

SUTTON, GEORGE MIKSEL. Eskimo year. New York 1934. — 911 a.

SWEETSER, MOSES FOSTER. The Maritime Provinces. A handbook for travellers, a guide to maritime provinces of Canada, also Newfoundland and the Labrador Coast. Boston 1875. — 912.

SÖDERBERG, SVEN. Vinland. Föredrag i Filologiska sällskapet i Lund, maj 1898. Tryckt i Sydsvenska Dagbladet Snällposten N:r 295, 30 okt. 1910. Malmö 1910. — 912 a.

SØRENSEN, THORVALD. Bodenformen und Pflanzendecke in Nordostgrönland. Beiträge zur Theorie der polaren Bodenversetzungen auf Grund von Beobachtungen über deren Einfluss auf die Vegetation in Nordostgrönland. Medd. om Grønland, Bd. 93, No 4. København 1935. — 912 b.

TAIT, R. H. Newfoundland. A Summary of the History and Development of Britain's Oldest Colony from 1497 to 1939. New York 1939. — 913.

TANNER, V. Naturförhållanden på Labrador. Iakttagelser under den finländska expeditionens färder och forskningar år 1937. Societas Scientiarum Fennica, Årsbok XVII B., no. 1. Helsingfors 1938. — 914.

—»— Folk och kulturer på Labrador. Ett bidrag till belysande av konvergensföreteelserna. Societas Scientiarum Fennica, Årsbok XVII B., no. 2. 1939. — 915.

—»— Folkrörelser och kulturväxlingar på Labrador-halvön. De vita pälsjägarnas framryckande på montagnaisindianernas jaktmarker på halvöns sydöstra del. Résumé. Migrations et changements de culture au Labrador de la Terre-Neuve. L'invasion des trappeurs blancs sur les terrains de chasse des Indiens dans le Sud-Est de la presqu'île. Svensk Geografisk Årsbok, 1939, p. 80. Lund 1939. — 916.

TANNER, V. Om de blockrika strandgördlarna (Boulder Barricades) vid subarktiska ocean-kuster. Förekomstsätt och uppkomst. Summary. On the Boulder Barricades at Subarctic Ocean coasts. Terra, Bd. 51, p. 157. Helsingfors 1939. — **917.**

—»— Über die Glaziation und die Peneplain von Labrador. Comptes Rendus du Congr. Int. de Géographie Amsterdam 1938. T. I, p. 197. Leiden 1938. — **918.**

—»— Suggestions regarding colonization in Newfoundland-Labrador based on utilization of the soil for mixed farming. The manuscript dated September 4 th 1939, St. John's, and written at the request of Sir Wilfred Woods, K. C. M. G., Commissioner for Public Utilities, is in the archives of the Department of Public Utilities at St. John's. — **918 a.**

—»— Ett blivande storindustricentrum på Labrador-halvön. Tekn. Fören. i Finland Förhandl., Årg. 60, p. 103. Helsingfors 1940. — **919.**

—»— The Glaciation of the Long Range of Western Newfoundland. A brief contribution. Geolog. Fören. Förhandl., Bd. 62, p. 361. Stockholm 1940. — **919 a.**

—»— Ruinerna på Sculpin Island (Kanayoktot) i Nain's skärgård, Newfoundland-Labrador. Ett förmodat nordboviste från medeltiden. Geografisk Tidskr., Bd. 44, p. 129. København 1941. — **920.**

—»— Finnländische Forschungen in Labrador. Mitteil. d. Naturforschenden Gesellsch. Schaffhausen (Schweiz), Bd. XVII, p. 16. Schaffhausen 1941. — **920 a.**

—»— De gamla nordbornas Helluland, Markland och Vinland. Ett försök till lokali-sering av huvudetapperna i de isländska Vinlandsagorna. Budkaveln, No. 1, 1941. Åbo 1941. — **920 b.**

—»— Folkrörelser och kulturväxlingar på Labradorhalvön. II. Den vita befolkningen med ständigt hemvist på Newfonndland-Labrador: liveyeres och trappers. Svensk Georgr. Årsb., 1942, p. 355. Lund 1942. — **920 c.**

TARDUCCI, F. John and Sebastian Cabot. Detroit 1893. — **921.**

TARR, RALPH, S. Evidence of Glaciation in Labrador and Baffins land. Amer. Geologist, Vol. XIX, p. 191. Minneapolis 1897. — **922.**

—»— Climate in Labrador, Baffin- and Greenland. Amer. Journ. of Sc., IVth ser., Vol. III, p. 315. New Haven 1897. — **928.**

—»— The Arctic Sea Ice as a Geological Agent. Ibid., p. 223. 1897. — **924.**

—»— Rapidity of Weathering and Stream Erosion in the Arctic Latitudes. Amer. Geologist, Vol. XIX, No. 3, p. 131. Minneapolis 1897. — **925.**

TAYLOR, E. G. R. Hudson's Strait and the Oblique Meridian. Imago Mundi, Vol. III, p. 48. London 1939. — **926.**

TEACHOUT, FLOYD S. The Interior of Northern Newfoundland. Among the Deep Sea Fishers, Vol. 27, No. 2, p. 60. New York 1929. — **927.**

THALBITZER, W. Skraelingene i Markland og Grønland, deres Sprog og Nationalitet. Overs. over Det. Kongl. Danske Vidensk. Selsk. Forh., 1905, p. 185. København 1905—06. — **928.**

—»— Grönlandske sagn om Eskimoernes fortid. Stockholm 1913. — **929.**

—»— Four Skraeling words from Markland (Newfoundland) in the Saga of Erik the Red. Bull. 18th Congress of Americanists. London (1912) 1913. — **980.**

—»— Parallels within the culture of the arctic peoples. XX Congresso Internacional de Americanistas, p. 283. Rio de Janeiro 1924. — **981.**

—»— The Cultic deities of the Innuit. Congr. Int. degli Americanisti, Roma, Vol. II, Roma 1926. — **982.**

Thalbitzer, W. Die kultischen Gottheiten der Eskimos. Arch. f. Religionswissensch., Vol. XXVI, Leipzig 1928. — **933.**

—»— Les magiciens esquimaux. Journ. d. l. Soc. d. Américanistes, N. S., Vol. XXII. Paris 1930. — **934.**

Thomas, J. G. Journeys in Labrador. Geographical Journ., Vol. 66, p. 79. London 1925. — **935.**

Thomas, Lowell. Kabluk of the Eskimo. London 1932. — **936.**

Thomas, Olive J. see Whitbeck, Ray H.

Thompson, Bram. Who owns Labrador? Canadian Law Times, Vol. 41, p. 724. Toronto 1921. — **937.**

Thompson, Laura Nels. The Nurse Answers a Call by Komatik. Among the Deep Sea Fishers, Vol. 31, No. 3, p. 121. New York 1933. — **938.**

Thordarson, Matthias. The Vinland voyages. Amer. Geogr. Soc., Research Ser. No. 18. New York 1930. — **939.**

Thorell, Tamerlane. Notice of some spiders from Labrador. Proc. Boston Soc. Nat. Hist., Vol. XVII, p. 490. Boston 1875. — **940.**

Thoulet, J. Sur un mode d'érosion des roches, par l'action combinée de la mer et de la gelée. Note prés. p. M. Berthelot. Comptes rendus hebd. d. séances de l'Acad. d. Sc., T. CIII, p. 1193. Paris 1886. — **941.**

—»— Sur le mode de formation des bancs de Terre-Neuve. Ibid., 1042. — **942.**

Thwaites, Reuben Gold. The Jesuit Relations and Allied Documents. Travels and Explorations of the Jesuit Missionaries in New France 1610—1791. The original French, Latin, and Italian texts, with English translations and notes; illustrated by portraits, maps and facsimiles. Vols. I—LXVIII. Cleveland 1896—1901. — **943.**

Tolman, C. See Gill, J. E.

Toop, John R. An Incident at Northwest River Recalled. Among the Deep Sea Fishers, Vol. 31, No. 3, p. 126. New York 1933. — **944.**

Torfaeus, Tormod. Historia Vinlandiæ Antiquæ. Havniæ 1705. — **945.**

—»— Ancient Wineland. Translation by Chas. G. Herberman and John G. Shea. New York 1891. — **946.**

Townsend, Charles W. Along the Labrador Coast. Boston 1907. — **947.**

—»— A Labrador Spring. Boston 1910. — **948.**

—»— Captain Cartwright and his Labrador Journal, edited by Chas. W. Townsend, with an introduction by Dr. Wilfred Grenfell. Vol. I. Labrador, Description and Travels. Vol. II. Frontier and Pioneer Life, Labrador. Vol. III. Hunting. Labrador. London 1911. — **949.**

—»— A Short Trip into the Labrador Peninsula by Way of the Natashquan River. Bull. of the Geogr. Soc. of Philadelphia, Vol. XI, No. 3, p. 38. Philadelphia 1913. — **950.**

—»— Some More Labrador Notes. The Auk, Vol. 30, p. 1. New York 1913. — **951.**

—»— Bird Conservation in Labrador. Canad. Comm. of Conservation. Ottawa 1916.—**952.**

—»— In Audubon's Labrador. The Auk, Vol. 34, p. 133. New York 1917. — **953.**

—»— In Audubon's Labrador. Boston & New York 1918. — **954.**

—»— The Birds. In Grenfell's Labrador. The Country and the People, p. 374. New York 1922. — **955.**

Townsend, C. W. See Cartwright, George.

Townsend, Charles W. and Allen, Glover M. List of the Birds of Labrador. In Grenfell's Labrador. The Country and the People, p. 495. New York 1922. — **956.**

TOWNSEND, CHARLES W. and ALLEN, GLOVER M. Birds of Labrador. Proceed. Boston Soc. Nat. Hist., Vol. XXXIII, p. 277. Boston 1907. — **957.**

—»— Ibid. Boston & New York 1918. — **958.**

TRAP, F. H. The Cartography of Greenland. In »Greenland», published by the Commission for the direction of the geological and geographical investigations in Greenland, Vol. I, p. 136. Copenhagen 1928. — **959.**

TRASK, PARKER D. See RICKETTS, NOBLE G.

TROBE, [BENJAMIN] DE LA. Extract of two Meteorological Journals of the Weather, observed at Nain, in 57° North Latitude, and at Okak, in 57°30′, North Latitude, both on the Coast of Labradore. Communicated by Mr. De la Trobe to the President, and by him to the Society. Philosoph. Transactions of the Roy. Soc. London, Vol. LXIX, Part I, p. 657. London 1779. — **960.**

—»— cf. Latrobe.

TUCKER, EPHRAIM W. Five months in Labrador and Newfoundland during the summer of 1838. Concord. 1839. — **961.**

TURNER, LUCIEN M. List of the birds of Labrador, including Ungava, East Main, Moose and Gulf districts of the Hudson's Bay Company, together with the island of Anticosti. Proc. U. S. National Mus., Vol. VIII, p. 233. Washington 1885. — **962.**

—»— On the Indians and Eskimos of the Ungava District, Labrador. Transact. of the Roy. Soc. of Canada, T. V, sect. 2, p. 99. Montreal 1887. — **963.**

—»— The Physical and Zoological Character of the Ungava District, Labrador. Transact. of the Roy. Soc. Canad. Proceed., Vol. 5, Sect. IV, p. 79. Montreal 1887. — **964.**

—»— Ethnology of the Ungava-district, Hudson Bay Territory. Smithsonian Institution, Bureau of Amer. Ethnology, Eleventh Ann. Rep.. 1889—90, p. 159. Washington 1894. — **965.**

TWENHOFEL, WILLIAM H. Physiography of Newfoundland. Amer. Journ. of Science. Fourth Ser., Vol. XXXIII, p. 1. New Haven 1912. — **966.**

—»— Geology of the Mingan Islands. Bull. Geol. Soc. Am., Vol. 37, p. 535. New York 1926. — **967.**

—»— Geology of Anticosti Island. Geol. Surv. Canada, Mem. 154. Ottawa 1928. — **967 a.**

—»— See Schuchert, C.

TWENHOFEL, W. H. and CONINE, W. H. Postglacial Terraces of Anticosti Island. Amer. Journ. of Sc., 5th Ser., Vol. I, p. 268. New Haven 1921. — **967 b.**

TWOMEY, ARTHUR C. Ungava Expedition. The Beaver, Outfit 270, p. 44. Winnipeg 1939. — **968.**

—»— Walrus off the Sleepers. Ibid., Outf. 269, p. 6. 1939. — **969.**

TYRREL, J. BURR. Is the Land around Hudson Bay at present rising? Amer. Journ. of Sc., IVth ser., Vol. II, p. 200. New Haven 1896. — **970.**

UEBE, RICHARD. Labrador. Eine physiographische und kulturgeographische Skizze. Halle a. S. 1909. — **971.**

UHLENBECK, C. C. Uralische Anklänge in den Eskimosprachen. Zeitschr. d. deutschen morgenländ. Gesellsch., Vol. LIX. Berlin 1905. — **972.**

UHLIG, JOHANNES. Untersuchung einiger Gesteine aus dem nordöstlichsten Labrador. Mitt. d. Ver. f. Erdk. Dresden, H. 8, p. 230. Dresden 1909. — **973.**

UMFREVILLE, EDUARD. Über den gegenwärtigen Zustand der Hudsonbay, der dortigen Etablissements und ihres Handels. Herausgg. von E. A. W. Zimmermann. Helmstädt 1791. — **974.**

UNGAVA. Magazine of Labrador. Publ. in Cartwright [Labrador]. **974 a.**

UNITED STATES HYDROGRAPHIC OFFICE. Newfoundland and Labrador Pilot. Advance chapters of Hydrographic Office. Publication 73. Washington 1917. — **975.**

U. S. COAST GUARD. International Ice Observations and Ice Patrol Service in the North Atlantic Ocean, Season of 1926. U. S. Coast Guard Bull. No. 15. Washington 1927. — **976.**

U. S. HYDROGRAPHIC OFFICE. Newfoundland and the Labrador coast. Pilot guide. 3rd. ed. Washington 1909. — **977.**

UPHAM, WARREN. The Fishing Banks between Cape Cod and Newfoundland. Amer. Journ. of Sci., 3rd Ser., Vol. 47, p. 123. New Haven 1894. — **977 a.**

VAILLANCOURT, E. Au Labrador canadien. Bull. Soc. Géogr. Québec, Vol. 13, p. 287. Québec 1929. — **978.**

VASSEUR, N. LE. Au Labrador par la rivière Natashquan. Bull. Soc. Géogr. de Québec, T. 7, p. 276. Québec 1913. — **979.**

VEATCH, A. C. and SMITH, PAUL A. Atlantic Submarine Valleys of the United States and the Congo Submarine Valley. Geol. Soc. of America, Spec. Paper No. 7. Washington 1939. — **979 a.**

VERRILL, ADDISON EMERY. List of plants collected at Anticosti and the Mingan Islands during the summer of 1861. Proceed. Boston Soc. Nat. Hist., Vol. IX, p. 146. Boston 1862. — **980.**

—»— Occurrence of fossiliferous Tertiary rocks on the Grand Bank and George Bank. Amer. Journ. of Sc., 3rd Ser., Vol. 16, p. 323. New Haven 1878. — **980 a.**

VIERKANDT, A. Die Erforschung des Inneren der Halbinsel Labrador. Globus, Bd. 69, p. 24. Berlin 1896. — **981.**

VINE, A. C. See EWING, MAURICE.

VIRCHOW, R. Eskimos aus Labrador. Verhdl. d. Berliner Gesellsch. f. Anthropologie, Ethnologie und Urgeschichte, 1880, p. 253. Also in Zeitschr. f. Ethnologie, Bd. XII, p. 253. Berlin 1880. — **982.**

VOGELSANG, H. Sur le Labradorite coloré de la côte de Labrador. Verdl. de Geolog. Reichsanst.. 1868, p. 107. Wien 1869. — **983.**

VOORHIS, E. Historic forts and trading posts of the French régime and of the English fur-trading companies. Ottawa 1930. — **984.**

WAGHORNE, ARTHUR C. A Summary Account of the Wild Berries and other Edible Fruits of Newfoundland and Labrador. New Harbour (Nfld) 1888. — **985.**

—»— The flora of Newfoundland, Labrador, and Saint-Pierre and Miquelon. I—III. Proc. and Transact. Nova Scotian Instit. of Sc. Halifax 1895, 1896, 1898. — **986.**

WAKEFIELD, A. W. Child Life in Labrador. Among the Deep Sea Fishers, Vol. 9, No 1, p. 22. Toronto 1911. — **987.**

—»— A Winter's Work at Mud Lake, Hamilton Inlet. Among the Deep Sea Fishers, Vol. 11, No. 3, p. 7. Toronto 1913. — **988.**

WALDMANN, S. Les Esquimaux du Nord du Labrador. Bull. d. l. Soc. Neuchateloise de Géographie, T. XX, p. 430. Neuchatel 1910. — **989.**

WALDO, FULLERTON LEONARD. With Grenfell on the Labrador. New York 1920. — **990.**

—»— Grenfell: Knight errant of the North. Philadelphia 1924. — **991.**

WALLACE, DILLON. Hubbard's terrible death in Labrador. Letter from D. Wallace tells of starvation. New York Times, March 24, 1904. New York 1904. — **992.**

WALLACE, DILLON. Hunters of the Labrador. Their Life, Their Ken of the Outside World, and their Hardships. Field and Stream, July 1905. New York 1905. — **993.**

—»— The Lure of the Labrador Wild. The Story of the Exploring Expedition conducted by Leonidas Hubbard Jr. London 1905. — **994.**

—»— The long Labrador Trail. Toronto 1907. — **995.**

WALLACE, FREDERICK WILLIAM. Life on the Grand Banks. An account of the Sailor-Fishermen who Harvest the Shoal Waters of North American Eastern Coasts. Nat. Geogr. Magz., Vol. XL, No. 1, p. 1. Washington 1921. — **996.**

WALLACE, W. S. Labrador. Historical Introduction. In Greenfell's Labrador. The Country and the People. New York 1922. — **997.**

WALLACE, W. S. See McLEAN, JOHN.

WARD, ROBERT DE C. See BROOKS, CHARLES, F.

—»— A Cruise with the International Ice Patrol. Geogr. Review, Vol. 14, p. 50. New York 1924. — **998.**

WARMOW. Journal. Missionsblatt aus der Brüdergemeinde. Gnadau 1859. — **999.**

WASSON, DAVID A. Ice and Esquimaux. Atlantic Monthly, Vol. XIV, p. 728; XV, pp. 39, 201, 437, 564. Boston 1864, 1865. — **999 a.**

WATKINS, H. G. River Exploration in Labrador by Canoe and Dog Sledge. Geogr. Journ., Vol. LXXV, p. 97. London 1930. — **1000.**

WATT, J. S. C. Labrador Year. The Beaver, June 1937, p. 20. Winnipeg 1937. — **1001.**

WAUGH, T. W. Notes 1921 concerning the Barren Ground-Davis Inlet Naskaupis. Manuscript in the National Mus. of Canada. — **1002.**

—»— Naskapi Indians of Labrador and their neighbors. Wom. Can. Hist. Soc. Ott., p. 126. Ottawa 1925. — **1003.**

WAYFARER, THE. The Wayfarer in Labrador. Cambridge 1934. — **1004.**

WAYLING, THOMAS. Eskimo Exodus. Canad. Geogr. Journ., Vol. XIII, p. 518. Montreal 1936—37. — **1005.**

WAYMOUTH, GEORGE. *The Voyage of Captaine* George Waymouth, with two Fly-*boates*, *one of 70, th'other of 60 Tonnes; 35 men, victualied for 18 monethes; set forth by the* Moscovia and Turkie. See Voyages of Captain Luke Fox . . . ed. Miller Christy, Hakluyt Soc., p. 80. London 1894. — **1006.**

WEARE, G. E. Cabot's Discovery of North America. London 1897. — **1007.**

WEEKES, MARY. How to Build a Model Cree Tipi. The Beaver, Outfit 269, p. 19. Winnipeg 1938. — **1008.**

—»— [CORNWALLIS KING, W.] Caribou Hunt. The Beaver, Outfit 269, p. 45. Winnipeg 1939. — **1009.**

WEISE, A. J. Discoveries of America to 1525. London 1884. — **1010.**

WEIZ, SAMUEL & PACKARD, A. S. Map of Labrador, compiled by J. Leuthner, from British Admiralty maps, and an unpublished Moravian map (prepared by Rev. Samuel Weiz). Bull. Amer. Geogr. Soc., No. 4. New York 1887. — **1011.**

WEIZ, SAMUEL. List of Vertebrates Observed at Okak, Labrador, with Annotations by A. S. Packard, Jr. Proceed. Nat. Hist. Soc. Boston, Vol. X, p. 264. **Boston** 1864—66. — **1012.**

WENNER, CARL-GÖSTA. En kultur mot undergång. Labradors eskimåer av i dag. Jorden Runt. 1940, p. 213. Stockholm 1940. — **1012 a.**

WERLAUFF, E. CHR. Symbolæ ad geographiam medii ævi, ex monumentis islandicis. Havniæ 1821. — **1013.**

WESTMAN, GEORGIUS A. Itinera priscorum Scandianorum in Americam. Dissertatione graduali ... Aboe 1757. — **1014.**

WESTON, T. C. Reminiscences among the rocks in connection with the Geological Survey of Canada. Toronto 1899. — **1015.**

WETMORE, R. H. Plants of Labrador. Rhodora, Vol. XXV, p. 4. Boston 1923. — **1016.**

WEYER, EDWARD MOFFAT, JR. The Eskimo, their environment and folkways. New Haven & London 1932. — **1017.**

WHEELER, 2ND, E. P. Journeys about Nain. Winter Hunting with the Labrador Eskimo. The Geogr. Review, Vol. XX, No. 3, p. 454. New York 1930. — **1018.**

—»— A Study of some Diabase Dikes on the Labrador Coast. Journ. of Geology, Vol. XLI, No. 4, p. 418. Chicago 1933. — **1019.**

—»— An Amazonite Aplite Dike from Labrador. American Mineralogist, Vol. 20, No. 1, p. 44. Lancaster, Pa. 1935. — **1020.**

—»— The Nain-Okak section of Labrador. The Geogr. Review, Vol. XXV, Nos. 2, p. 240 New York 1935. — **1021.**

—»— Topographical Notes on a Journey across Labrador. Ibid., Vol. XXVIII, No. 3. p. 475. 1938. — **1022.**

WHITBECK, RAY H. and THOMAS, OLIVE J. The Geographic Factor. New York 1932. — **1023.**

WHITBOURNE, RICHARD. A Discourse and Discovery of the New-found-land. London 1622. — **1024.**

WHITE, JAMES. Labrador boundary. University Mag., Vol. 8, p. 215. 1909. — **1025.**

—»— Boundary Disputes and Treaties. Canada and Its Provinces. Vol. VIII, p. 751. Toronto 1913. — **1026.**

—»— Forts and Trading Posts in Labrador Peninsula and adjoining Territory. Ottawa 1926. — **1027.**

WHITELEY, GEORGE G. The New World's Oldest Industry. The Beaver, Outfit 266, No. 3, p. 8. Winnipeg 1935. — **1028.**

WHITNEY, CASPAR. The Leonidas Hubbard, Jun. Expedition to Labrador. Outing. The Outdoor Magazine of Human Interest, Vol. XLV, p. 643. New York 1905. — **1029.**

WICHMAN, ARTHUR. Ueber Gesteine von Labrador. Deutsch. Geol. Gesellsch. Zeitschr., Bd. 36, p. 485. Berlin 1884. — **1030.**

WILKINS, D. F. H. Note on the Geology of the Labrador Coast. Canadian Naturalist, N. S., Vol. VIII. Ottawa 1878. — **1030 a.**

WILLIAMS, GEORGE D. The Birds of Labrador. Among the Deep Sea Fishers, Vol. 12, No. 1, p. 33. New York 1914. — **1031.**

—»— A Vacation on the Labrador. Ibid., Vol. 23, No. 4, p. 169. New York 1926. — **1032.**

—»— A Crisis in the Labrador Fishery. Ibid., Vol. 25, No. 4, p. 158. New York 1928. — **1033.**

WILLIAMS, G. H. Porphyritic Diabase or Diabase Porphyrite from Nachvak, and Hornblendic Pyroxenite from near Skynner's Cove, Nachvak, Labrador. Ann. Rep., Geol. Surv. Canada, Vol. V, Part I, 1890—91, Part F, Appendix I, Nos. 38 and 43. Ottawa. — **1033 a.**

WILLIAMS, M. Y. Palaeozoic geology of Mattagami and Abitibi rivers. Ont. Bur. Mines Rept., Vol. 29, Part 2. Toronto 1920. — **1033 b.**

WILLIAMSON, J. A. The Cabots and the discovery of North America. London 1929. — **1034.**

—»— The Voyages of John and Sebastian Cabot. Histor. Assoc. Pamphlet No. 106. London 1937. — **1035.**

WILSON, A. W. G. The Laurentian peneplain. Journ. of Geology, Vol. 2, p. 615. Chicago 1903. — **1036.**

WILLSON, BECKLES. The Great Company 1667—1871. Toronto 1899. London 1900. —**1037.**

WILSON, M. E. [Review of the Canadian shield. In 'Geologie der Erde'. In the press.] Quoted by SNELGROVE in **864.** — **1038.**

WILTSIE, JAMES W. Another Problem Solved. Among the Deep Sea Fishers, Vol. 10, No. 4, p. 13. Toronto 1913. — **1039.**

WINSOR, JUSTIN. Narrative and critical history of America. Vol. I. Aboriginal America. Boston and New York 1889. — **1040.**

WINTEMBERG, W. J. Preliminary Report on Field Work in 1927. Ann. Rep. for 1927. Bull. Nat. Mus. Canada, No. 56. Ottawa 1929. — **1041.**

—»— Preliminary Report on Field Work in 1928. Ann. Rep. for 1928. Ibid. Mus. Canada, No 62. Ottawa 1929. — **1042.**

—»— A Stone Culture from Northern Labrador and its Relation to the Eskimo-like Cultures of the Northeast. Geogr. Review, Vol. 20, p. 673. New York 1930. — **1043.**

—»— Shell-beads of the Beothuk Indians. Transact. Roy. Soc. Canada, Ser. 3rd, Vol. XXX, Sect II, p. 23. Ottawa 1936. — **1044.**

WINTEMBERG, in JENNESS: The Problem of the Eskimo. The American aborigines. Toronto 1933. — **1045.**

WINTER, HEINRICH. Das falsche Labrador und der Schiefe Meridian. Forschungen u. Fortschritte, Jahrg. 12, No. 9, p. 117. Berlin 1936. — **1046.**

—»— The Pseudo-Labrador and the Oblique Meridian. Facsimiles. Imago Mundi, Bd. 2, p. 61. Bath, London 1937. — **1047.**

WISSLER, CLARK. The American Indian. New York 1922. — **1048.**

—»— The relation of man to nature in aboriginal America. New York 1926. — **1049.**

—»— The effect of civilization upon the length of life of the American Indian. The Scientific Monthly, T. XLIII, p. 5. New York 1936. — **1050.**

WOOD, RUTH KEDZIE. The Tourist's Maritime provinces, with chapters on the Gaspé shore, Newfoundland and Labrador and the Miquelon Islands. New York 1915. — **1051.**

WOOD, W. Animal Sanctuaries in Labrador. Canada Comm. of Conservation. Ottawa 1911. — **1052.**

—»— Supplement to an address on animal sanctuaries in Labrador. Ottawa 1912.— **1052 a.**

—»— Draft of a plan for beginning animal sanctuaries in Labrador. Ottawa 1913. — **1052 b.**

WOODWORTH, R. H. Interesting Plants of Northern Labrador. Rhodora, Vol. XXIX, p. 54. Boston 1927. — **1053.**

—»— Notes on the Torngat Region of Northern Labrador. Geogr. Review, Vol. XVII, p. 632. New York 1927. — **1054.**

WOOLLARD, G. P. See EWING, MAURICE.

WORMSKIOLD, M. Gammelt og Nyt om Grønlands, Viinlands og nogle flere af Forfædrene kiendte Landes formeentlige Beliggende. Det skandinav. Litter. Selsk. Skr. 1814, p. 387. Kjøbenhavn 1814. — **1054 a.**

WRIGHT, G. FREDERIK. Observation upon the Glacial Phenomena of Newfoundland, Labrador and Southern Greenland. Amer. Journ. of Sc., 3rd Ser., Vol. XLIX, p. 86. New Haven 1895. — **1055.**

—»— Ibid., Abstr. American Geologist. Vol. 15, p. 198. (1895). — **1056.**

—»— Ibid., Abstr. Science, N. S., Vol. 1, p. 60. (1895). — **1057.**

WYATT, A. G. N. Survey on the Labrador Coast. Geogr. Review., Vol. LXXXI, p. 59. New York 1933. — **1058.**

—»— Surveying cruises of H. M. S. »Challenger» off the coast of Labrador in 1932 and 1933. The Hydrographic Review, Vol. XI, p. 70. Cannes 1934. — **1059.**

—»— Surveying cruises of H. M. S. CHALLENGER off the coast of Labrador in 1932 and 1933. Geogr. Journ., Vol. LXXXIV, p. 33. London 1934. — **1060.**

WYMAN, J. On boulder accumulations on the coast of Labrador. Boston Soc. Nat. Hist, Proceed. 3, p. 182. Boston 1850. — **1061.**

WÜST, GEORG. See STOCKS, THEODOR.

YOUNG, ARMINUS. A Methodist Missionary in Labrador. Toronto 1916. — **1062.**

—»— One hundred years of mission work in the wilds of Labrador. London 1931. — **1063.**

ZECHLIN, EGMONT. Die angebliche deutsche Entdeckung Amerikas. Vergangenh. u. Gegenwart 1935, nr 7/8. Leipzig 1935. — **1064.**

—»— Zur Frage einer deutsch-dänisch-portugiesischer Vorentdeckung Amerikas. Forschungen und Fortschritte, No. 14, Berlin 1935. — **1065.**

—»— Das Problem der vorkolumbischen Entdeckung Amerikas und die Kolumbusforschung. Histor. Zeitschr., Bd. CLII, p. 1. München 1935. — **1066.**

ZIMMERMANN, E. A. W. See UMFREVILLE, EDUARD.

ANONYMOUS. Relation De Ce Qui S'Est Passé De Plus Remarquable Aux Missions Des Pères de la Compagnie de Iesus En La Nouvelle France és années 1661 & 1662. Paris MDCLXIII. — **1067.**

—»— Ibid. 1668 & 1669. Paris MDCLXX. — **1068.**

—»— Mémoire touchant la découverte, les établissemens, et les possessions de l'Isle de Terre-Neuve pour les sujets de Labour ... Mars 1710. See Leland No. 20088. — **1069.**

—»— Mémoire pour les mers et les costes du Nord Hudson Bay, Labrador, Acadia and the St. Lawrence. See Leland No. 848. — **1070.**

—»— Copy of journal by Louis Jolliet of his exploration of Labrador and the Eskimo country 1694, Apr. 28—Aug. 1. See Leland No. 9275. — **1071.**

—»— Bernou. Series of sketch maps, including Labrador ... See Leland No. 1017. — **1072.**

—»— Documents relating to Hudson Bay and Labrador 1656—1697, 1713—1744. See Leland No. 9284. — **1073.**

—»— A Description of the Coast, Tides and Currents, in Button's Bay and in the Welcome: ... Shewing a Probability, that there is a Passage from thence to the Western Ocean of America. London [1745?]. — **1074.**

—»— Articles of Agreement. For carrying on an Expedition by Hudson's Streights, for the Discovery of a North West Passage to the Western and Southern Ocean of America. London 1745. — **1075.**

—»— Histoire générale de voïages, ... enrichi de cartes géographiques, Nouvellement composées sur les Observations les plus authentiques ... Tome XIV. Paris MDCCLVII. — **1076.**

ANONYMOUS. A brief account established among the Esquimaux, on the coast of Labrador. London 1774. — **1076 a.**

—»— The North American Pilot for Newfoundland and Labrador. Vol. I. London 1806. — **1077.**

—»— Moravian explorations in northern Labrador. London 1814. — **1078.**

—»— Voyage en Amérique: Terre du Labrador. Eyriès: Abrégé des voyages modernes, Vol. 8, p. 199. 1823. — **1078 a.**

—»— Missions in Labrador from their commencement to the present time. 2nd Ed. Dublin 1832. — **1078 b.**

—»— Religious Tract Society. Dangers on the ice off the coast of Labrador with some particulars concerning the natives. London 1830 (?) — **1079.**

—»— Missions in Labrador from their Commencement to the present time. Dublin 1831. — **1080.**

—»— [LEIGH, J.] The Moravians in Labrador. Edinburgh 1833. — **1081.**

—»— Eskimos zu Nain in Labrador. Journ. für die neuesten Land- und Seereisen, Vol. LXXXVIII, p. 373. Berlin 1838. — **1082.**

—»— Report of the Special Commissioners appointed by the Government of Canada to investigate Indian affairs. Appendix to Journals of the Legislative Assembly of the Province of Canada. 16th volume. Ottawa 1858. — **1082 a.**

—»— Sketches of Newfoundland and Labrador. Printed and Published by S. H. COWELL, Ipswich [about 1860]. — **1083.**

—»— Retour d'une expédition canadienne au Labrador. Nouv. Annales des Voyages, 1861, p. 375. Paris 1886. — **1083 a.**

—»— Aus den Aufzeichnungen eines Kabeljaufishers in Labrador. Globus, Bd. II, pp. 281, 314. Hildburghausen 1862. — **1083 b.**

—»— Labrador School Returns; 1871—74. St. John's 1872—75. — **1083 c.**

—»— Labrador. Bull. Soc. de Géogr. de Genève, T. I, p. 113. Genève 1860. — **1084.**

—»— Die Robbenschlägerei in Labrador. Ausland, Bd. XXXIV, p. 1171. Augsburg 1861. — **1085.**

—»— Observations météorologiques au Labrador. Bull. Soc. de Géographie de Genève, T. II, p. 163. Genève 1861. — **1086.**

—»— Report of the Labrador Mission, for 1865, 1866, 1867, pp. 1—46. — **1087.**

—»— Die Missionen der Brüder-Unität in Labrador. Gnadau, Pemsel 1871. — **1088.**

—»— Die Missionen der mährischen Brüder unter den Eskimos in Labrador. Ausland, Bd. XLII, p. 788. Augsburg 1869. — **1089.**

—»— Positionsbestimmungen an der Küste von Labrador. Petermann's Geogr. Mitth., Bd. XV, p. 230. Gotha 1869. — **1090.**

—»— Kurzer Abriss der Geschichte unserer Mission in Labrador. Missionsblatt aus der Brüdergemeinde, April 1871. Gnadau 1871. — **1091.**

—»— Beschreibung der Küste von Labrador von K. St. Charles bis zur Sandwich Bay. Hydrograph. Mitteil., Bd. I, p. 175. Berlin 1873. — **1092.**

—»— The Grand Falls of Labrador. Goldthwaite's Geographical Magazine, 1891, Vol. 1, No. 2, p. 117. New York 1891. — **1093.**

—»— The Name of Labrador. Bull. Amer. Geogr. Soc., Vol. 26, p. 77. New York 1894. — **1094.**

—»— »Livyeres» of Labrador. Miserable Condition of Coast Folk in the Far North. New York Times, 15 Dec., p. 10, col. 3 f. New York 1901. — **1095.**

ANONYMOUS. A Labrador Expedition, Nat. Geogr. Mag., Vol. XV, p. 183. Washington 1904. — **1096.**

—»— Origin and meaning of the word »Labrador». Ibid., Vol. XVII, p. 364. Washington 1906. **1097.**

—»— Introducing reindeer into Labrador. Ibid., Vol. XVIII, p. 686. Washington 1907. — **1098.**

—»— Ethnology of Canada and Newfoundland. Annual archæolog. report 1905, being part of appendix to the report of the Minister of Education, Ontario, p. 71. Toronto 1906. — **1099.**

—»— Plants from Labrador. Kew Bull. of Miscell. Inform., No. 3, p. 76. London 1907. — **1099 a.**

—»— The Charting of Northern Labrador. Bull. Amer. Geogr. Soc., Vol. 41, p. 224. New York 1909. — **1100.**

—»— [GIFFORT PINCHOT?] Vast areas of virgin forests in Newfoundland and Labrador. Journ. of the Soc. of Arts., 8 Apr. 1910. London 1910. — **1101.**

—»— Die Häfen der britischen Kolonie Neufundland. 1. Die Häfen an der Küste Labradors. Annal. d. Hydrogr. u. Marit. Meteorolog., Jahrg. 42, p. 350. Berlin 1914, — **1102.**

—»— The Glaciers of the Torngat Mountains, Labrador. Geogr. Review, Vol. 4, p. 58. New York 1917. — **1103.**

—»— The Labrador Eskimos. Geogr. Review, Vol. 5, p. 232. New York 1918. — **1104.**

—»— Plants collected in Labrador in the summer of 1915. Canad. Geol. Surv. Mem. 124, p. 53. Ottawa 1921. — **1105.**

—»— Relation Between Land and Sea on the Northeastern Coast of Labrador. Geogr. Review, Vol. 11, p. 611. New York 1921. — **1106.**

—»— Reported discovery of placer gold in Labrador. Canad. Min. Journ. Vol. 44, No. 21, p. 396. Toronto 1923. — **1107.**

—»— Historical sketch. Governmental and commercial relations with the Labrador Indians, and ecclesiastical relations. Labrador Appeal, Vol. 6, p. 3655. Cfr **764.** London 1926. — **1108.**

—»— La frontière du Labrador, décision du Comité judiciaire du Conseil privé en cause de la frontière entre le dominion du Canada et la Colonie de Terre-Neuve dans la presqu'île de Labrador. Internat. Intermediary Institute Bull., T. 16, p. 246. Harlem 1927. — **1109.**

—»— Newly Discovered Falls in the Hamilton Basin of Labrador. Geogr. Review, Vol. 17, p. 667. New York 1927. — **1110.**

—»— The Valley River of Labrador. Geogr. Journ., Vol. LXX, p. 289. London 1927. — **1111.**

—»— Lake Melville, True Lake or Inlet of the Sea? Edit. Note. Geogr. Review, Vol. 17, p. 666. New York 1927. — **1112.**

—»— Recent Work on Iceberg Drift and Control in the North Atlantic. Geogr. Review, Vol. 17, p. 678. New York 1927. — **1113.**

—»— Unitas Fratrum, the Moravian Mission of Labrador. The Beaver, Outfit 258, p. 8. Winnipeg 1927. — **1114.**

—»— The Labrador Survey 1932. Geogr. Review, Vol. 23, p. 315. New York 1933 — **1115.**

ANONYMOUS. La famille Foucher de Labrador. Bull. d. rech. histor., Vol. XI, April 1934,
    p. 250. Paris 1934. — **1116.**

—»— Labradorite from Nepoktulegatsuk [Taber's Island]. Rocks and Minerals, Vol. 10,
    No. 10, p. 150. Peekskill, N. Y. 1935. — **1117.**

—»— Voyage au Canada. La mission Jacques Cartier. Préface de G. Hanotaux. Paris
    1935. — **1118.**

—»— Meteorology of a Newfoundland Forest Fire. Geogr. Review, Vol. 27, p. 142.
    New York 1937. — **1119.**

—»— Forty-fifth Anniversary. Among the Deep Sea Fishers, Vol. XXXV, p. 65. New
    York 1937. — **1120.**

—»— »Good Custom Is Better Than Law». Stern Code of Roving Furriers in the Un-
    written Law in the Northern Wilds. The Daily News, October 19, 1938. St. John's,
    1938. — **1121.**

—»— Facts And Figures Concerning the People of Labrador. Ibid. — **1122.**

—»— Isolated Communities—Blanc Sablon and St. Pièrre and Miquelon. Geogr. Review,
    Vol. XXIX, No. 4, p. 668. New York 1939. — **1123.**

—»— Facsimile of part of M.S. found on coast of Labrador. In. Roy. Geogr. Soc. Libr.
    London. — **1124.**

—»— »Assada». La découverte de l'Amérique par les Normands. La Géographic. T. XXIV.
    Paris 1912. — **1125.**

—»— [Fr. T.] Grenzregelung in Labrador. Petermann's Mitt., Bd. 73, p. 353. Gotha 1927.
    — **1126.**

—»— [R. J. S.] History of Missions in Labrador. Dublin 1832. — **1127.**

—»— Inventaire de pièces sur la côte du Labrador. Vol. I. Québec. [Publ. Pierre-George
    Roy?] Not available.

CARTOGRAPHY.

Contemporaneous maps and charts.

Maps:—

The newest and most complete map for an explorer's practical use within the whole area of Newfoundland—Labrador is

Newfoundland—Labrador. Scale 1 : 1.267,200. Draughted by W. B. TITFORD. Published by the Department of Natural Resources, Survey Division, St. John's 1937.

This map, however, only indicates positions: it gives a fairly general view of the coastal trend and the river systems, but the morphography is lacking.

A map which is topographic in the strict sense of the word is available only for the northernmost part of the northern triangular peninsula between the Atlantic and Ungava Bay; it is:

Map of Northernmost Labrador. Four sheets of the scale 1 : 100,000 with contour intervals at 50 meters, and

Map of Northernmost Labrador. Scale 1 : 300,000 and contour intervals at every 250 meters.

Both these maps were constructed for the *American Geographical Society* by O. M. MILLER, assisted by WALTER A. WOOD and CHARLES B. HITCHCOCK, from oblique aerial pictures, taken by the *Forbes-Grenfell Expedition* of 1931 and the *Forbes Expedition* of 1935 and published in 1938 in the American Geographical Society's Special Publication No. 22. Annexed to these is a diagrammatic chart, showing the distribution of control (ground stations, resected air stations, etc.) in relation to areas covered by the sheets mentioned above.

For the interior, especially the northern part of the 'Lake Plateau', a topographical map may probably be expected as soon as the present extensive iron ore prospecting (261) is concluded and the cartographical material obtained has been compiled.

Charts:—

The British and Canadian up-to-date charts available for the different districts are to be found through the *British Admiralty:* Index to Admiralty Charts of Newfoundland and parts of Labrador and the St. Lawrence, Sheets V. and V 2., London 1929 (cf. bibliogr. 680) to which are added the Admiralty Charts published later: Chart No. 251, Cape Creep to Cape Porcupine, 1936, and Chart No. 265, Approaches to Nain and Port Manvers, 1936.

A series of up-to-date charts of Newfoundland—Labrador is also published by the *U.S.A. Hydrographic Office.*

A map has been made from aero-photographs of the outer coast by the *British War Office*, south-east from a line ten miles east of Nain to a point just east of Davis Inlet (275, p. 133) but the writer has not seen it.

A general survey of the coasts of the Labrador Peninsula is given by

British North America. Hudson Bay and Strait. Published by the [British] Admiralty 1884, Edition 3rd June 1927. Large New corrections 23rd Sept. 1932.

North America: Canada, Hudson Bay and Strait, Second Edition, [Washington], May 1930. Published at the Hydrographic Office under the authority of the Secretary of the Navy. No. 5380. Corrected April 17, 22, 36; or

Sailing directions and descriptions of ports and anchorages are to be obtained from the *British Admiralty's* Newfoundland and Labrador Pilot, Ed. 1929 (680, cf. 977); this information is essentially supplemented by FORBES' Navigational Notes on the Labrador Coast (273, 275) with 16 sketch-maps of courses and harbours. The existing lights and foghorns are to be found in (677). Information about the tides can be obtained from the *Canadian Hydrographic and Map Office*'s tables (479). The *U.S. Hydrographic Office*'s monthly Pilot Chart of the North Atlantic Ocean gives reliable information about the ice conditions; (cf. also 204 b).

Older maps and charts.

As regards the older maps and charts of Newfoundland-Labrador, there is every reason to repeat what has already been stated in the preface to the bibliography, that it was scarcely possible to make a nearly complete register of the maps referring to this lonely part of the world by means of my more or less fortuitous visits to some libraries. However, in order to give a general picture of the main development of the cartography, and especially to amplify the notes to be found in the literary sources of the different topics, lists have been made of the older and the more modern maps which the author found in the libraries mentioned (p. 829 sq.) and in the literature; the first list is in alphabetical order, the second in chronological order.

Two atlases of fundamental value give a preparatory orientation for the older chartography:

Atlas to accompany the case of the Colony of Newfoundland in regard to the Labrador Boundaries (London 1926) and

Labrador Boundary Canadian Atlas (Ottawa 1926).

Valuable information of a chartographic nature also concerning Labrador is to be found in the works of BJØRNBO (83, 84), CORTESÃO (178), DIONNE (225), GANONG (293—301), HARRISSE (409—413, 415), KUNSTMANN (541), LARSEN 553), PACKARD (716), STEVENSON (896), TAYLOR (926), TRAP (959), WINTER 1047) and others.

ALPHABETICAL LIST.

ABBE, ERNEST C. Map of Northeastern Labrador compiled from data supplied by the American Geographical Society. Rhodora, No. 448, p. 103. Boston 1936. — **I.**
—»— See American Geographical Society, 1936.
AMERICAN GEOGRAPHICAL SOCIETY. Labrador, showing the boundary award of 1927. Geogr. Rev., Vol. XVII, p. 657. New York 1927. — **I a.**
—»— Map of the Torngat Region of Northern Labrador. 1 : 500,000. June 1931. — **II.**
—»— Sketch map of northeastern Labrador compiled from data supplied by the American Society. With paper by Ernest C. Abbe. Rhodora, Vol. 38, No. 448. Lancaster, Pa. 1936. — **III.**
—»— Map of northernmost Labrador 1 : 100,000. Constructed at the American Geographical Society from the oblique aerial surveys of the Forbes—Grenfell Expedition of 1931 and the Forbes Expedition of 1935. Geogr. Rev., Vol. XXVI, No. 1. New York 1936. — **IV.**
—»— Map of northernmost Labrador [in six sheets]. Constructed at the American Geographical Society from oblique aerial photographs taken by the Forbes—Grenfell expedition of 1931 and the Forbes expedition of 1935. In charge of survey and mapping operations: O. M. Miller. Scale 1 : 100,000. Annexed to Forbes, Northernmost Labrador . . . 1938. — **V.**
—»— Map of Torngat Region compiled from Sir Wilfred Grenfell's sketch maps, Canadian charts and other data 1931. Cfr. **235**, Fig. 17. — **V a.**
—»— Coast line in the vicinity of Cape Harrison. 1931. Cfr. **235**, Fig. 27. — **V b.**
ANDERSON, F. and OTHERS. Canada, East Coast. Lake Melville surveyed by Captain Anderson . . . 1921. Naval Service of Canada. 1 : 146,666. Ottawa 1922. — **VI.**
ANSPACH, C. A. Charte der Insel Newfoundland und eines Teils der Küste von Labrador. See **15**. Weimar 1821. — **VII.**
D'ANVILLE. Amérique Septentrionale 1746. — **VIII.**
ARROWSMITH, A. Chart of Labrador and Greenland including the Northwest Passages

of Hudson, Frobisher. and Davis, by A. Arrowsmith, Hydrographer to His Majesty. London 1809 [additions 1866 in Amer. Geogr. Soc. collection]. — **IX.**

BAKER, E. H. B. Approaches to Nain. Hydrographic Rev., Vol. XIII, No. 1. Monaco 1936. — **X.**

BANNERMAN, HAROLD M. See GILL, JAMES E.

BARTHOLOMEW, J. G. Map showing portion of Labrador claimed by Newfoundland. Annexed to Gosling: Labrador. 1910 (**825**). — **XI.**

—»— Labrador from the most recent surveys. [Hypsography indicated]. Annexed to Gosling: Labrador 1910. — **XII.**

—»— Labrador. Quebec province. In Handy Reference Atlas of the World. Scale about 360 miles = 1 inch. London 1912. **XIII.**

BARTHOLOMEW, JOHN. Labrador Boundary. 1 : 10,000,000. Statesman's Year Book. 1928. — **XIV.**

BEAUVILLIERS. Carte depuis la Rivière des Hommes jusqu'à la petite baye des Rochers, Coste de Labrador 1715. — **XV.**

—»— Carte depuis la petite baye des Rochers jusqu' à la baye St. Louis 1715. — **XVI.**

BELL, ROBERT. Map of the Labrador Peninsula, compiled from the latest expeditions. The Edinburgh Geogr. Instit. 1895. Scott. Geogr. Mag., Vol. XI, p. 335, cf. **56.** — 1895. — **XVII.**

BELLIN, M. Carte de l'Amérique septentrionale Depuis le 28 Degré de Latitude jusqu'au 72. MDCCLV. — **XVIII.**

—»— Suite de la **Carte Reduite** du Golphe de St. Laurent contenant Les Costes de Labrador depuis Mecantina jusqu'à la Baye des Esquimaux . . . MDCCLIII — **XIX.**

BOZARDI, JOHANNES. Estotilandia et Laboratoris terra. Lovanii 1597. — **XX.**

BRITISH ADMIRALTY. Newfoundland and Labrador Pilot. Sixth Edition. London 1929 [Descriptions of courses, harbours and views]. — **XXI.**

[BRITISH] GEOGRAPHICAL SOCIETY, THE ROYAL. The Labrador Peninsula. Scale 1 : 7,500,000. In Hutton, S.K. 1912. — **XXI a.**

BRITISH GOVERNMENT. Atlas to accompany the case of the Colony of Newfoundland in regard to the Labrador Boundary [London 1926]. — **XXII.**

Index of maps to accompany the Case of the colony of Newfoundland [46 maps]

[British Museum Press Marks.]

1	SANSON (N.) ..	..	Canada	.. 1656	B.M. 39. e. 2. (32.)
2	CORONELLI (M. V.) ..		Do.	.. 1689	B.M. 70620. (4.)
3	MORTIER. (P.) ..	..	Do.	.. 1693	B.M. 147. d. 25.
4	L'Isle (G. de)		Amérique Septentrionale	.. 1708	Not in B.M.
5	Do.	..	Do.	[undated]	B.M. 69915. (17.)
6	Do.		Do.	.. 1700	B.M. 37. f. 13. (60.)
7	Do.		Canada	[undated]	B.M. 70615. (12.)
8	Do.		Do.	.. 1703	B.M. 37. e. 13. (61.;
9	SENEX (J.) ..	..	North America ..	.. 1710	B.M. 148. e. 2. (12.)
10	BELLIN (N.) ..	..	Amérique Septentrionale	.. 1743	B.M. 118. e. 17. (1.)
11	Do. ..	..	Nouvelle France ..	.. 1744	B.M. 118. e.17. (8.)
12	Do. ..	..	Amérique Septentrionale	.. 1755	B.M. 143. e. 14.
13	BOWEN (E.) ..	..	North America ..	.. 1763	B.M. 69915. (85.)

14	GIBSON (J.) ..	..	America ..	..	.. [1763]	B.M. 69810. (101.)
15	ROCQUE (J.) ..	..	North America	..	[undated]	B.M. K118. (32.)
16	CARVER (J.) ..	..	Quebec ..	..	.. 1776	B.M. 31. e. 11. (18.)
17	ARROWSMITH (A.)	..	British North America	..	1814	B.M. 45. f. 7. (46.)
18	BRUÉ (A. H.) ..	..	Amérique Septentrionale	..	1815	B.M. 69915. (36.)
19	Do. ..	..	Do	..	1821	B.M. S.T.A. (1.)
20	TANNER (H. S.)	..	North America	..	1822	B.M. 32. e. 17.
21	ARROWSMITH (J.)	..	British North America	..	1834	B.M. 70410. (10.)
22	Do.	..	North America ..	..	1839	B.M. 69917. (16.)
23	BRUÉ (A. H.) ..	..	Amérique Septentrionale	..	1840	B.M. 38. f. 14.
24	ARROWSMITH (J.)	..	British North America	..	1842	B.M. 7040. (11.)
25	KEEFER (T. C.)	..	North America ..	..	1855	B.M. 10470. d. 23.
26	ARROWSMITH (J.)	..	Do. ..	..	1857	B.M. N.R. 15.
27	COLTON (J. H.)	..	Do. ..	..	1862	B.M. 69915. (46.)
28	BERGHAUS (H. C. W.)		World ..	..	.. 1863	B.M. 950. (55.)
29	BRUÉ (A. H.) ..	..	Amérique Septentrionale	..	1863	B.M. 69915. (48.)
30	ANON ..	..	America ..	..	1864	B.M. 69915. (50.)
31	RUSSELL (A.) and MARA (E. A.)	}	Canada ..	..	.. 1871	Not in B.M.
32	ANON ..	..	Do. ..	..	1872	B.M. 70615. (31.)
32A	G. E. DESBARATS	..	Montreal ..	..	1873	
33	BAUR (C. F.) ..	..	America ..	..	1873	B.M. 69810. (83.)
34	JOHNSTON (J.)	..	Canada ..	..	1874	B.M. 70615. (33.)
35	Do.	..	Do. ..	..	1878	B.M. 70619. (11.)
36	JOHNSON (E. V.) and EDMONDS (A. M.)	}	Do. ..	..	1882	Not in B.M.
37	JOHNSTON (W.) and (A. K.)	}	British North America	..	1882	B.M. 69917. (110.)
38	TACHE (J.) and GENEST (F. X.)	}	Canada ..	..	.. 1883	B.M. 70615. (37.)
39	JOHNSTON (J.)	..	Labrador ..	..	1890	B.M. 70912. (1.)
40	STANFORD	..	North America	..	1891	B.M. 69915. (68.)
41	JOHNSON (E. V.) and EDMONDS (A. M.)	}	Canada ..	..	1891	B.M. 70619. (5.)
42	LOW (A. P.) ..	..	Labrador ..	..	1896	B.M. C.S. 16.
43	ANON ..	..	Canada ..	..	1898	B.M. 70615. (44.)
44	STANFORD	..	North America	..	1899	B.M. 69915. (78.)
45	WHITE (J.) ..	..	Canada ..	..	1908	B.M. 31. a. 59. (1.)
46	STANFORD	..	North America	..	1912	B.M. 69915. (91.)

Index of maps to accompany the counter-case of the Colony of Newfoundland. Labrador Nos 49—52.

49	HAMILTON INLET	..	The Narrows, Lake Melville.
50	Do.		Ivucktoke Inlet.
52	WOODY ISLAND.		

BRYANT, HENRY G. Map of Grand Falls. Cf. **105.** 1894. — **XXII a.**

—»— Die Halbinsel Labrador mit den Ergebnissen der Forschungen von Henry G. Bryant 1891. Deutsche Rundschau, Vol. 16, Pl. 10. Wien 1891. — **XXIII.**

—»— St. Augustine River 1912. See PORTER, W.

CABOT, WILLIAM BROOKS. Map of Assiwaban R. and Notakvanon R. 1911. In Cabot: Labrador, 1920. — **XXIV.**

CABOT, WILLIAM B, cf. No. 119. 1912. — **XXIVa.**

CANADIAN ARCHIVES. Labrador, Hudson's Bay and the Arctic sea. [Maps in the British Museum] [In Canada Archives. Publications of the Canadian archives. Index to reports, from 1872 to 1908. 8. Pp. 160—162]. Ottawa 1909. — **XXV.**

CANADIAN DEPARTMENT OF THE INTERIOR, NATIONAL DEVELOPMENT BUREAU. Dominion of Canada. 1 : 2,217,600. Ottawa 1930. — **XXVI.**

CANADIAN GEOLOGICAL SURVEY. Eskimo Tribes of the Labrador Peninsula. In E. W. HAWKES 1916. — **XXVI a.**

CANADIAN GOVERNMENT. Labrador Boundary Canadian Atlas [Ottawa 1926]. — **XXVII.**

Index of maps to accompany the case:

Case maps.

1. Map of Labrador Peninsula. Scale 35 miles to 1 inch. [Ottawa, 1923].

2 a. Map of Eastern Portion of Labrador Peninsula, North-western and North-eastern sheets. Department of Justice, Ottawa, 1923. Scale 12 miles to 1 inch.

2 b. Map of Eastern Portion of Labrador Peninsula. South-western and South-eastern sheets. Department of Justice, Ottawa, 1923. Scale 12 miles to 1 inch.

3. Maps illustrating the Evolution of Labrador:

   A. Part of anonymous Portuguese map of the World. [»King» map], 1502?. After Nordenskiöld.

   B. Part of map of the World. By Salvat de Palestrina. [»Kunstmann Map No 3»], 1504?

   C. Part of map of the World. By Diego Ribero, 1529. After W. Griggs.

   D. Part of map of the World. By Descelière, 1550. After Earl of Crawford.

   E. »Carta da Navegar» of Nicolo and Antonio Zeni [the fictitious »Zeno» map]. Published in Zeno's Commentaries, 1558.

   F. Part of Mercator's »Weltkarte», 1569.

4. Septentrio America. By Jodocus Hondius, Amsterdam. (In Mercator's atlas, Amsterdam, 1630). Scale 660 miles to 1 inch.

5. Amérique Septentrionale. Par N. Sanson. Revuë et changée en plusieurs endroits. Par G. Sanson. Paris, chez Pierre Mariette, 1669. Scale 350 miles to 1 inch.

6. Le Canada ou Partie de la Nouvelle France dans l'Amérique Septentrionale, contenant la Terre de Labrador, la Nouvelle France, etc. Par H. Jaillot, Paris, 1696. Scale 99 miles to 1 inch.

7. A. Coste du Canada depuis Quégasca, Pays des Esquimaux, jusqu'à la Rivière Quesesasquiou, Côte de Labrador. Dessinée sur le Memoire de M. de Courtemanche. Par le Sieur Deshaise, 1704. Scale 4 1/2 leagues to 1 inch.

   B. Nouvelle découverte depuis la Rivière des François qui est à dix lieues au dessous de Monsieur de Courtemanche du coste du Nordest jusqu'à Chechaosquismas et Quessesoikou. Par Pierre Constantin, 1715. Scale 4 leagues to 1 inch.

8. Coste des Eskimaux. By Louis Fornel, 1748. Reproduced from manuscript plan in the Archives du Service Hydrographique de la Marine, Paris. Scale 6 $^1/_2$ leagues to 1 inch.

9. Americæ, Mappa generalis. D.I.M. Hasii......concinnata et delineata ab. Aug. Gottl. Boehmio. In lucem proferentibus Homannianis Heredibus, 1746. [Nurnberg]. Scale 550 miles to 1 inch.

10. Carte des Possessions Angloises & Françoises du Continent de l'Amérique Septentrionale. Par Jean Palairet. Londres, 1763. Scale 110 miles to 1 inch.

11. A Map of the British Colonies in North America with the roads, distances, limits and extent of the Settlements. Humbly inscribed to the......Lords Commissioners for Trade and Plantations. By John Mitchell. London, for Jefferys & Faden [1775]. Scale 44 miles to 1 inch.

12. Map of part of Labrador illustrating the proposed Northern Boundary of the Province of Quebec. To accompany the report of the Deputy Minister of the Interior, dated the 29th January, 1896. [Department of the Interior, Ottawa, 1896]. Scale 41 miles to 1 inch.

13. Canada, East Coast, Lake Melville. Published by the Naval Service of Canada. W. J. Stewart, Chief Hydrographer, Ottawa, March, 1922. Scale 2 geographical miles to 1 inch.

14. Slope in the Hamilton, St. Lawrence and Hudson Rivers. [Dept. of Justice, Ottawa, 1924]. Horizontal scale, 6 $^2/_3$ miles to 1 inch. Vertical scale 1 foot to 1 inch.

15. Maps and Diagrams, Hamilton River, vicinity of Rigolet. (These maps and diagrams have been inserted in Vol. V [cfr 764]. where they form pp. 2323, 2334A, 2342A and 2358B.)

Counter-case maps.

16. Tabula Novæ Franciæ (Père Creuxius). Anno 1660. Scale 125 miles to 1 inch (approx.)

17. [New France between Trois-Rivières and Mingan. Par le Père P. Laure, 1730 (?)].

18. Carte du Domaine en Canada. Dediée a Monseigneur le Dauphin. Par le Père P. Laure, 23 Aoust, 1731.

19. Suite de la Carte Reduite du Golphe de St. Laurent, contenant les costes de Labrador depuis Mecatina jusqu'à la Baye des Esquimaux le Detroit de Belle-Isle et partie des Costes de l'Isle de Terre Neuve connues sous le nom du Petit Nord. Dréssée au Dépôt des Cartes, Plans et Journaux de la Marine. Paris, chez Mr. Bellin, Ingénieur de la Marine, 1743.

20. Seigniories of Terre Ferme de Mingan, Isles et Islets de Mingan and Isle d'Anticosti. [By James White. Toronto, 1925.] Scale 16 miles to 1 inch.

21. Concessions in Labrador and Newfoundland granted by French. [By James White. Toronto, 1925.] Scale 35 miles to 1 inch.

22. Concessions below Anticosti Island granted by French. [By James White. Toronto, 1925.] Scale 13.8 miles to 1 inch.

Supplementary maps.

23. L'Amérique Septentrionale dressée sur les observations de Mrs. de l'Académie Royale des Sciences & quelques autres. Par G. de l'Isle, géographe. Paris, chez l'auteur, 1700. Scale 300 miles to 1 inch.

24. Carte du Canada ou de la Nouvelle France et des Découvertes qui y ont été faites. Dressée sur plusieurs observations. Par Guillaume de l'Isle. Paris, chez l'auteur,

1703. [With lines indicating British and French claims according to d'Auteuil, 1719.] Scale 160 miles to 1 inch.

24 a. Hudson's Bay and Labrador (Treaty of Utrecht). By Samuel Thornton, 1709. [From manuscript map in Hudson's Bay Company's records.] Scale 76 miles to 1 inch.

25. Carte de la Partie Orientale de la Nouvelle France ou du Canada. Par N. Bellin, Ingénieur de la Marine. [Paris.] 1744. Scale 55 miles to 1 inch.

26. Carte de l'Amérique Septentrionale depuis le 28 degré de latitude jusqu'au 72. Par N. Bellin [Paris], 1755. Scale 110 miles to 1 inch.

27. An accurate map of North America describing and distinguishing the British, Spanish and French Dominions on this great Continent, according to the Definitive Treaty concluded at Paris 10th February, 1763. Also all the West India Islands. By Eman. Bowen, geographer and John Gibson, engraver. London [1673]. (Accompanied the Report of the Lords of Trade, 8 June, 1763). Scale 85 miles to 1 inch.

28. A general map of North America. By the late John Rocque, Topographer to His Majesty. — Cartes generales de l'Amérique Septentrionale. London, N. A. Rocque & A. Dury [1763]. Scale 150 miles to 1 inch.

29. Chart of Labrador and Greenland, including the North-West Passages of Hudson, Frobishers and Davis. By A. Arrowsmith, Hydrographer to His Majesty. London Published 1 June, 1809. By A. Arrowsmith. Corrections to 1825.

30. British North America. Dedicated to the Honble. Hudson's Bay Co., By J. Arrowsmith. London, 15 February, 1832. Scale 164 miles to 1 inch.

31. Map of North America. By John Arrowsmith. Papers, Hudson's Bay Co. London, 1850. Scale 185 miles to 1 inch.

32. Map of North America. Select Committee on the Hudson's Bay Company. Drawn by J. Arrowsmith. London, J. Arrowsmith, 1857. Scale 185 miles to 1 inch.

33. Aboriginal map of North America denoting the Boundaries and the Locations of various Indian Tribes. Select Committee on the Hudson's Bay Company. By John Arrowsmith, London, 1857. Scale 185 miles to 1 inch.

34. The Dominion of Canada and Northwest and Hudson's Bay Territories. By A. L. Russell. Copp, Clark & Co., lith., Toronto, 1868. Scale 190 miles to 1 inch.

35. Chart No. 1422. North America. East Coast. Labrador. Hydrographic Office, 1871. Published at the Admiralty, 30 March, 1871. Rear-Admiral G. H. Richards, Hydrographer.

36. The Dominion of Canada. A reduction of the map prepared and issued under the direction of the Minister of Railways and Canals (1882). McCorquodale & Co., London. [1886.] Scale 140 miles to 1 inch. [In Canadian Pictures. By the Marquis of Lorne. 1886.]

37. Map of the Dominion of Canada. 1886. Showing the location of some of the Principal Products, etc. Indian and Colonial Exhibition, London, 1886. Department of Agriculture, April, 1886. Canada Bank Note Co., photo-lith., Montreal, 1886. Scale 90 miles to 1 inch.

38. Map showing Provisional Districts of Canada. Department of the Interior. By John Johnston, geographer. Ottawa [1897]. Accompanies Order in Council No. 3388, 18 December, 1897. Scale 230 miles to 1 inch.

39. Map of the Dominion of Canada, 1900. Department of the Interior. By James White. [Ottawa, 1900.] Preliminary edition. Scale 100 miles to 1 inch.

40. North America. From Map 15 in Indexed Atlas of the World, Historical, Descriptive, Statistical. Vol. 1, United States. Rand, McNally & Co., Chicago, 1902. Scale 235 miles to 1 inch.
41. Dominion of Canada. From Plate 124 in XXth Century Citizen's Atlas of the World. 2nd Edition. By John Bartholomew, Edinburgh [1901]. Scale 215 miles to 1 inch.
42. Sketch map of the Labrador Peninsula to illustrate a paper by Dr. Wilfred T. Grenfell, London, Royal Geographical Society. 1911. [In Geographical Journal, April, 1911.] Scale 118 miles to 1 inch.
43. Map of North Atlantic Ocean. Plate 1, Philip's Mercantile Marine Atlas, 3rd Edition, 1908. By George Philip & Son, London. 1908.
44. Map of Arctic Regions, Plate 21, Philip's Mercantile Marine Atlas, 3rd Edition, 1908. By George Philip & Son, London. Scale 540 miles to 1 inch.
45. Inset map in Scarborough's New Map of Newfoundland. Scarborough Company of Canada, Hamilton, Ont. 1916. Scale 86 miles to 1 inch.
46. Standard Map of Dominion of Canada. From Plates 468—469 in Commercial Atlas of America. Rand, McNally & Co., Chicago, 1922. Scale 130 miles to 1 inch.
47. North-eastern Canada. Plate 83, Times Survey Atlas of the World. By J. G. Bartholomew. The Times, London, 1922. Scale 79 miles to 1 inch.
48. Dominion of Canada with Newfoundland—Commercial Development. Plates 103—104 in Philip's Chamber of Commerce Atlas, 1st edition, 1925. George Philip & Son, London [1925]. Scale 160 miles to 1 inch.
49. Ontario—Manitoba Boundary. [Department of Justice, Ottawa, 1926.] Scale 50 miles to 1 inch.
50. South-eastern Alaska and part of British Columbia, showing Award of Alaska Boundary Tribunal, 20 October, 1903. [Department of Justice, Ottawa, 1926.] Scale 17 miles to 1 inch.
51. Boundary Claims, 1700—1763, in the Labrador Peninsula and in North-eastern Ontario and Manitoba. By James White [Ottawa, 1923]. Scale 70 miles to 1 inch.
52. Coast of Labrador. By John Arrowsmith. Scale 55 miles to 1 inch. [Map No. III in The Colonial Church Atlas. Third Edition. Society for Promoting Christian Knowledge. London, 1853.]

CANADA-LABRADOR BOUNDARY. Reprint from Canada and its Provinces, a History ... Toronto 1914. — **XXVIII.**
CARTWRIGHT, GEORGE. See LANE, MICHAEL.
CHIMMO, W. Chart of the North East Coast of Labrador to accompany the Paper by Commander W. Chimmo. Geogr. Journ, Vol. 68, opp. p. 258. London 1868. — **XXIX.**
COLEMAN, A. P. Northeastern portion of Labrador and New Québec. Geol. Surv. of Canada. Publ. no. 1701. Ottawa 1916. — **XXX.**
—»— Nachvak Fiord, Labrador. Ibid. — **XXXI.**
COLONIAL OFFICE. Labrador, a dependency of Newfoundland. Colonial Office List, 1905, p. 292. London 1925. — **XXXII.**
CONSTANTIN. See COURTEMANCHE.
COURTEMANCHE, M. DE. Coste du Canada depuis Quegasca Pays des Esquèmaux Iusqu à la Rivière Quesesasquiou Côte de Labrador. Dessinée sur le Mémoire de M. de Courtemanche par le sieur Deshaire 1704. — **XXXIII.**
—»— Nouelle découverte depuis la rivierre des francois [Pinwari. R.] qui est à dix lieües

au dessous de Monsieur de Courtemanche du Côsté du Nord'est iusqu'à Chechasos-
quimas et Quessesoikou [Hamilton River], faite par Pierre Constantin En l'année
1715. — **XXXIV.**

CURTIS, ROGER. A chart of part of the Country of Labrador, taken by order of Commodore Shuldham in a Tour up the Coast in the Year 1773. Scale 1 : 500,000.
— **XXXV.**

DALY, REGINALD A. Sketch Map of Nachvak Bay. See **200.** Cambridge 1902. — **XXXV a.**

DECOÜAGNE. Carte du Canada tirée sur un très grand nombre de memoires des plus
recents augmentée et corrigée sur touttes celles qui ont été faittes avant 1711.
Presentée a Monseigneur le Comte de Ponchartrain, commandeur des ordres du Roy,
secretaire et ministre destat par son très humble et très obedient seviteur Decoüagne.
— **XXXVI.**

DEE, JOHN. Atlantidis, vulgariter Indiae Occidentalis nominatae, emendatior descriptio
quam adhuc est divulgata. 1580. See E. G. R. Taylor. — **XXXVI a.**

DELABARRE, EDMOND BURKE. Cf. No. **218.** 1901. — **XXXVII.**

[DE] l'ISLE, G. DE. L'Amérique Septentrionale Dressée sur les Observations de Mrs de
l'Académie Royale des Sciences & quelques autres, & sur les Memoires les plus
recens. Paris 1700. — **XXXVIII.**

DEL' ISLE, GUILLAUME. Carte du Canada ou de la Nouvelle France at des Découvertes
qui y ont été faites. Paris 1703. — **XXXIX.**

DEPARTMENT OF JUSTICE CANADA. Labrador and Adjoining Portion of Quebec. Scale
1 : 760,320. Ottawa 1922. — **XXXIX a.**

DEPARTMENT OF MINES [HAWKES]. Eskimo Tribes of the Labrador Peninsula. Mem. 91.
Ottawa 1916. — **XL.**

DESBARATS, G. E. Canada and part of United States. Montreal 1874. — **XLI.**

DESHAIRE, See COURTEMANCHE.

EATON, D. J. V. Map of Labrador Peninsula. Compiled by D. J. V. Eaton. Sheet Nos
585 So. West: 586 So. East: 587 N. West: 588 N. East. Accompanying Low's Report
Geol. Survey Canada, Pt L, Vol. 8 (N.S.) 1896. — **XLII.**

EATON, D. J. V. [LOW, A. P.]. Map of Labrador Peninsula [geological], 4 Sheets. See Rept.
Part L, Vol. VIII, N.S. 1 : 1,584,000. — **XLIII.**

ELLIS, HENRY. A New Chart of the parts where a North West Passage was sought in the
Years 1746 and 1747 exhibiting the Track of the Ships throughout that Expedition. London 1748. — **XLIV.**

FLAHERTY, Robt. J. Sketch Maps. Ungava Explorations, May 1911 — November 1913, to
accompany Report of Expedition to Hudson Bay. Report of ROBT. J. FLAHERTY.
29 sheets, bound as atlas. Amer. Geogr. Soc. Library, sign. 63— D. — **XLV.**

—»— Two Route Surveys across Ungava Peninsula, Labrador, by ROBERT J. FLAHERTY
(1) Via the Povungnituk and Payne Rivers, westward by canoe, June—July 1912.
(2) Via Lake Minto and Leaf River, Gartword, by dog sledge, March—April 1912.
Geogr. Rev., Vol. VI, No. 2. 1 : 506,880. New York 1918. — **XLVI.**

—»— Sir WM. MACKENZIE'S Expeditions, 1910—1916. Previously unexplored areas of
Hudson Bay and Northern Ungava, compiled from plans and sketches of R. J.
Flaherty, F.R.G.S. Cfr. Geogr. Rev., Vol. VI, No. 2. New York 1918. — **XLVI a.**

FORBES, ALEXANDER. Navigational Notes on the Labrador Coast. Contains 16 charts.
Amer. Geogr. Soc., Special Publ. No. 22. New York 1938. — **XLVII.**

—»— See AMERICAN GEOGRAPHICAL SOCIETY, 1938.

FRISSELL, VARICK. Sketch map . . . to show the results of Mr. Frissell's exploration [Vicinity of Hamilton River, Grand Falls and Grenfell Falls]. Geogr. Journ., Vol. LXIX, p. 338. London 1927. — **XLVIII.**

GENEST, P. M. A. Carte De La Nouvelle France Pour servir à l'Etude de L'Histoire du Canada. Depuis sa Découverte Jusqu'en 1760. — **XLIX.**

GILBERT, JOSEPH. Côte de Labradore depuis le Détroit de Belle Isle Jusqu'au Cap Bluff Levée par Joseph Gilbert en 1767. [Chez le Rouge]. London 1770. — **L.**

—»— A Chart of part of the Coast of . . . from the Straights of Belle Isle, to Cape Bluff. London 1770 — **LI.**

—»— A Chart of part of the Coast of Labradore, from the Straights of Bell Isle, to Cape Bluff. Surveyed by Joseph Gilbert in 1767. In North American Pilot. London 1786. — **LII.**

GILL, JAMES E., BANNERMAN, HAROLD M., TOLMAN, CARL. Geological map of the Wapussakatoo district [Headwaters of Hamilton River]. Bull. Geol. Soc. Amer., Vol. 48, p. 567. New York 1937. — **LIII.**

GRENFELL, W. T. and OTHERS. Map of Labrador, cfr **346**, New York 1910 & 1922. — **LIII a.**

GRENFELL, SIR WILFRED. Sketch maps of the Labrador coast. Cfr. **235**, Figs 15 & 16. — **LIII b.**

L. M. H. Pictorial Grenfell stations 193—? — **LIV.**

HAMMOND [for Dr Wilfred T. Grenfell] Labrador and Newfoundland. In Among the Deep Sea Fishers, p. 95. — **LV.**

HANTSCH, BERNHARD. Nordost-spitze von Labrador. 1 : 250,000. Mitt. d. Vereins f. Erdk. zu Dresden, 1909, p. 241. Dresden 1908. — **LVI.**

[HANTSCH, BERNHARD]. Nordostspitze von Labrador. Scale 1 : 1,250,000. Mitt. d. Vereins für Erdk. zu Dresden, H. 8. Dresden 1909. — **LVII.**

[HARPE, M. DE LA.] Carte de la Baye de Hudson, par M. B., Ingr de la Marine 1757. — In Histoire Générale des Voyages, T XIV, opposite p. 640 (cf **405**) — **LVIII.**

HAVEN, JENS. Stück der Nördlichen Küste von Labrador vom 53—56. Grade aufgenommen durch Jens Haven im August u. Septb. 1765. — V. Bossart fecit. — **LIX.**

HAWKES, E. W. Eskimo Tribes of the Labrador Peninsula. Annexed to Hawkes' The Labrador Eskimo. Canada Geol. Survey. Mem. 91, No. 14, Anthropolog. Series. Ottawa 1916. — **LX.**

HIND, H. Y. Map of Labrador. Shewing the Canoe Route from Seven Islands to Hamilton Inlet. Annexed to Hind's Explorations . . . of the Labrador Peninsula. London 1864. — **LXII.**

—»— Map of the River Moisie and Adjoining Country Shewing the Route followed by the Labrador Expedition. Ibid. 1864. — **LXIII.**

HOLME, RANDLE F. Map of the Peninsula of Labrador, compiled from information supplied by Mr. Randle F. Holme, collated with the most recent Amiralty Charts . . . Roy. Geogr. Soc. Proceed., Vol. X, opp. p. 260. London 1888. — **LXIV.**

HOMANNIANORUM HEREDUM. America Septentrionalis a Domino d'Anville in Galliis edita . . . Noribergae 1756. — **LXIV a.**

HOWLEY, J. P. Geological map of Newfoundland. Published by the Department of Mines and Agriculture. St. John's 1919. — **LXV.**

HUBBARD, JR. MRS. LEONIDAS. Map of Eastern Labrador showing Grand Lake and the courses of the Nascaupee and George Rivers as surveyed and mapped, June 27 to

August 27, 1905. With the Susan and Big Rivers showing the route of Mr Leonidas Hubbard, Jr. in the Summer of 1903. Scale 25 statute miles to an inch. Amer. Geogr. Soc. Bull. 1906. Cfr **458**. — **LXVI**.

HUTTON, S. K. See **XXI a.**

HYDROGRAPHIC OFFICE [U.S.A.]. Pilot Chart of the North Atlantic Ocean, Washington D. C. Issued monthly. — **LXVI a.**

ISELIN, COLUMBUS. Ekortiarsuk Fiord, sketch map. Cfr **235**, Fig. 94. — **LXVII**.

JAILLOT, HUBERT. Partie de la Nouvelle France . . . Paris 1695. — **LXVIII**.

JANNASCH. See LINDER.

JEFFERYS, THOMAS. The North American Pilot for Newfoundland, Labrador. London 1775. — **LXIX**.

JOHNSTON, J. British North America. Labrador Boundary Map. Ottawa 1890. — **LXX**.

KEULEN, JOANNES VAN. Pascaarte vande Norder Zee custen van America vande West-hoeck van Ysland doorde Straet Davis en Hudson, tot aen Terra Neuf. 't Amsterdam 1681. — **LXXI**.

KINDLE, E. M. Sketch Map of waterways connected with Hamilton Inlet . . . Cf. **523**. 1924. — **LXXI a.**

KRANCK, E. H. Geological Sketch Map of the Seaboard of Newfoundland — Labrador. St. John's 1938. — **LXXI b.**

LANE, MICHAEL. A Chart of part of the Coast of Labrador, from Grand Point to Shecatica. London 1770. — **LXXII**.

—»— Partie des Côtes de Labrador Depuis le Cap Charles à la Baye de Sandwich; Levée . . . par M. Lane. Publié à [Chez Le Rouge] Londres en 1777. — **LXXIII**.

—»— Côte de Labrador depuis Grande Pointe à Shecatica. Levée par M. Lane. Chez Le Rouge. 1768. Paris, 1778. — **LXXIV**.

—»— The Island of Newfoundland, laid down from surveys taken By Order of the Right Honorable the Lords Commissioners of Admiralty, by Lieut. Michael Lane. Principal Surveyor of the said Island 1790. Ibid. — **LXXV**.

—»— A Chart of Part of the Coast of Labrador from Cape Charles to Sandwich Bay, Surveyed by Order of the Hon.[ble] Commodore Byron, Governor of Newfoundland, Labrador &c. in the Years 1770 and 1771 by Michael Lane. Annexed to Cartwright's A Journal . . . Newark 1792. — **LXXVI**.

LAURE, FATHER. Map of the Moisie River system and Lake Ashwanipi. Dated Checoutimi, August 23. 1731. Manuscr. in the Canadian Library of Parliament. — **LXXVI a.**

LESLIE, L. A. D. See WATKINS, H. G.

LEUTHNER, F. Map of [Newfoundland] Labrador compiled by F. Leuthner from British Admiralty Maps and an Unpublished Moravian Map. Amer. Geogr. Soc. Bull., Vol. 19, No. 4. New York 1887. — **LXXVII**.

LEVERETT. See TAYLOR.

LIEBER, OSCAR MONTGOMERY. Sketch showing the Geology of the Coast of Labrador. Balto 1860. — **LXXVIII**.

—»— Die Küste von Labrador. Nach den bisherigen Englischen Aufnahmen . . . und handschriftlichen Mitteilungen von Oscar Montgomery Lieber, Geolog der Amerikanischen Expedition . . . 1860 . . . Sheet 9, Petermanns Mitteil, Vol. 7. Gotha 1861. — **LXXIX**.

LINDER, WEIZ und JANNASCH. Küste von Labrador. Masstab. 1 : 1 $1/4$ Million. Entworfen und gezeichnet von Capitain LINKLATER aus England und den Missionaren der Brüderkirche LINDER, WEIZ und JANNASCH in Labrador. Dresden 1896. — L X X X.

LINKLATER. See LINDER.

LOTTER, T. CONRAD. Partie Orientale de la Nouvelle France ou du Canada avec l'Isle de Terre-Neuve et de Nouvelle Escosse, Acadie et Nouv. Angleterre. Augsbourg 1740 (?). — L X X X I.

LOW, A. P. Map of Labrador Peninsula, Three sheets: No. 585, Southwest; 587, Northwest; 588, Northeast. Scale 1 : 1,584,000. Geol. Surv. Canada, Part L. Vol. VIII (N. S.). Ottawa 1896. — L X X X II.

—»— Sketch Map of the northern portion of the Labrador Peninsula from exploratory surveys by A. P. Low. B.Sc. and Eskimo sketches collected in 1899—1900. Scale 50 miles to inch. Geolog. Surv. of Canada. To accompany Part A, Vol. XIV. Ottawa 1901. — L X X X III.

—»— Sketch Map of the northern portion of the Labrador from exploratory surveys by A. P. Low, B. Sc. and Eskimo sketches collected in 1899. Sheet No. 785, Vol. 13. Canada Geol. Survey Report 1900. Ottawa 1902. — L X X X IV.

LOW, A. P. See EATON, D. I. W.

MILLER, O. M. See AMERICAN GEOGRAPHICAL SOCIETY, 1938.

MISSIONARIES OF THE UNITAS FRATRUM. The Northern Extremity of Labrador with Ungava Bay Explored by the Missionaries of the Unitas Fratrum in 1811. Append. in KÖHLMEISTER and KMOCH. — L X X X V.

MITCHELL, W. A Map of the British Colonies in North America . . . London (about 1755). — L X X X VI.

MORTIER, PIERRE. Carte particulière de L'Amérique Septentrionale, où sont Compris les Destroit de Davids, les Destroits de Hudson, &c. Amsterdam 1700. — — L X X X VII.

MOSBY, OLAV. See SMITH, EDWARD.

MURRAY, ALEX. A Diocesan Map of the Church of England in Newfoundland and Labrador, showing the Deaneries and Missions. London 1877. — L X X X VIII.

PACKARD, A. S. Map of Labrador. New York 1891. — L X X X VIII a.

PORTER, RUSSELL W. Map of Saint Augustine River, Province of Quebec, Canadian Labrador, showing explorations of Henry G. Bryant's Expedition, 1912. Bull. Geogr. Soc. Philadelphia, Vol. XI, p. 62. Philadelphia 1913. — L X X X IX.

PRICHARD, H. Map to illustrate the route of Mr H. Hesket Prichard and Mr Gathorne-Hardy in Labrador 1910. In Hesket Prichard 1911. — L X X X IX a.

REICHEL, L. T. Chart of the Coast of Labrador sketched for the use of the Captain of the Harmony by L. T. Reichel 1862. — XC.

—»— L. T. Reichel's Aufnahmen der Umgegend von Okak und Nain in Labrador. Sheet 5, Petermann's Mitteil, Vol. 9., p. 160. Gotha 1863. — XCI.

RICHARDS, G. M. Map of Canoe Route from Lake Michikamau to Ungava Bay and Sledge Route from Fort Chimo to Nachvak Bay, 1905, Ibid. — XCII.

—»— Map of Portage Route from Hamilton Inlet to Lake Mishikamau, Labrador, 1905. Annexed to Dillon Wallace 1907. — XCIII.

ROBYN, I. Niewe warsende graadt kaert Van d'noorder Zee Custen van America Van Fero door d'Straet Davis. En kutson tat aen Terra Neuf. 168. — XCIV.

SANSON, G. Amérique Septentrionale. 1669. — XCV.

[SCHOOLCRAFT, HENRY R.] Ethnographical Map of the Indian Tribes A.D. 1600. In Information respecting the history, condition and prospects of the Indian Tribes . . ., Vol. II, opposite p. 28, 1852 (cfr **880**). — **XCVI.**

SCOTT, J. M. See WATKINS, H. G.

SEUTTER, MATH. Partie Orientale de la Nouvelle France ou du Canada avec . . . et Nouvelle Angleterre [ca: 1740?]. — **XCVII.**

SMITH, EDWARD, SOLUE, FLOYD M. and MOSBY, OLAV. Bathymetrical Map of the Northwestern North Atlantic. Washington 1937. — **XCVIII.**

SNELGROVE, A. K. Newfoundland Labrador showing location of mineral concession to Labrador Mining & Exploration Co. Ltd. (Weaver Minerals Ltd. In Snelgrove: Mines and mineral resources of Newfoundland. St. John's 1938. — **XCIX.**

SOULE, FLOYD M. See SMITH, EDWARD.

SPECK, FRANK G. Distribution of the Montagnais-Naskapi Bands in the Labrador peninsula. Cf. **873** and **877**, 1931 & 1935. — **XCIX a.**

SPIESS, F. [Contour map of Eastern Labrador.] Hamburg 1940. In **880 a.** — **XCIX b.**

TANNER, V. Principal vegetation regions on Newfoundland-Labrador. Soc. Scient. Fennica, Årsbok XVII B, No. 1. Helsingfors 1938. — **C.**

—»— Indians and white trappers in Newfoundland-Labrador. Ibid., No. 2. 1939. — **CI.**

—»— The approximate distribution of the aborigines on Labrador nowadays. Svensk Geogr. Årsbok 1939, p. 90. Lund 1939. — **CII.**

—»— The penetration of the white trappers on the hunting grounds of the Indians. Ibid., p. 96. 1939. — **CIII.**

TAYLOR et LEVERETT. Début de la formation des lacs glaciaires sur le front de la nappe labradorienne. In Haug, Emile, Traité de Géologie, p. 1874. Paris 1911. — **CIV.**

THOULET, J. Carte générale bathymétrique des Océans: par ordre de . . . Prince de Monaco. 1904. — **CV.**

TITFORD, W. B. Newfoundland-Labrador. Published by the Department of Natural Resources. Scale 1: 1.267.200. St. John's 1937 — **CV a.**

TOLMAN, CARL. See GILL, JAMES E.

VANDERMAELEN, PH. Labrador. Nos 28 & 29, Vol. 4, Atlas Universel de Géographie. Bruxelles 1827. — **CVI.**

WALLACE, DILLON. Map of Susan R. and Beaver River. 1904. In Wallace: The Lure of the Labrador Wild, 1905. — **CVII.**

WATKINS, H. G., SCOTT, J. M. and LESLIE, L. A. D. River exploration in Labrador. Geogr. Journ., Vol. LXXV, No. 2. London 1930. — **CVIII.**

WEIZ. See LINDER.

WEIZ, S. Nordspitze von Labrador. 1868. [Enlarged photost. copy of map facing p. 226 in Packard's The Labrador Coast.] [Scale 1 : 1.000.000] [also appears in Science. Vol. XI, p. 78]. New York 1888. — **CVIII a.**

—»— Nordspitze von Labrador. See Map of the Northern Extremity of Labrador, Science, Outp. 78, Vol. XI, No. 263. 1888. — **CVIII b.**

—»— Cfr. **285**, Fig. 177.

WHEELER, E. P.2ND. Reconnaissance route survey of Labrador. 12 sheets, across country from the Fraser River, up the Whale River and back to Saglek Bay. Scale 1 : 125.000. Manuscript in the American Geographical Society's Archives. — **CIX.**

—»— Reconnaissance map of the Labrador Coast in the vicinity of Nain. Ibid., Vol. XX, p. 468. New York. 1930. — **CX.**

WHEELER, E. P. 2nd. Reconnaissance map of Labrador in the vicinity of Okak Bay. Geogr. Rev., Vol. XXV, p. 254. New York 1935. — **CXI.**

—»— Reconnaissance route survey of part of Labrador. 19 sheets, across country from Bell Island, up the Whale River and back to Mugford Tickle. 1 : 126,720. Man. of Amer. Geogr. Soc., New York. 1937. — **CXII.**

—»— Changes on the map of a part of the Labrador peninsula made as a result of E. P. WHEELER's reconnaissance survey of 1935. Geogr. Review, Vol. XXVIII, p. 477. New York 1938. — **CXII a.**

WHITE, JAMES. Canada. 35 miles 1 inch. Ottawa 1902. A later issue 1908. — **CXIII.**

ANONYMOUS. Ethnological Map of the Indian Tribes A. D. 1600. SCHOOLCRAFT, HENRY R. Vol. II, 1852. — **CXIV.**

—»— M. B. Ingr. de la Marine. Carte de la Baye de Hudson 1757. See HARPE. M. de la. 1780. — **CXV.**

—»— Labrador showing the boundary award, 1927. In McGRATH. — **CXVI.**

CHRONOLOGICAL LIST.

1502?	KING',	sub **XXVII.**	1715	COURTEMANCHE [CONSTANTIN],	
1504?	PALESTRINA, »	»		sub **XXIV.**	
1529	RIBERO, »	»	»	BEAVILLIERS, sub **XVI & XVII.**	
1550	DESCELIERS, »	»	1730	LAURE, sub **XXVII & LXXVIa.**	
1558	ZENO, »	»	1731	» »	»
1569	MERCATOR, »	»	1740 [?]	LOTTER, sub **LXXXI.**	
1580	DEE, »	**XXXVII.**	»	SEUTTER, »	**XCVII.**
1597	BOZARDI, »	**XX.**	1743	BELLIN, »	**XXVII.**
1630	HONDIUS »	**XXVII.**	1744	» »	»
1656	SANSON, »	**XXII.**	1746	D'ANVILLE, »	**VIII.**
1660	CREUXIUS, »	**XXVII.**	»	BOEHMIO, »	**XXVII.**
1669	SANSON, »	**XCV.**	1748	ELLIS, »	**XLIV.**
1681	KEULEN, »	**LXXI.**	»	FORNEL, »	**XXVII.**
168-	ROBYN, »	**XCIV.**	1753	BELLIN, »	**XVIII.**
1689	CORONELLI, »	**XXII.**	1755	» »	**XIX & XXVII.**
1693	MORTIER, »	»	»	MITCHELL, »	**LXXXVI.**
1695	JAILLOT, »	**LXVIII.**	1756	HOMANNIANORUM HEREDUM,	
1696	» »	**XXVII, 6.**		sub **LXV.**	
1700	MORTIER, »	**LXXXVII.**	1757	M. B., sub **LVIII.**	
»	[DE] L'ISLE, »	**XXXVIII.**	1760	GENEST, »	**XLIX.**
1703	DEL'ISLE, »	**XXXIX.**	1763	BOWEN, »	**XXII & XXVII.**
1704	COURTEMANCHE, [DESHAIRE],		»	GIBSON, »	**L.**
	sub **XXXIII.**		»	PALAIRET, »	**XXVII.**
1708	L'IISLE, G. DE, sub **XXII.**		»	ROCQUE, »	»
1709	THORNTON, sub **XXVII.**		1765	HAVEN, »	**LIX.**
1710	SENEX, »	**XXII.**	1767	GILBERT, »	**LII.**
1711	DECOÜAGNE, »	**XXXVI.**	1770	LANE, »	**LXXII.**

1770 GILBERT, sub **L.**
»   »   » **LI.**
1773 CURTIS,   » **XXXV.**
1775 JEFFERYS, » **LXIX.**
»   MITCHELL, » **XXVII.**
1776 CARVER,   » **XXII.**
1777 LANE, sub **LXXIII.**
1778 »   » **LXXIV.**
1790 »   » **LXXV.**
1792 »   » **LXXVI.**

1809 ARROWSMITH, sub **IX.**
1811 KOHLMEISTER & KMOCH,
      sub **LXXXV.**
1814 ARROWSMITH, sub **XXII.**
1815 BRUÉ   »   »
1821 »   »   »
1821 ANSPACH,   » **VII.**
1822 TANNER,   » **XXII.**
1827 VANDERMAELEN, sub **CVI.**
1832 ARROWSMITH, sub. **XXVII.**
1834 »   » **XXII.**
1839 »   »   »
1840 BRUÉ,   »   »
1842 ARROWSMITH, »   »
1850 »   » **XXVII.**
1852 SCHOOLCRAFT, » **XCVI.**
»   ANON.,   » **CXVI.**
1853 ARROWSMITH sub. **XXVII.**
1855 KEEFER,   » **XXII &**
      XXVII.
1857 ARROWSMITH, » **XXVII.**
»   »   » **XXVII.**
1860 LIEBER, sub **LXXVIII.**
1861 »   » **LXXIX.**
1862 REICHEL, » **XC.**
»   COLTON,   » **XXII.**
1863 REICHEL, » **XCI.**
»   BERGHAUS, sub **XXII.**
»   BRUÉ,   »   »
1864 HIND, sub. **LXII & LXIII.**
»   ANON.,   » **XXII.**
1866 ARROWSMITH, sub **IX.**
1868 RUSSELL, sub **XXVII.**
»   WEIZ,   » **CXIV.**
»   CHIMMO,   » **XXIX.**
1871 RICHARDS, » **XXVII.**

1871 RUSSELL & MARA, sub **XXII.**
1872 ANON.,   »   »
1873 DESBARATS,   »   »
»   BAUR,   »   »
1874 DESBARATS, sub **XLI.**
»   JOHNSTON,   » **XXII.**
1877 MURRAY,   » **LXXXVIII.**
1878 JOHNSTON,   »   **XXII.**
1882 JOHNSON & EDMONDS, sub »
»   JOHNSTON,   »   »
1883 TACHE & GENEST   »   »
1886 LORNE, sub **XXVII.**
»   DEPARTMENT OF AGRICULTURE,
      sub **XXVII.**
1887 LEUTHNER, sub **LXXVII.**
1888 WEIZ,   » **CXV.**
»   HOLME,   » **LXIV.**
1890 JOHNSTON, » **LXX.**
1891 BRYANT,   » **XXIII.**
»   PACKARD,   » **LXXXVIII a.**
»   STANFORD, » **XXII.**
»   JOHNSON & EDMONDS, sub **XXII.**
1894 BRYANT, sub **XXII a.**
1895 BELL,   » **XVII a.**
1896 EATON,   » **XLII.**
»   »   » **XLIII.**
»   Department of the Interior, sub.
      XXVII.
»   LINDER, sub **LXXX.**
»   LOW,   » **LXXXII.**
1897 JOHNSTON, » **XXVII.**
1898 ANON.,   » **XXII.**
1899 STANFORD, »   »
1900 WHITE,   » **XXVII.**
1901 DELABARRE, sub **XXXVI a.**
»   LOW, sub **LXXXIII.**
»   BARTHOLOMEW, sub **XXVII.**
1902 DALY, sub **XXXV a.**
»   LOW,   » **LXXXIV.**
»   ANON.,   » **XXVII.**
1904 THOULET, sub **CV.**
1905 RICHARDS, » **XCII, XCIII &**
      CVII.
»   WALLACE,   » **XCII, XCIII &**
      CVII.
1906 HUBBARD, sub **LXVI.**
1908 WHITE,   » **CXIII.**

1908 HANTSCH, sub **LVI.**	1924 KINDLE, sub **LXXIa.**
» WHITE, » **XXII.**	1925 COLONIAL OFFICE, sub. **XXXII.**
1909 Hantsch » **LVII.**	» WHITE, sub **XXVII.**
» CANADIAN ARCHIVES, sub **XXV.**	» » » »
1910 GRENFELL, sub **LIIIa.**	» » » »
» BARTHOLOMEW, sub **XI & XII.**	1926 BRITISH GOVERNMENT, sub **XXII.**
» GRENFELL, sub **XXVII.**	» CANADIAN GOVERNMENT, » **XXVII.**
1911 TAYLOR & LEVERETT, sub **CIV.**	1927 AMERICAN GEOGR. SOCIETY, sub **I &**
» PRICHARD, sub **LXXXIXa.**	**CXVIII.**
1912 BRITISH GEOGRAPHICAL SOCIETY,	» FRISSELL, sub **XLVIII.**
sub **XXIa.**	1928 BARTHOLOMEW, sub **XV.**
» BRYANT, sub **LXXXIX.**	1929 BRITISH ADMIRALTY, sub **XXI.**
» CABOT, » **XXIV.**	1930 WATKINS, sub **CVIII.**
» HUTTON, » **XXIa.**	» WHEELER, » **CX.**
» PORTER, » **LXXXIX.**	1931 SPECK, » **XCIXa.**
» STANFORD, » **XXII.**	» AMERICAN GEOGRAPHICAL SOCIETY,
» BARTHOLOMEW, sub **XIV.**	sub **II & Va.**
1914 CANADA LABRADOR BOUNDARY,	1932 ISELIN, sub **LXVIa.**
sub **XXVIII.**	1935 WHEELER, » **CXI.**
1916 HAWKES [CANADIAN GEOLOGICAL	1936 ABBE, » **I.**
SURVEY], sub **XXVI & LX.**	» AMERICAN GEOGRAPHICAL SOCIETY,
» COLEMAN sub **XXX.**	sub **IV.**
» » » **XXXI.**	1936 BAKER, sub **X.**
1918 FLAHERTY, » **XLV & LVI &**	1937 WHEELER, » **CXII & CIX.**
XLVIa.	» TITFORD, » **CVa.**
1919 HOWLEY, sub **LXIVa.**	» GILL, » **LIII.**
1920 CABOT, » **XXIV.**	» SMITH, » **XCVIII.**
1922 ANDERSON, » **VI.**	1938 TANNER, » **C.**
» STEWART, » **XXVII.**	» FORBES, » **XLVII.**
» DEPARTMENT OF JUSTICE CANADA,	» WHEELER, » **CXIIa.**
sub **XXXIXa.**	» SNELGROVE, » **XCIX.**
» BARTHOLOMEW, sub **XXVII.**	» AMERICAN GEOGRAPHICAL SOCIETY,
1923 ANON., sub **XXVII.**	sub **V.**
» DEPARTMENT OF JUSTICE,	1939 KRANCK, sub **LXXIb.**
sub **XXII.**	» TANNER, » **CI, CII & CIII.**
» DEPARTMENT OF JUSTICE,	» U.S. HYDROGRAPHIC OFFICE,
sub **XXII.**	sub **LXVII.**
1924 DEPARTMENT OF JUSTICE,	
sub **XXVII.**	

ADDENDA:

P. 234, 1. 13 WRIGHT, W. B. (The Quaternary Ice Age, London 1914, and Geol. Magaz.,
Vol. 57, p. 382; Ibid. Vol. 62, p. 227, and Rapp. d.1. Commission des Terrasses, Congr.
Int. de Géographie Cambridge 1928.)

* * *

Printed in the United States
By Bookmasters